PSpice® Simulation of Power Electronics Circuits

JOIN US ON THE INTERNET VIA WWW, GOPHER, FTP OR EMAIL:

WWW: http://www.thomson.com
GOPHER: gopher.thomson.com
FTP: ftp.thomson.com
EMAIL: findit@kiosk.thomson.com

A service of I(T)P®

PSpice Simulation of Power Electronics Circuits

An introductory guide

Raymond Ramshaw and Derek Schuurman

Department of Electrical Engineering
and Computer Engineering
The University of Waterloo
Ontario
Canada

CHAPMAN & HALL
London · Weinheim · New York · Tokyo · Melbourne · Madras

Published by Chapman & Hall, 2–6 Boundary Row, London SE1 8HN, UK

Chapman & Hall, 2–6 Boundary Row, London SE1 8HN, UK

Chapman & Hall GmbH, Pappelallee 3, 69469 Weinheim, Germany

Chapman & Hall USA, 115 Fifth Avenue, New York, NY 10003, USA

Chapman & Hall Japan, ITP-Japan, Kyowa Building, 3F, 2-2-1 Hirakawacho, Chiyoda-ku, Tokyo 102, Japan

Chapman & Hall Australia, 102 Dodds Street, South Melbourne, Victoria 3205, Australia

Chapman & Hall India, R. Seshadri, 32 Second Main Road, CIT East, Madras 600 035, India

First edition 1997

Printed in Great Britain by St Edmundsbury Press, Bury St Edmunds, Suffolk

ISBN 0 412 75140 2

A Catalogue record for this book is available from the British Library

∞ Printed on permanent acid-free text paper, manufactured in accordance with ANSI/NISO Z39.48-1992 and ANSI/NISO Z39.48-1984 (Permanence of Paper).

CONTENTS

PREFACE

The aim of this book is to provide instruction in the use of a computer program that simulates power electronics circuits. This program is called MicroSim™ PSpice®.

Power electronics circuits are used to control the amount of power that is transmitted from an energy source to a load. The purpose of the control is to modify the load characteristics to specified requirements. The requirements may be to control a lighting or heating level or to control the acceleration, speed, or position of a motor shaft.

The control by power electronics circuits is accomplished by the use of semiconductor switches that are turned on and off sequentially. There are many circuit configurations whose designs must usually satisfy certain performance criteria. In the design stage analytical studies are carried out, where possible, to check the performance of the control circuit or system. Many designs are too complex to analyze by hand. Prototype testing at the early stages is not usually economic. However, with all the computer tools and aids available to us today, it is possible to carry out a computer-aided circuit simulation. The simulation results can be checked to see if the performance requirements have been satisfied. If necessary, circuit parameters can have values modified, a computer simulation rerun, and performance rechecked. This can take virtually no time compared with either a hand analysis, or a practical construction and testing of the circuit. Simulation has become very important.

There is a number of circuit simulation packages on the market. One of the most popular is the generic SPICE program that is used world wide. This text describes how to use the commercial version PSpice® to simulate power electronics circuits. Some knowledge of power electronics is of use to be able to interpret simulations. In some cases, it is helpful to have a book on power electronics available as a reference. We have followed the general sequence found in "Power Electronics Semiconductor Switches" by Raymond Ramshaw (Chapman & Hall 1993) (abbreviated to PESS). However, this text on PSpice is written to be a stand-alone guide for both students and practising engineers.

The format we have used is that power electronics circuits are described very briefly; the generation of a computer file that becomes the input for the PSpice program is developed in detail, and the output of the program is discussed. The development of simulation expertise is created by tutorial examples, drill exercises and problems for practice. The statements and commands for PSpice input files are summarized in Appendix C.

In this book the semiconductor switches are modelled for PSpice in two ways. The first way is to treat the semiconductor switch as close as possible to that of an ideal switch for system simulation, if the study of the load characteristics is of paramount importance. The second way is to use a detailed model of the semiconductor switch in order to study the switch or the converter performance alone. Some attention is given to the details of modelling power semiconductor devices for PSpice.

We have used a particular commercial program PSpice® from MicroSim™ Corp. There is an evaluation version of PSpice that can be freely copied for use on a PC, a MAC or a workstation. All the examples, drills and problems in this text can be simulated using this version.

It has not been possible to cover all the material associated with MicroSim™ PSpice®. However, what has been covered will give a good working knowledge about how to simulate power electronics circuits. Armed with this knowledge, you will find MicroSim PSpice manuals invaluable aids for future work.

This guide is for tutorial purposes only. The results of any simulation using circuit files associated with this text must not be construed as useful design data. This text is intended as instruction only in the use of the PSpice program.

Our aim is to have you sit at the computer with this text as a guide and to progressively work through the EXAMPLES, the Drill Exercises and the Problems chapter by chapter. Accordingly, you will need to load the free evaluation version of MicroSim™ PSpice® program into a subdirectory of your hard drive.

If you have access to the Internet, we suggest you view the World Wide WEB (WWW) home page of the authors by using the URL

http://www.ece.uwaterloo.ca/~pwr_elec

From this home page you can download circuit files and library files which are relevant to this text (see Appendix E). Also, this home page allows you to download the "zipped" solutions of the Drill Exercises and the Problems. To unzip these solutions files instructors can obtain the password from the publishers, Chapman & Hall. You can also access MicroSim's evaluation version of PSpice for downloading. There are additional links to PSpice and SPICE-related materials that you may find useful.

If you cannot obtain a copy of MicroSim's evaluation version of PSpice from the Internet, you may obtain your free copy using any of the following.

1. You are allowed to copy anyone's PSpice evaluation version.
2. Phone MicroSim Corporation directly at (714) 770-3022, ext. 351 and place your order.
3. Fax MicroSim Corporation with your address and your request for the most recent evaluation version of PSpice at (714) 455-0554.
4. Dial MicroSim's BBS at (714) 830-1550.
5. Write to MicroSim Corporation, 20 Fairbanks, Irvine, California 92718.

Acknowledgement and gratitude are due to Elizabeth Ramshaw who has accepted once again scribbles and sketches and transformed them into the typeset text and figures set before you.

Thanks are due to MicroSim Corporation. PSpice is a friendly program. We have really enjoyed composing this guide. We hope that you have as much enjoyment learning about the program.

Raymond S. Ramshaw
Derek C. Schuurman

Waterloo.

Chapter 1

PSpice and Electric Circuits

1.1 INTRODUCTION

Power electronics circuits comprise semiconductor switches. The switched circuits can be complex in terms of both the power circuit and the signal control circuits. The general layout of a power electronics circuit is shown in Fig. 1.1. One aim during a design process of a power electronics system is to predict the performance, either at the system level or at the device level. A common way to make a performance prediction is by using a computer-aided, electric circuit simulator. As models of semiconductor switches are being made available or are being enhanced, so the use of circuit simulators is increasing in the field of power electronics.

An important feature in today's circuit simulation tools is that it is not necessary to write down the sets of algebraic and integro-differential equations to describe the circuit behaviour. We have only to describe the circuit configuration and parameter values. The simulation program creates the equations, solves them and gives us the results in terms of the circuit-variable responses.

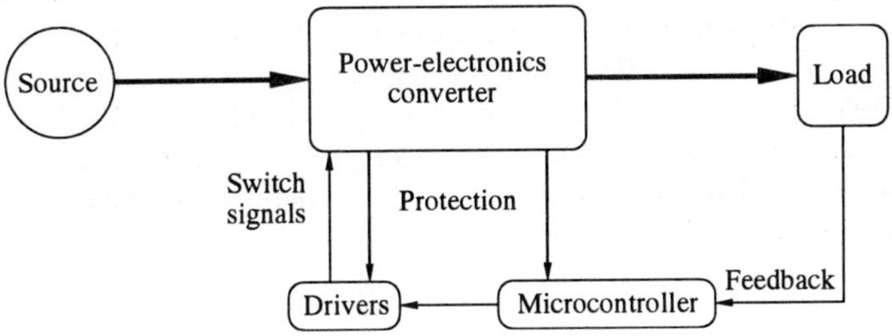

Fig. 1.1 Block diagram of a power-electronics circuit.

One of the industry standards for circuit simulation is the public-domain program SPICE (Simulation Program with Integrated Circuit Emphasis). SPICE has been developed at the University of California, Berkley, over the last twenty years and development continues. Although it was created as an aid in the design of integrated circuits, SPICE is used to solve a variety of non-IC circuit problems including power-electronics circuits.

In this text, we are going to give an introductory guide to the use of a MicroSim™ proprietary package called PSpice® based on a version of SPICE. PSpice is popular, the evaluation version is free, it can be used on PCs, MACs and workstations, and it has friendly preprocessors and a postprocessor. There is a number of other simulation packages. However, these packages are similar in many ways and incorporate many of the generic attributes of SPICE. So this guide is appropriate for them too.

In this chapter we introduce the method of describing the configuration of simple linear circuits. An important part of the simulation of power electronics circuits is to be able to determine the performance factors associated with the supply, the converter and the load. This requires the *measurement* of average values, rms values, harmonic values of the variables and the effects of nonlinear elements. The postprocessor (PROBE) of PSpice is useful in this respect.

1.2 PSPICE

PSpice® by MicroSim Corp. is a popular commercial version of SPICE. This software package includes utilities which help the simulation process to be a pleasant task. The various utilities and their association with PSpice are illustrated in Fig. 1.2.

The flow of the simulation process, depicted in Fig. 1.2 is from left to right. First, a text file containing the circuit netlist and analysis commands must be made available. This *circuit* file can be created using your text editor or it can be created using a graphical schematic editor† which is available for the Microsoft™ WINDOWS version of PSpice. MicroSim has also developed a graphical stimulus editor which is an optional utility for defining dependent voltage or current sources for the circuit file.

† We do not use the schematic-capture facility in this guide. An introduction to circuit files is best accomplished through the personal creation of the ASCII files.

The circuit file becomes the input for the actual PSpice simulation. If the circuit file contains any external references to component models, then PSpice will obtain the appropriate details from a component library file. The evaluation version of PSpice includes some component models in its library, whereas the full, professional version of PSpice contains libraries with the descriptions of thousands of different components. In addition, MicroSim offers a utility called PARTS which can be used to define and characterize additional component models to augment the existing libraries.

Once a PSpice simulation has been run, a text file containing an output listing is always created. This file can echo the circuit file as it is read into PSpice, it flags any errors if they occurred, and it reports any requested results. Usually, a data file is also created for use with PROBE, which is a utility that formulates simulation results graphically. PROBE is a highly useful tool for viewing and manipulating simulation results. It is likened to an oscilloscope.

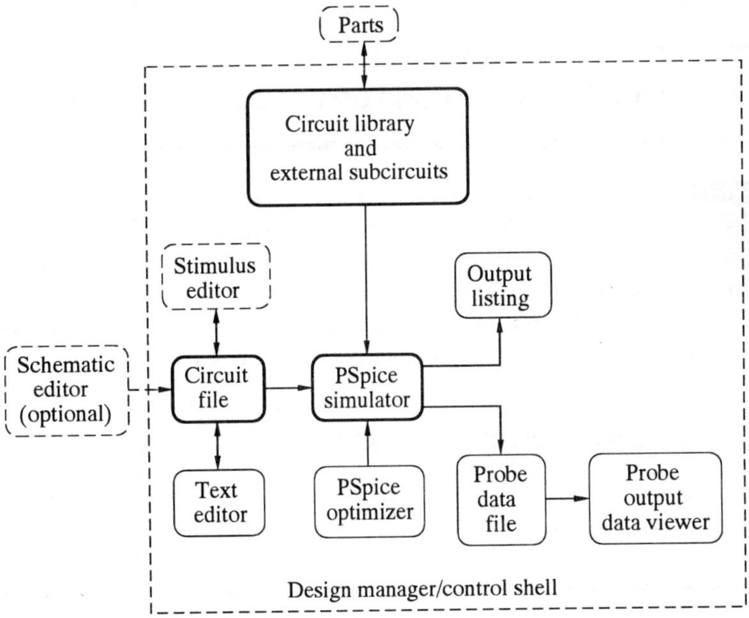

Fig. 1.2 Flow diagram of the simulation process.

All the programs and utilities associated with PSpice can be orchestrated from within the *Control Shell*,† a supervisory program, which allows the user to *seamlessly* move between the various programs as required.

PSpice performs circuit analysis. Therefore, we must have a circuit diagram that includes the element values if a simulation is to be carried out. PSpice does not design a circuit; it helps you to realize a circuit that operates well by allowing you to modify parameter values and view the results of an analysis.

In terms of limitations, PSpice is not interactive. You cannot specify a required output of a circuit and expect the simulation process to determine element values. If the evaluation version of PSpice is used, there is a circuit limit of 10 transistors; this is not a restriction in this book. If speed of operation is of concern, a math coprocessor and the DOS version of PSpice is faster than most WINDOWS versions. New PSpice releases are WINDOWS based.

If you learn to use PSpice, it is not difficult to adapt to other SPICE variants. MicroSim maintains PSpice compatibility with SPICE.

1.3 CIRCUIT DESCRIPTION FOR PSPICE SIMULATION

We design a circuit and we wish to determine its efficacy. This is the stage at which we do a computer simulation because it is the quickest and easiest way to check a circuit's responses. In this section we will discuss some of the PSpice statements used to set up a program for the simulation of linear electric circuits. Three examples will prepare us to start investigating the principles of power-electronics power modulation that is described in the next chapter.

1.3.1 DC ANALYSIS (DC SOURCE)

A dc source supplying a resistive load is easy to check for responses and is a good introduction to PSpice.

EXAMPLE 1.3.1

The circuit diagram in Fig. 1.3a is a simple example, but it is useful to describe how a computer simulation can be carried out. The circuit comprises a direct-

† In WINDOWS all programs are orchestrated by the PSpice design manger.

voltage source V_s and a load resistance $R_l = 5.0\,\Omega$. Determine the values of the load current I_l for different values of source voltage.

Solution

The accomplishment of the task in this example is carried out in four steps.

STEP 1

Number the nodes on a circuit diagram with non-negative integer values for all elements of the circuit. See Fig. 1.3b. Give names to all elements, as shown. The first letter of the names follows PSpice rules. Such a diagram is called a PSpice configuration.

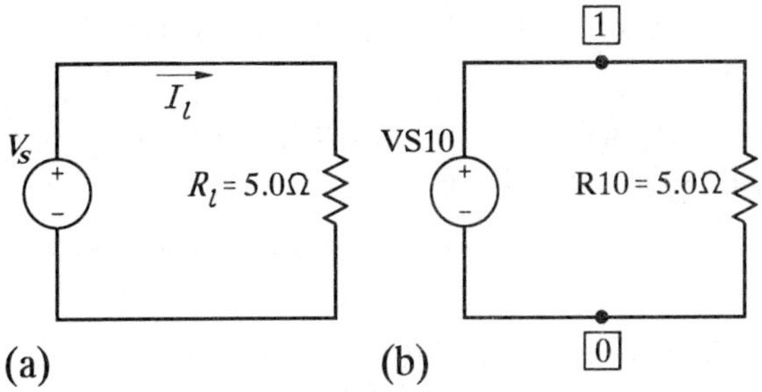

Fig. 1.3 A simple dc circuit.
(a) Circuit diagram, (b) node numbering and element naming.

STEP 2

A computer circuit file (also known as a source file or an input file or a netlist) is composed by you with your text editor of choice†. The ASCII file must have a particular format that can be interpreted by PSpice. For convenience, the file is divided into sections that are entitled SOURCES, CIRCUIT ELEMENTS, ANALYSIS and OUTPUT. Each section includes statements that can be recognized by PSpice. At the right-hand side of the lines in the file listing a number is written for reference. With the PSpice circuit configuration in Fig. 1.3b as reference, the contents of a circuit file, describing the circuit, are shown on the next page.

† Within the Control Shell of PSpice, the **files** menu contains a text editor. For WINDOWS, you can use NOTEPAD which is with the WINDOWS' ACCESSORIES.

```
»»»»» EX1_3_1.CIR »»»»»»»»»»»»»»»»»»»»»»»»»»»»»»»»»»»»»»»»
DC SWEEP OF A LINEAR RESISTIVE CIRCUIT; .................. (1)
* PSpice file name EX1_3_1.CIR.    DC analysis. .................. (2)
* Determine the current responses. .............................. (3)
* SOURCE ........................................................ (4)
VS10  1   0   DC   1.0V; ........................................ (5)
* CIRCUIT ELEMENT ............................................... (6)
R10   1   0   5.0ohms; .......................................... (7)
* ANALYSIS ...................................................... (8)
.DC   LIN    VS10    4V   12V   2V;  DC sweep. ............ (9)
* OUTPUT ........................................................ (10)
.PRINT    DC  V(1)  I(R10);  Writes results to file EX1_3_1.OUT. ... (11)
.END;  Signifies the end of the file. ........................ (12)
»»»»»»»»»»»»»»»»»»»»»»»»»»»»»»»»»»»»»»»»»»»»»»»»»»»»»»»»»»»»
```

STEP 3 The file EX1_3_1.CIR can be written to the subdirectory where the executable program PSPICE resides. For example, if we are in the subdirectory C:\PSPICE, at the DOS prompt, we type in

C:\PSPICE> PSPICE EX1_3_1.CIR EX1_3_1.OUT <ENTER>

or just type in C:\PSPICE> PSPICE EX1_3_1 <ENTER>

We do not need to type the extension .CIR in the command argument; it is understood by PSpice to be a circuit file. The name EX1_3_1.OUT can be the name of your choice for the output file generated by PSpice. It can also be directed to any drive or directory. If no indication of the name or directory is given, the file EX1_3_1.OUT is automatically written to the same directory where the circuit file is stored. The circuit file does not have to be in the PSpice directory. If it is not there, its full address must be typed in the command line. In the Windows' environment we choose from the menu FILE/OPEN and select the circuit file we wish to run.

After you have hit carriage return (<CR> or <ENTER> or <↵>) the PSpice program will provide information on the screen during the execution of the program. If there is an error you will be told to view the output file. The error message in the output file should help you to correct the error so that you can rerun the program successfully. PSpice is user friendly.

STEP 4 Examine the output of the program. Resulting from the successful execution of the program EX1_3_1.CIR, the pertinent output data from the file EX1_3_1.OUT are as follows.

»»»»» EX1_3_1.OUT »»»»»»»»»»»»»»»»»»»»»»»»»»»»»»»»»»

```
DC SWEEP OF A LINEAR RESISTIVE CIRCUIT .................... (13)
DC TRANSFER CURVES         TEMPERATURE 27DEG C ........... (14)
VS10            V(1)            I(R10) ..................... (15)
4.000E+00       4.000E+00       8.000E-01 .................. (16)
6.000E+00       6.000E+00       1.200E+00
8.000E+00       8.000E+00       1.600E+00
1.000E+01       1.000E+01       2.000E+00
1.200E+01       1.200E+01       2.400E+00

TOTAL JOB TIME 2.63 ....................................... (17)
```

»»»

EXPLANATION OF THE FILES EX1_3_1.CIR and EX1_3_1.OUT

. . . .

Line (1) of circuit file EX1_3_1.CIR

DC SWEEP OF A LINEAR RESISTIVE CIRCUIT; (1)

- This title statement always takes the first line in the file. You can include in this line whatever is relevant to you. We described this example in the first line as DC SWEEP OF A LINEAR RESISTIVE CIRCUIT. It must not have more than 80 characters.
- The semicolon (;) following a PSpice statement and any comment that follows is ignored by PSpice. The comment is for a reader's benefit.
- The statements that are to be recognized by PSpice are independent of case. We have used a sans-serif font in upper case in the text in order to distinguish the PSpice statements from comments, but it is not done in an actual circuit file.

. . . .

Line (2) of circuit file EX1_3_1.CIR

* PSpice file name EX1_3_1.CIR. DC analysis. (2)

- The asterisk at the beginning of the line and anything that follows on the line is a comment. It is ignored by PSpice. The comment here is to remind us of the name that we have given to the circuit file and that a dc analysis is to be performed on the circuit.

- All circuit-file names have the extension .CIR. Otherwise PSpice will not recognize them. Here, we chose the name EX1_3_1.CIR to provide easy reference in this book. The name of the file can have any number of characters and any character that is recognized by DOS.

. . . .

Line (3) of circuit file EX1_3_1.CIR

```
* Determine the current responses.                                  (3)
```

- This is another comment line ignored by PSpice.

. . . .

Line (4) of circuit file EX1_3_1.CIR

```
* SOURCE                                                            (4)
```

- This comment informs us that what follows will be data statements regarding voltage and current sources.

. . . .

Line (5) of circuit file EX1_3_1.CIR

```
VS10   1     0   DC   1.0V;                                         (5)
```

- This is a data statement for PSpice. It can appear anywhere between the title statement and the end statement of the file.
- There are four fields in this data statement. The fields are for a voltage-source name, node designation, the type of source and the numerical value of the source voltage, respectively. The fields are separated by spaces.
- The first field contains the source name that must begin with the letter V, since it represents an independent voltage source. The source-name characters that follow V can be letters, numbers and characters $, %, * and /. Any reasonable number of characters can be used. Table 1.1 gives a list of the coding of the source names. Here, the source name is VS10.
- The second field gives the numbers (1, 0) of the two nodes to which the source is connected. In PSpice the node label can have letters, or any text string. The first node is understood to have a positive polarity with respect to the second node†. Convention prevails that current direction in a circuit is from high potential to low potential. There is a space between the node numbers which can be values between 0 and 9999. There must be one

† A current source has its current direction from the first node to the second node.

node in the main circuit that is designated the reference (ground) node and must be given the number 0. Every element must have a dc path to ground. Every node must have at least two connections; a node that is left *hanging* will be detected as an error and the program will not execute.

- The third field is for the type of the source output, in this case DC (a constant value). If there is nothing in this field, it is, by default, treated as a dc source. Table 1.2 gives a list of PSpice source outputs.

Table 1.1 First letter of source names for PSpice

Type of source	First
Voltage-controlled voltage source (VCVS)	E
Current-controlled current source (CCCS)	F
Voltage-controlled current source (VCCS)	G
Current-controlled voltage source (CCVS)	H
Independent current source	I
Independent voltage source	V

Table 1.2 Type of source outputs

Source output	PSpice description
Alternating current	AC
Direct current	DC
Exponential	EXP
Piecewise linear	PWL
Polynomial	POLY (n)
Pulse	PULSE
Single-frequency modulation	SFFM
Sinusoid	SIN

- The fourth field is for the numerical value of the source voltage. In this example, it is given as 1.0V. For this example, the value of the voltage is a nominal value, because the analysis calls for other values. The value can be an integer, or a real number. The number can be in the exponential form or it can have a symbolic scaling factor. Table 1.3, shown on the next page, lists the exponential forms and the scaling factors. Whatever letters come after the value of the voltage are ignored by PSpice. The SI units such as V or amperes or ohms have no meaning to PSpice. There is a default unit of volts assumed by PSpice, since the initial-letter V in the source type describes a voltage source.

- The general data statement for a dc-voltage source is
  {source_name} {positive_node, negative_node} {source_type} {value}

  Vxxx N+ N- <DC> value

  The angular brackets <> surround an optional field.
  There can be other fields. See the appendix on data statements.

. . . .

Table 1.3 PSpice number scaling.

Multiple	Prefix	Abbreviation	Exponential form	PSpice symbol
10^{12}	tera	T	1E12	T (or t)
10^{9}	giga	G	1E9	G (or g)
10^{6}	mega	M	1E6	MEG (or meg)
10^{3}	kilo	k	1E3	K (or k)
10^{-3}	milli	m	1E-3	M (or m)
10^{-6}	micro	μ	1E-6	U (or u)
10^{-9}	nano	n	1E-9	N (or n)
10^{-12}	pico	p	1E-12	P (or p)
10^{-15}	femto	f	1E-15	F (or f)

. . . .

Line (6) of circuit file EX1_3_1.CIR
* CIRCUIT ELEMENT (6)

- This line is a comment line. The data statements concerning the elements follow this comment line in the circuit file.

. . . .

Line (7) of circuit file EX1_3_1.CIR
R10 1 0 5.0ohms; (7)

- This is a data statement regarding the load resistance.
- There are three fields separated by spaces in this statement.
- The first field contains the name R10 given to the resistor in the circuit configuration in Fig. 1.3b. Its name has to begin with R so that PSpice

recognizes it as a resistor. What follows to complete the name can be anything that is convenient. Table 1.4 lists the initial letters that are used in PSpice for the circuit elements.

- The second field of the statement contains the numbers (1, 0) of the nodes to which the resistor is connected.
- The third field has the resistance value, given as 5.0 ohms. The unit ohms in the field is ignored. It would have been sufficient to have written 5.0.

Table 1.4 First letter of element names for PSpice

Element	First letter of PSpice name
Capacitor	C
Inductor	L
Mutual inductance (transformer)	K
Resistor	R

- The general form of the data statement for a resistor is
  {element_name} {positive_node, negative_node} {value}

```
Rxxx          N+          N-          value
```

It is possible to have a fourth field for modelling nonlinear resistors†.

. . . .

Line (8) of circuit file EX1_3_1.CIR
* ANALYSIS (8)

- This line is a comment that what follows in the circuit file is a control statement regarding the type of analysis to be carried out.

. . . .

Line (9) of circuit file EX1_3_1.CIR
```
.DC  LIN  VS10  4V  12V  2V; DC sweep.
```
(9)

- This is a control statement for PSpice to perform a dc analysis for a range of different values of the source voltage. This statement has six fields of information.
- Field one is the **dot** command specifying the type of analysis. .DC requests a dc analysis that is more than a simple dc analysis. If this statement were

† See MicroSim's manuals.

omitted PSpice would carry out a simple dc analysis with a source value of 1.0V, as designated in line (5). The output file would automatically have the dc operating point written to it. The .DC command signifies that a dc analysis is requested for the dc transfer curves. The analysis will involve a dc **sweep**, a range of source-voltage values, for each of which the node voltages and branch currents are calculated. Table 1.5 lists the dot commands for different analyses.

Table 1.5 Dot commands for analyses

Analysis	Dot command
DC analysis	
DC sweep	.DC
DC operating point	.OP
Node set	.NODESET
Small-signal sensitivity	.SENS
Small-signal transfer function	.TF
AC analysis	
Frequency response	.AC
Transient analysis	
Transient analysis	.TRAN
Initial transient conditions	.IC
Fourier analysis	.FOUR
Noise analysis	.NOISE
Monte-Carlo analysis	.MC
Distortion analysis	.DISTO

- Field two contains the coding for the way the dc sweep is to be performed. LIN signifies that the sweep is to be linear. There are other forms of sweep involving logarithmic scaling. See the Appendix C about .DC.
- Field three contains the name of the source (VS10) that is to undergo a change of voltage value that forms the sweep. Two sources could be included in a control statement for dc sweeps. This is useful if the circuit contains a device such as a transistor.
- Field four contains the starting value (START) of the source voltage (4V) for the first dc analysis.

- Field five contains the final value (STOP value) of the source voltage (12V) for the last dc analysis.
- Field six contains the increment value (INC=2V) for the sweep. In EXAMPLE 1.3.1 the dc analysis is performed sequentially for values 4V, 6V, 8V, 10V and 12V of the source voltage VS10.
- The general form for this control statement is

```
.DC    LIN    Vxxx    START    STOP    INC
```

. . . .

Line (10) of circuit file EX1_3_1.CIR
* OUTPUT (10)

- This is a comment line to indicate that what follows in the circuit file is a request for a particular form of recording output results.

. . . .

Line (11) of circuit file EX1_3_1.CIR
.PRINT DC V(1) I(R10); Writes results to file EX1_3_1.OUT (11)

- This is an output statement describing the format of the analysis results that are to be written to the output file (EX1_3_1.OUT). There are four fields.
- Field one contains the dot command .PRINT. This specifies that data from the analysis is to be recorded in the output file.
- Field two specifies the type of analysis (DC) from which the output data is to be generated.
- Fields three and four contain circuit variables whose values are to be printed. A list of variables is generally added after the analysis-type field. If the list exceeds the line length of 80 characters, PSpice recognizes the continuation on following lines, if the first character of each new line is a plus (+) sign.
- Field three contains a voltage variable V(1). This particular voltage is the load voltage, the voltage across nodes 1 and 0. It could have been written as V(1, 0). The numbers separated by a comma in parentheses are the two node numbers. V(0, 1) is the negative of V(1, 0). The designation V(1) by default is a voltage at node 1 with respect to ground. Since the output voltage V(1) is equal to the source voltage, we could have specified the voltage by name as VS10. Equally well, we could have specified the output voltage as V(R10), the voltage across the element named R10.
- Field four contains a current variable I(R10). This represents the load current. Current variables are not specified by node numbers because this

would be misleading. The symbol I means a current variable to PSpice and the coding within the parentheses represents the name of an element that conducts the current. The current direction is indicated by the first node in the resistor data statement. For this circuit of EXAMPLE 1.3.1 we could have written -I(VS10), because the same currents are in elements R10 and VS10. If it is not convenient to measure current in a branch element, a dummy voltage source VX=0volt can be connected in the branch. The current I(VX) through the source would give the required value of the branch current. The dummy source does not affect the circuit operation.

- The general output statement for printing is

  . PRINT DC [LIST OF OUTPUT VARIABLES]

. . . .

Line (12) of circuit file EX1_3_1.CIR

. END; Signifies the end of the file. (12)

- The last line of a PSpice circuit file is a control statement. It has the dot command .END so that PSpice is aware that the entire circuit-file description has been loaded and the program is ready to be executed.

EXPLANATION OF LINES (13) TO (17) IN THE OUTPUT FILE

. . . .

Line (13) of output file EX1_3_1.OUT

DC SWEEP OF A LINEAR RESISTIVE CIRCUIT (13)

- The title of the source file is repeated for information.

. . . .

Line (14) of output file EX1_3_1.OUT

DC TRANSFER CURVES TEMPERATURE 27DEG C (14)

- This line reminds us about the type of analysis that was performed and the steady-state temperature of the elements in the circuit. If no temperature is specified in the input circuit file the default value is 27° C.

. . . .

Line (15) of output file EX1_3_1.OUT

VS10 V(1) I(R10) (15)

- This line gives the column headers of the data requested in the .PRINT statement. Always included is the source voltage (VS10) upon which the

DC sweep has been exercised. The variables, load voltage V(1) and load current I(R10) were called for in line (11) of the input circuit file.

. . . .

Line (16) of output file EX1_3_1.OUT
4.000E+00 4.000E+00 8.000E-01 (16)

- The data values of VS10, V(1) and I(R10) are given as real numbers scaled by the integer exponential factor, as shown in this line.

. . . .

Line (17) of output file EX1_3_1.OUT
TOTAL JOB TIME 2.63 (17)

- The last line gives the time of program execution in seconds.

This ends the explanation of the files EX1_3_1.CIR and EX1_3_1.OUT.

. . . .

ERRORS It might arise that you do not compile the circuit file in a way that is satisfactory for PSpice. Should such a state occur, you will be informed that there is an error. PSpice is friendly. The error will be flagged in the output file to help you make the correction. Some common errors are as follows.

A space in a statement is missing. There is a dot missing from a DOT command.
There is no asterisk or semicolon in front of a comment.
A command is spelt incorrectly. The solution does not converge.
A node may have only one element connected to it.
A loop may contain only voltage sources and inductors.

The circuit file EX1_3_1.CIR could have been written without comments. The listing would be as follows.

```
DC SWEEP OF A LINEAR RESISTIVE CIRCUIT
VS10     1    0      DC    1.0
R10      1    0      5
.DC      LIN  VS10   4     12     2
.PRINT   DC   V(1)   I(R10)
.END
```

If the circuit file is to be used more than once, or passed to others, it is convenient to have comments in the file.
Create this circuit file, do a PSpice simulation and confirm the results.

END OF EXAMPLE 1.3.1

Drill Exercise D1.3.1
Create a circuit file for the circuit diagram in Fig. 1.3. Make PSpice do a dc sweep of the source voltage from 5V to 30V in steps of 5V. Do a hand calculation and check the results in the output file.

Drill Exercise D1.3.2
Consider the PSpice configuration in Fig. 1.4. Using PSpice, print the output voltage V(3, 0) and source current I(VS10) for a DC sweep of the source voltage from 50V to 200V in steps of 25V.

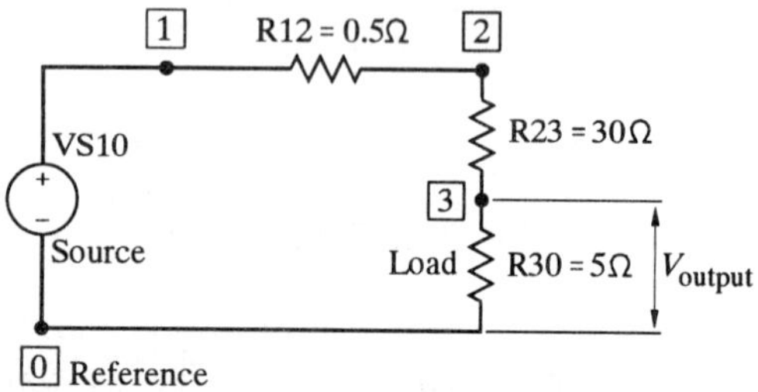

Fig. 1.4

1.3.2 TRANSIENT ANALYSIS (AC SOURCE)

We will choose a simple ac circuit to exemplify the use of new statements.

EXAMPLE 1.3.2

Consider the circuit diagram in Fig. 1.5a. Determine the load current i as a function of time using PSpice.

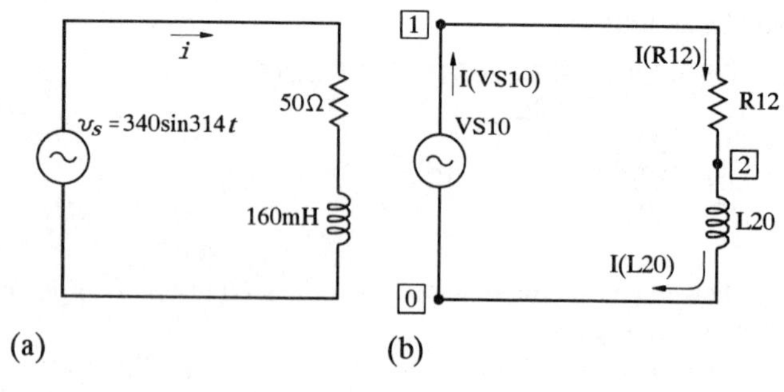

Fig. 1.5 (a) An ac circuit, (b) PSpice circuit configuration.

Solution

This example is described in four steps, configuration, file, run and check.

STEP 1

Create a PSpice circuit configuration, as shown in Fig. 1.5b. Names are given to the elements according to the rules. First letters are V for independent voltage source, R for resistance and L for inductance. The other characters in the names are a matter of choice. Number the two nodes for each circuit element.

STEP 2

The PSpice circuit file, named EX1_3_2.CIR is composed in a way that is equivalent to the circuit diagrams in Fig. 1.5. Account is taken of the independent, sinusoidal-voltage source. The requirement is to determine the instantaneous value of the current. Consequently, a transient analysis (also known as a time-varying analysis) must be performed.

```
»»»»» EX1_3_2.CIR »»»»»»»»»»»»»»»»»»»»»»»»»»»»»»»»»»»»»»»»»»
RESPONSES IN AN AC CIRCUIT WITH RL ELEMENTS; ........... (1)
* PSpice file name EX1_3_2.CIR. Transient analysis. ............... (2)
* Determine the current response in this ac circuit. .................. (3)

* SOURCE ................................................ (4)
VS10    1   0   SIN(0  340V  50Hz  0); ...................... (5)

* CIRCUIT ELEMENTS
R12     1   2   50; ..................................... (6)
L20     2   0   160mH; .................................. (7)

* ANALYSIS
.TRAN   0.1ms   60ms   0    50us; ....................... (8)

* OUTPUT
. PLOT  TRAN   V(1)     I(R12); ......................... (9)
. PROBE;   Graphics postprocessor. ....................... (10)
. END;     End of the file.
»»»»»»»»»»»»»»»»»»»»»»»»»»»»»»»»»»»»»»»»»»»»»»»»»»»»»»»»»»»»»
```

EXPLANATION OF THE CIRCUIT FILE EX1_3_2.CIR

. . . .

Line (1) of circuit file EX1_3_2.CIR

RESPONSES IN AN AC CIRCUIT WITH RL ELEMENTS; (1)

- The first line is the title. It does not have to be there. It could be a comment line with an asterisk (*) in column 1. If the first line is a PSpice statement of any kind, it will be lost to the program and the results will be in error.

. . . .

Line (2) of circuit file EX1_3_2.CIR

* PSpice file name EX1_3_2.CIR. Transient analysis. (2)

- This comment line is useful but not necessary.

. . . .

Line (3) of circuit file EX1_3_2.CIR

* Determine the current response in this ac circuit. (3)

- This is a comment line to remind us about the aim of the problem.

. . . .

Line (5) of circuit file EX1_3_2.CIR

```
VS10 1 0  SIN  (0 340V 50Hz 0);
```
(5)

- This is a PSpice data statement that describes an independent voltage source (V).
- There are three fields in this statement.
- The first field is for the name of the voltage source VS10.
- The second field is for the two node numbers (1, 0) of the nodes to which the source is connected. There is a space or a comma between numbers.
- The third field is for the source waveform (SIN represents a sinusoidal source). See Table 1.2. The parameter values within the parentheses in the statement describe the waveform. Refer to Fig. EX1.6.
  The first parameter value VO = 0 is the initial value of the voltage (the dc offset level).
  The second parameter value VA = 340 is the amplitude in volts.
  The third parameter value FREQ = 50 is the frequency in hertz.
  The fourth parameter TD = 0 is the Time Delay in seconds of the sinewave.
  There are 6 possible parameters. The last two parameter values are omitted here, so the default values of zero are assumed. The extra parameters are the damping factor ALPHA of the sinewave and the phase angle THETA of the phasor†. The first two parameter values are required, but the last four parameter values are optional. Omissions can be in any combination of the last parameter, or the last two parameters, or the last three parameters, or the last four parameters (as shown by nested angle brackets below).

---

† Every phasor has a phase angle with respect to some reference. That is, $V = V\angle\,\theta$.

Fig. 1.6

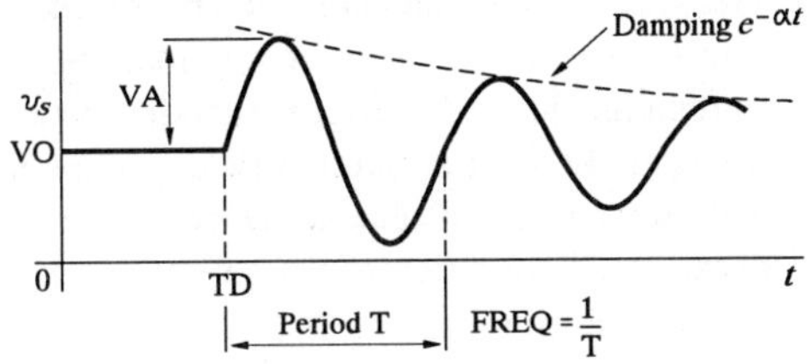

- The general form of this data statement for the voltage source is
  {source} {node_numbers} {source_type (parameters)}

  V*xxx* N+ N- SIN (VO VA <FREQ <TD <ALPHA <THETA>>>>)

  Table 1.6 lists the parameters of the sinewave.

. . . .

Table 1.6 Parameters of an ac-source sinewave

Parameter	Name	Unit	Default value
Initial value	VO	volts	none
Signal amplitude	VA	volts	none
Frequency	FREQ	hertz	1/TSTOP
Time delay	TD	seconds	0
Damping factor	ALPHA	inverse seconds	0
Phase angle	THETA	degrees	0

. . . .

Line (6) of circuit file EX1_3_2.CIR

R12 1 2 50; (6)

- This data statement is for the linear, resistive element in the circuit.

. . . .

Line (7) of circuit file EX1_3_2.CIR

L20 2 0 160mH; (7)

- This data statement is for the linear, inductive element in the circuit.
- The name of the inductor is L20.
- The inductor is connected to nodes (2, 0).

- The value of the inductance is 160 millihenrys.
- The format of the inductor statement is similar to the format of the resistor statement. The difference is that the inductor name must begin with L.
- There are two extra optional fields for the inductor data statement. One field is for modelling a nonlinear inductance. The second extra field is for the value of the initial condition of the current in the inductor. The default value of the initial condition is zero.

. . . .

Line (8) of circuit file EX1_3_2.CIR

```
.TRAN   0.1ms   60ms   0    50us;
```
(8)

- This control statement has five fields and can have more if needed.
- The first field contains the dot command .TRAN. This is the PSpice command for carrying out a transient analysis.
- The second field in this control statement contains 0.1 ms. This is the time interval separating the values of the variables that are printed or plotted. In general, this is called TSTEP, but is not associated with the integration time step in the numerical solution.
- The third field contains 60 ms. This field is called TSTOP and signifies the ending time of the transient analysis.
- The fourth field contains 0. This field is called TSTART. It is optional with a default value equal to 0. From time $t=0$ to TSTART no variable values are written to the output file. That is, printing starts at TSTART.
- The fifth field contains 50μs. This is called TMAX and is optional. The value of TMAX sets the maximum computational step size. There is a default value of TMAX equal to TSTOP/50.
- Other optional fields include an operating point (OP) command. A command UIC (Use Initial Conditions) must be used if initial conditions are specified for the inductor and capacitor data statements. UIC will bypass the calculation of the operating point.
- The general statement for this particular control statement for transient analysis is

  ```
  .TRAN</OP>  TSTEP  TSTOP  <TSTART  <TMAX>>  <UIC>
  ```

  Remember that the angular brackets enclose options that, in some cases, are nested. For example, TMAX must be used with TSTART, but TSTART can be used without TMAX. Both TSTART and TMAX can be omitted, and UIC can be included without them.

. . . .

. . . .

Line (9) of circuit file EX1_3_2.CIR

```
.PLOT  TRAN   V(1)  I(R12);                                    (9)
```

- This line is an output statement to describe how the results of simulation are to be presented. There are three fields.
- The first field contains the dot command .PLOT. PSpice reacts to this command by sending the specific results requested to the output file.
- The second field contains TRAN. This identifies the type of analysis, which must be one of TRAN, DC, or AC.
- The third field contains a list of the output variables to be plotted. Here, we have listed the output voltage and current, V(1) and I(R12) respectively, of the circuit. There is a space between each variable, and eight variables can be printed from one PLOT command. For more than eight variables more .PLOT output statements can be added to the circuit file. V(1) is the voltage variable across nodes 1 and 0. I(R12) is the current variable in the resistive element named R12.
- Other optional fields can be added to set the values of the lower limit and upper limit of any of the dependent variables listed. For the current we could have written I(R12) (-10A, 10A) to limit the plot range between -10 amperes and +10amperes.

. . . .

Line (10) of circuit file EX1_3_2.CIR

```
.PROBE;      Graphics postprocessor.                           (10)
```

- This dot command .PROBE is an order for PSpice to send the computed output data of all node voltages and branch currents to a file called EX1_3_2.DAT. The program PROBE will process the data for graphics display on the screen in an interactive manner with you, the user.
- All data files created by PROBE have the extension .DAT and have the same filename as the circuit file. PROBE data files are written to the circuit file subdirectory.
- This output statement can have another field that lists particular variables. Only those variables named will be made available to PROBE.
- A guide to the use of PROBE is given as we progress through the text.
- Because of the versatility of PROBE, we do not use the .PLOT facility much.

This ends the explanation of circuit file EX1_3_2.CIR

. . . .

STEP 3

For a simulation run, type in at the prompt (usually C:\PSPICE>)

PSPICE EX1_3_2 <ENTER>

Once PSpice has executed the simulation of the circuit, which is described in the file EX1_3_2.CIR, there will be two additional files in the same directory as the circuit file. File EX1_3_2.OUT is the output created by the PLOT statement and file EX1_3_2.DAT is the output created by PROBE. View the output file.

PROBE

In the DOS version of PSpice, immediately after the simulation, PSpice invokes PROBE automatically in version 5.0. In version 6.1 the command line PROBE EX1_3_2 <ENTER> must be typed in. Alternatively, if we wish PROBE to be invoked automatically after a simulation, we can type in SIM EX1_3_2 <ENTER>.

An empty graph appears on the screen. The *x*-axis is scaled with a time interval 0 to 60 ms. You have to choose the variables of the *y*-axis, either on the one graph or on multiple graphs (called **add_aXis** and **Add_plot** in the menus).

The menu on the bottom of the screen can be explored by the user, using the arrow keys or a mouse. You can learn much from a trial interaction. We used a mouse and carried out the following procedure.

- Click on **Add_trace**. Now we have the choice of entering variables or expressions, or hitting the function key **F4** to select variables from a list. We chose the latter.
- Click on V(1) and I(R12). Hit **ENTER** {that is, <↵>}.
  The voltage and current output appear on the graph as sinewaves. The voltage amplitude is 340V. The current amplitude cannot be read because the scale is too large.
- Click on the menu item **Remove_trace**.
- Click on the new menu item 1) V(1), or **Select** and follow instructions.
  Now we have the time variation of the output current plotted alone and to a new scale. The amplitude of the output current is about 4.8A and the current waveform has a delay of 2.5 ms with respect to the voltage waveform. This is made easier by using the **Zoom** facility or the **Cursor**. We could now analyze the circuit with respect to average power and power factor.
- Click on the menu until a graphical output appears like that shown in Fig. 1.7.
- Click on the **Exit** or hit **ESC** until you exit the program.

With keystrokes instead of using the mouse you could have typed in turn

A v(1) i(R12) <ENTER> **R** <SPACE> <ENTER> **E**

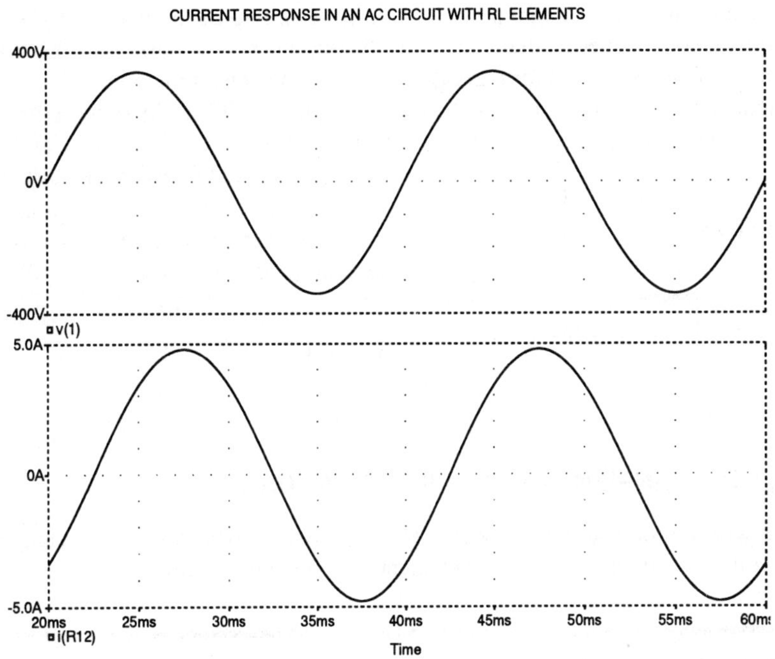

Fig. 1.7

STEP 4 The results of the simulation shown in Fig. 1.7 can be compared with a hand calculation (phase angle 45.1°).

END OF EXAMPLE 1.3.2

Drill Exercise D1.3.3
Consider the circuit diagram in Fig. 1.5a. Change the source frequency to 60Hz. Use PSpice to do a transient analysis over an interval of time equal to 50ms and use PROBE to plot the source voltage and current response. Determine the current amplitude and phase angle.

Drill Exercise D1.3.4

Consider the circuit diagram in Fig. 1.8. **(a)** Use PSpice to obtain plots of the instantaneous values of the load voltage v_s and the source current i. **(b)** If the simulation run is aborted for part (a), look at the contents of the output file. You might add to the circuit a small resistance (about $10^{-8}\,\Omega$) in series with the inductance. View the waveform of the source current over an interval of three periods. **(c)** Add a resistance of 30Ω in series with the inductance. From the source voltage and current waveforms determine the input power factor PF. (Ans: (c) PF = 0.944.)

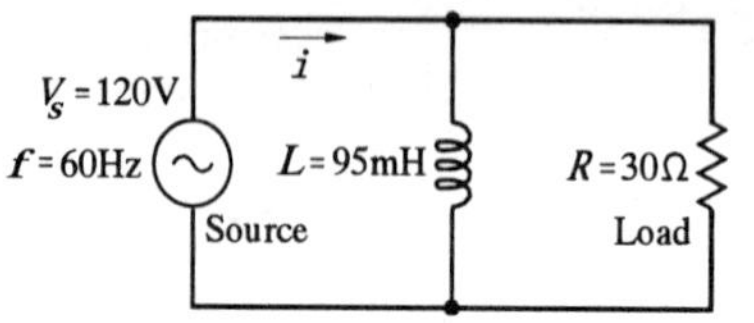

Fig. 1.8

1.3.3 TRANSIENT ANALYSIS (PWL SOURCE)

We will introduce a voltage source that represents a step input and a charging circuit that contains a capacitive element with finite initial conditions.

EXAMPLE 1.3.3

Consider the circuit diagram in Fig. 1.9a. The capacitor is initially charged to a voltage of 2 V, positive at the lower plate. At the time $t=0$, a voltage of 3V is applied across the RC circuit. Use PSpice to determine the time at which **(a)** the voltage across the capacitor is zero and **(b)** the voltage across the capacitor is 2.6V.

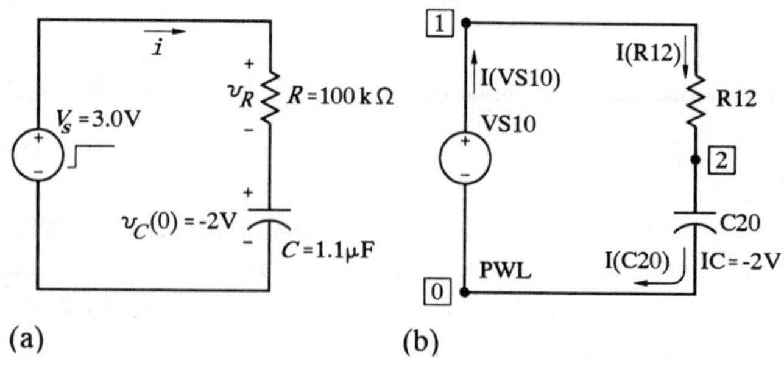

Fig. 1.9

Solution

This PSpice example is described in four steps, creating the PSpice circuit configuration, writing the circuit file, which we will call EX1_3_3.CIR, running the program and viewing the graphic results by the use of PROBE.

STEP 1 This first step involves node numbering and element naming. The outcome is depicted in Fig. 1.9b. The new PSpice element is the capacitor *C* which has the name C20; the name must begin with the letter C (or c). Note that every node is connected to two elements.

STEP 2 The PSpice circuit file, named EX1_3_3.CIR, that is equivalent to the circuit configuration in Fig. 1.9b, is listed below.

```
»»»»» EX1_3_3.CIR »»»»»»»»»»»»»»»»»»»»»»»»»»»»»»»»»»»»»»»
A FIRST-ORDER RC CIRCUIT WITH A STEP INPUT;          Title line.
* SOURCE, a piecewise linear (PWL) input voltage.
VS10 1  0   PWL(0s 0V, 1ns 3V, 0.7s 3V); ............... (1)
* CIRCUIT ELEMENTS
R12  1  2   1E+5;              100-kΩ resistor connected to nodes 1, 2.
C20  2  0   1.1uF   IC=-2V;    A capacitor. .................. (2)
* ANALYSIS, gives time-varying values of the capacitor voltage.
.TRAN   1ms  0.6s  0   0.1ms   UIC; ..................... (3)
* OUTPUT
.PROBE;         Output results to EX1_3_3.DAT for graphics processing.
* CONTROL
.END;           End of the circuit file.
»»»»»»»»»»»»»»»»»»»»»»»»»»»»»»»»»»»»»»»»»»»»»»»»»»»»»»»»»
```

This circuit file has a data statement that includes a new source PWL, a new data statement about the capacitive element *C* and a new optional field UIC in the control statement for analysis.

EXPLANATION OF THE CIRCUIT FILE EX1_3_3.CIR

. . . .

Line (1) of circuit file EX1_3_3.CIR

```
VS10   1   0   PWL(0s 0V, 1ns 3V, 0.7s 3V);
```
(1)

- There are three fields in this data statement. The first two fields are the same as in the two previous examples for voltage sources.
- The third field includes the type of source and its parameter values. See Table 1.2. In this case, we want a step input of magnitude 3 volts. A step input is represented by a piece-wise linear description PWL. Refer to Fig. 1.10. It depicts an approximation of a step-input waveform. PSpice cannot handle instantaneous changes. Accordingly, we allow the rise of voltage from 0V to 3V to take place in a short interval of time, such as one nanosecond. This is very close to an ideal step input of voltage.
  Within the parentheses, the waveform is described by pairs of coordinates (Ti Vi) that are arranged in time sequence (monotonic). The coordinates are time Ti and the value of the voltage Vi at time Ti. Each pair of coordinates describes one of the extremities of a linear portion of the waveform. In our example, there are two linear portions, the rise and the dc level. Accordingly, three sets of coordinates represent the waveform in the piecewise-linear manner.

Coordinates	Time	Voltage
T0 V0	0	0
T1 V1	1ns	3V
T2 V2	0.7s	3V

In the data statement we write

```
PWL(0 0, 1ns 3V, 0.7s 3V)
```

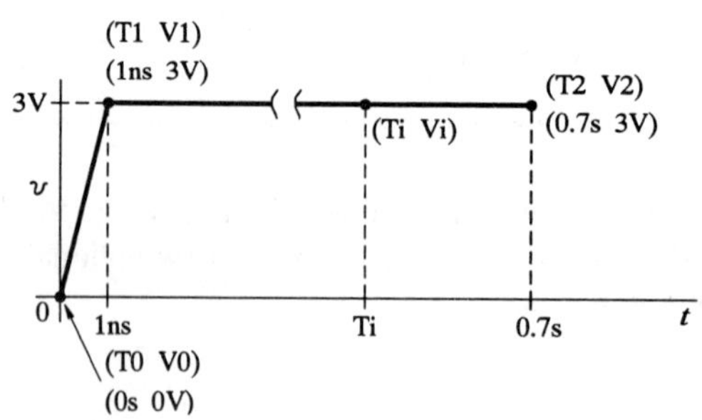

Fig. 1.10 An approximation for a step input of voltage (PWL).

- If the data statement takes more than one line, we continue on the next line, which must start with a plus (+) sign so that PSpice recognizes the continuation of data in the same statement. We have terminated the waveform at 0.7s because this time is greater than five time constants of the circuit ($\tau = RC$). Five time constants can be considered the time necessary for the variables, voltage and current, to reach the steady-state condition.
- The general statement for an independent voltage source that can be described in a piece-wise linear fashion is

  Vxxx N+ N- PWL(T0 V0, ... Ti Vi, ...)

  where Vi is the voltage at the time t=Ti. For more, see Chapter 8.

. . . .

Line (2) of circuit file EX1_3_3.CIR

```
C20   2   0   1.1uF  IC=-2V;          A capacitor.          (2)
```

- This is a data statement with four fields. It gives a PSpice description of the capacitor element in the circuit.
- The first field contains C20 which is the PSpice name given to the capacitor. The PSpice name of the ideal capacitive element must begin with C.
- The second field gives the numbers (2, 0) of the two nodes to which the capacitor is connected.
- The third field contains the value (1.1 μF) of the capacitance. PSpice reads u as 10^{-6}, assumes from the name C20 that the SI unit is the farad, and ignores the letter F after the letter u.
- The fourth field is optional. If it is absent, the voltage across the capacitor at time $t=0$ is taken by PSpice to be zero by default. Otherwise, the field contains the initial condition. In our example, IC= -2 V, the initial voltage across the capacitor is -2 V at node 2 with respect to node 0. It is important to note that, if initial conditions are included in the element data statement, there is an argument UIC† (Use Initial Conditions) in the control statement about the type of analysis.
- The general data statement for a capacitive element is

  Cxxx N+ N- [*value*] <initial_condition>

. . . .

Line (3) of circuit file EX1_3_3.CIR

```
.TRAN   1ms   0.6s   0   0.1ms   UIC;                      (3)
```

† If you use release 6.2 or later, refer to SKIPBP and NOICTRANSLATE in the IC= in the PSpice manual.

- This control statement has six fields, one more than in EXAMPLE 1.3.2. The extra field contains advice about initial conditions.
- The first field contains the dot command .TRAN and signifies that a transient analysis is to be performed by PSpice.
- The second field contains the value 1 ms, which is the time step TSTEP for printing data to the output file.
- The third field contains the value 0.6s, which is the limit of the time TSTOP over which the analysis is performed.
- The fourth field contains the time $t=0$ (TSTART) in the response, at which printing data in the file EX1_3_3.OUT is to begin. This is an option. By default it is TSTART = 0.
- The fifth field contains the value 0.1ms, which is the maximum time step TMAX for computation. This is an option. By default it is TSTOP/50.
- The sixth field in this particular statement contains UIC. With UIC included, the analysis takes into account any initial conditions IC that are written in the data statements for the elements. In the absence of UIC in the control statement all initial conditions for the elements are taken to be zero, no matter what IC values are given to the element data statements.
  In this example UIC means that the initial voltage $v_C(0) = -2$ V (IC = −2V) will be taken into account in the transient analysis.
- The general control statement for this type of analysis is

```
.TRAN  TSTEP  TSTOP  <TSTART  <TMAX>>  <UIC>
```

This ends the explanation of circuit file EX1_3_3.CIR

. . . .

STEP 3

Run the simulation with EX1_3_3.CIR. The results of running PSpice are stored in the file EX1_3_3.DAT. If we do not investigate the circuit responses of the simulation immediately we can view the results later by invoking PROBE.

STEP 4

We type C:\PSPICE> PROBE EX1_3_3 <ENTER>
Using the menus that appear on the screen, we can plot the time variation of the circuit input voltage v(1) and the capacitor voltage v(2). Fig. 1.11 shows the waveforms.

Part (a) of Solution. An inspection of Fig. 1.11 shows that the voltage across the capacitor becomes zero at time $t=56.2$ ms. If you ZOOM the plots over a small, specified region good resolution can be obtained. Otherwise, CURSOR can be used.

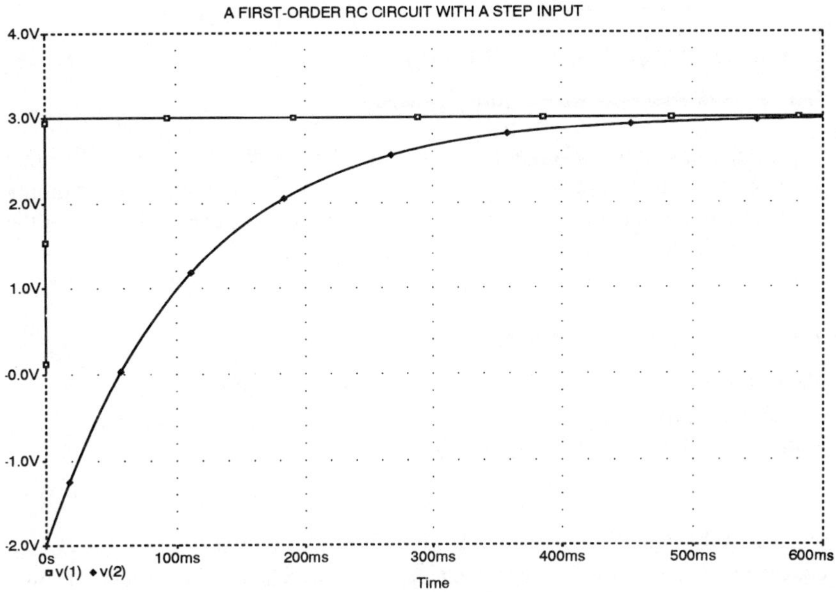

Fig. 1.11

Part (b) of Solution. The voltage across the capacitor is 2.6V at time $t = 278$ ms. These values can be checked by hand calculations from the equation for the capacitor voltage v_C. That is, $v_C = 3 - 5e^{-t/0.11}$ volts.

END OF EXAMPLE 1.3.3

Drill Exercise D1.3.5

Consider the circuit diagram in Fig. 1.9. Change the value of the capacitor to $C = 0.6\,\mu$F. Using PSpice and Probe determine the time at which the voltage across the capacitor is **(a)** zero and **(b)** 2.6V.

Drill Exercise D1.3.6

Consider the PSpice configuration in Fig. 1.12. From a simulation determine the value of the current in the resistor named R202 at time $t = 5.0$ ms, if the capacitor voltage is V(1) = 100V at time $t = 0$. (Ans: 70.9 mA.)

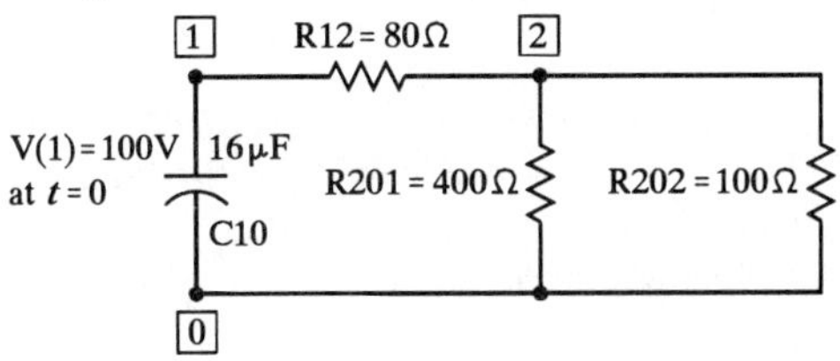

Fig. 1.12

1.4 OPERATIONAL AMPLIFIERS

The operational amplifier (op-amp) can be considered a device or a multiterminal element. It has many applications. In power electronics, the op-amp is commonly used in those circuits that provide signals controlling the on and off-states of the power-semiconductor switches.

1.4.1 THE IDEAL OP-AMP

Figure 1.13a illustrates an operational amplifier by means of the circuit symbol for the device. The characteristic of the ideal op-amp is that the output voltage v_o with respect to the input voltage v_i is given by

$$v_o = A v_i = A(v_2 - v_1) \tag{1.4.1}$$

where A is the amplification factor. Sources of ± 15V are connected to the device because it requires power to provide amplification. The negative input terminal is called the inverting terminal and the positive input terminal is called the noninverting terminal. Also, the ideal input resistance to the op-amp is infinite, so that the input currents i_+ and i_- are zero ideally.

In order to analyse an op-amp circuit we use two principles.

$$i_- = i_+ = 0 \quad \text{and} \quad v_- = v_+ . \tag{1.4.2}$$

In this way we are treating the op-amp as an ideal element and we can find the circuit gain $v_o/(v_2 - v_1)$ without being concerned about the amplifier gain A as long as it is high enough ($A \approx 100{,}000$).

Figure 1.13b shows the equivalent circuit diagram of a model of the op-amp. It incorporates a voltage-controlled voltage source (VCVS) whose dependent voltage is given by

$$e = A(v_2 - v_1) . \tag{1.4.3}$$

For an open-circuit output, eq. (1.4.1) is satisfied by

$$v_o = e = A(v_2 - v_1) . \tag{1.4.4}$$

The op-amp has an input resistance R_i and an output resistance R_o as shown in Fig. 1.13b. For ideal conditions R_i is infinite and R_o is zero. For practical near-ideal conditions R_i has a high value (100kΩ) and R_o has a low value (100Ω). It is not necessary to model the op-amp power supplies.

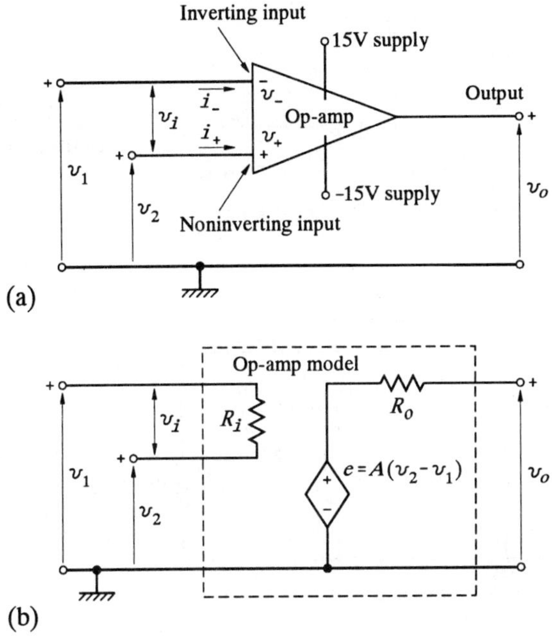

Fig. 1.13 The operational amplifier.
(a) The op-amp, (b) the equivalent circuit diagram.

In the following example we will introduce one of the forms of a data statement that describes the dependent source E. (See Table 1.1 First letter of source names for PSpice.)

EXAMPLE 1.4.1

An operational amplifier is to be used as an inverting voltage-controlled voltage source (VCVS). See Fig. 1.14 on the next page. Analysis using an ideal op-amp provides a circuit gain $V_o/V_i = -R_f/R_1 = 5$. If the op-amp is near-ideal (see Fig. 1.13b) with $R_i = 200\text{k}\Omega$, $R_o = 40\Omega$ and $A = 100{,}000$, determine from a PSpice simulation **(a)** the circuit gain V_o/V_i, **(b)** the input current I_1 and **(c)** the output current I_l. **(d)** Note the time taken for the simulation.

Solution

The four steps of the solution are to create a PSpice circuit configuration, to create a circuit file, to run the simulation and to view the results.

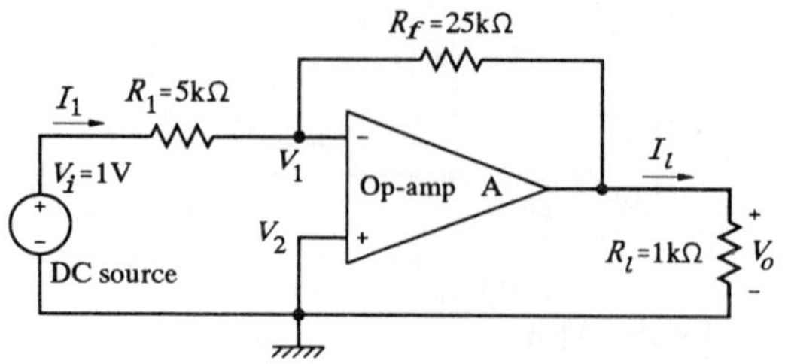

Fig. 1.14
An inverting op-amp.

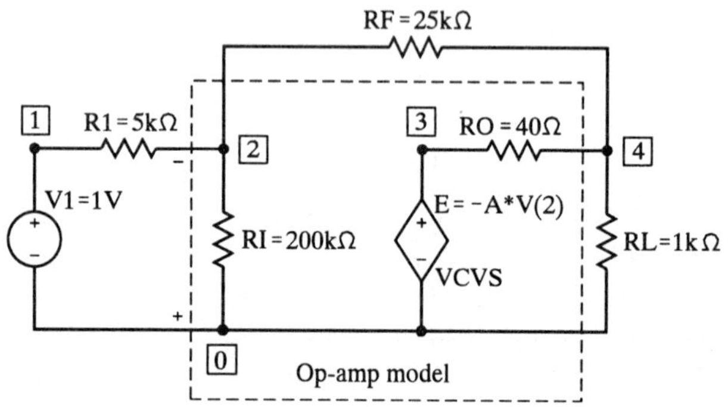

Fig. 1.15
PSpice configuration.

STEP 1 A PSpice configuration can be drawn using the circuit diagrams in Fig. 1.14 and Fig. 1.13b. The op-amp is to be represented by a voltage-controlled voltage source. The configuration is shown in Fig. 1.15. The VCVS voltage is $e=A(V_2-V_1)$ from Fig. 1.14 and this becomes E=-A*V(2) in Fig. 1.15.

STEP 2 We will use the PSpice configuration in Fig. 1.15 to write a circuit file EX1_4_1.CIR that describes the circuit.

»»»»» EX1_4_1.CIR »»»

```
AN INVERTING NEAR-IDEAL OP-AMP
* To determine the circuit gain, input current and output current.

* SOURCES
V1  1  0  DC  1V;        The input voltage source.
E   3  0  2   0  -1E5;   VCVS, a dependent source (op-amp)  .. (1)
```

```
* CIRCUIT ELEMENTS
R1  1  2  5kohms
RI  2  0  200kohms;     Op-amp input resistance.
RF  2  4  25kohms;      Feedback resistance.
RO  3  4  40ohms;       Op-amp output resistance.
RL  4  0  1kohms;       Load resistance.

* By default an operating-point analysis is carried out. The results, node
* voltages and source currents are written to EX1_4_1.OUT.

.END
```

» »

EXPLANATION OF CIRCUIT FILE EX1_4_1.CIR

The new statement in this file is line (1).

. . . .

Line (1) of circuit file EX1_4_1.CIR
E 3 0 2 0 -1E5; VCVS, a dependent source (op-amp) (1)

This statement describes a linear dependent source.
There are four fields.

- The first field contains the name of the voltage source. The name begins with E to signify that the source is a voltage-controlled voltage source.
- The second field contains the numbers (3, 0) of the nodes to which the voltage source is connected.
- The third field contains the numbers (2, 0) of the nodes, which are the reference nodes for the control voltage V(2) of the dependent source.
- The fourth field contains the negative of the voltage gain (-A = -100,000) of the op-amp, so that E = -1E5*V(2).
- The general form of this dependent source is

Source name	Source nodes	Control-voltage nodes	Voltage gain
E*xxx*	N+ N-	CN+ CN-	value

This is one of a number of forms of a dependent source. See the Appendix.

This ends the explanation of circuit file EX1_4_1.CIR.

. . . .

STEP 3 The PSpice simulation can now be run with EX1_4_1.CIR. An operating point analysis (OP) is carried out by default.

STEP 4 We can write down the required answers from an inspection of the contents of the file EX1_4_1.OUT.

(a) $V_l = -4.9997$V. Voltage gain $= V_o/V_i = -4.9997$. This is very close to the ideal value, -5. **(b)** $I_1 = 0.2$mA. This is the same as the ideal value V_1/R_1. **(c)** $I_l = V_l/R_l = 4.9997/10^3 = 4.9997$mA. **(d)** Total job time is 2.58s.

END OF EXAMPLE 1.4.1

Drill Exercise D1.4.1

Consider the circuit diagram in Fig. 1.16. The circuit is designed to have an ideal gain $V_l/V_s = 10$ with $R_1 = 3\text{k}\Omega$ and $R_f = 27\text{k}\Omega$. Do a PSpice simulation using the op-amp parameters input resistance $R_i = 100\text{k}\Omega$, the output resistance $R_o = 50\Omega$ and the op-amp gain A = 100,000. Determine the circuit gain V_l/V_s. (Ans: 9.9989.)

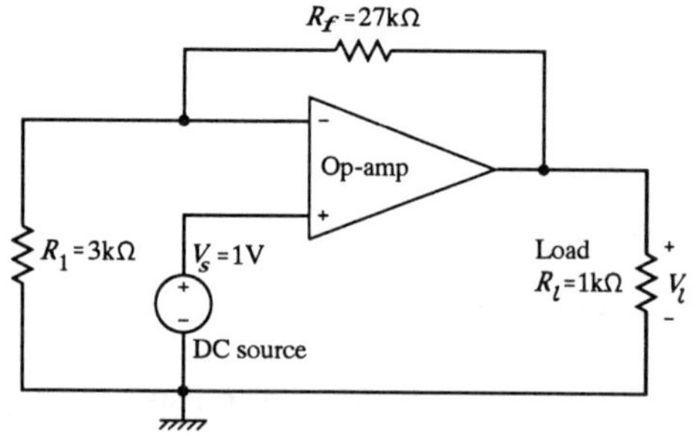

Fig. 1.16

Drill Exercise D1.4.2

The circuit diagram in Fig. 1.17 can be analyzed as though the op-amp were ideal. The result is that the output voltage across the load resistance appears ideally to be 1.5V. If the op-amp has the following characteristics, input resistance $R_i = 200\text{k}\Omega$, the output resistance $R_o = 100\Omega$ and the op-amp voltage gain A = 100,000, do a PSpice simulation to check the value of the load voltage V_l. Use the op-amp model that is depicted in Fig. 1.13. (Ans: 1.5000V.)

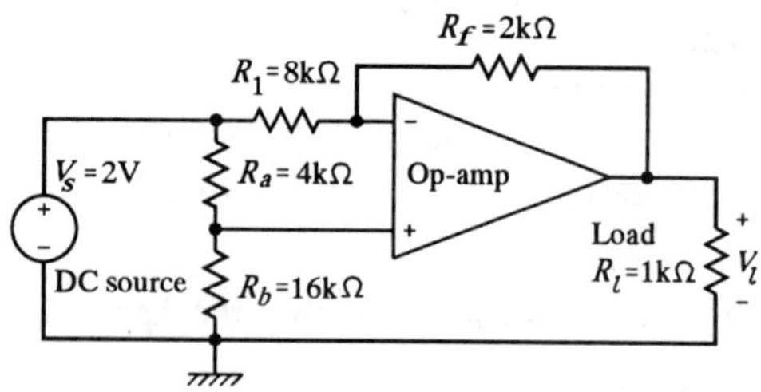

Fig. 1.17

For more Drill Exercises download from the WEB page
http://www.ece.uwaterloo.ca/~pwr-elec

1.4.2 BUILT-IN OP-AMP MACRO MODEL

The second way to simulate circuits with op-amps is to use a particular model of an op-amp that PSpice has in its library file, EVAL.LIB. We will use this model by *calling* its name in the circuit file. This is done in a special way. Figure 1.18 shows a connection diagram of a uA741 op-amp. If we want to include the PSpice model of the op-amp uA741 in any circuit file, we use two statements which have comments below.

```
* PSpice knows the library file from the following statement.
.LIB      EVAL.LIB
* The next statement calls the op-amp into the circuit file.
Xopamp   1    2    3    4    5      uA741 ; X signifies a subcircuit.
* |      |    |    |    |    |        |
* |      |    |    |    |    |      Subcircuit name in library.
* |      |    |    |    |   Output node connection in circuit.
* |      |    |    |   Node connection for negative supply, VEE.
* |      |    |   Node connection for positive supply VCC.
* |      |   Main-circuit node for inverting input connection.
* |     Main-circuit node for noninverting input connection.
*  Device name in main circuit.
*  The node numbers are not important. The correct sequencing of the
*  connections is important.
```

A simulation can be carried out without knowing anything about the internal elements of the op-amp.

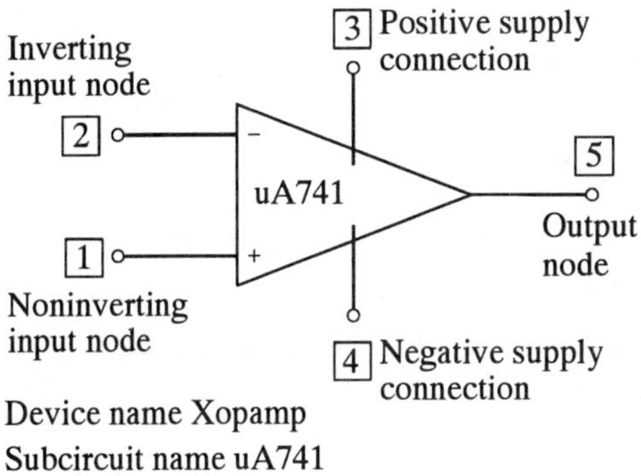

Fig. 1.18 Connection diagram of uA741 in PSpice configuration.

EXAMPLE 1.4.2

Repeat EXAMPLE 1.4.1 using the PSpice op-amp, model uA741. From the simulation determine (**a**) the circuit voltage gain V_l/V_1, (**b**) the input current I_1, (**c**) the output current I_l.

Solution

There are four steps.

STEP 1 From Fig. 1.14 we can draw the PSpice configuration using the subcircuit uA741 to represent the op-amp. This is shown in Fig. 1.19.

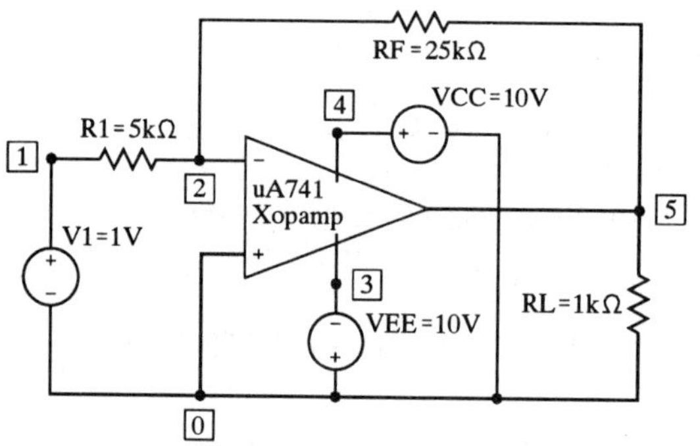

Fig. 1.19
PSpice configuration.

STEP 2 From the PSpice configuration of the circuit in Fig. 1.19 we can write the circuit file.

```
»»»»» EX1_4_2.CIR »»»»»»»»»»»»»»»»»»»»»»»»»»»»»»»»»»»»»»»»»

INVERTING AMPLIFIER WITH OP-AMP SUBCIRCUIT
* To determine the circuit voltage gain.

* Name the library filename which describes the subcircuit uA741.
.LIB   EVAL.LIB
* SOURCES
V1    1   0   DC   1V;     The input-voltage source.
VCC   4   0   DC   10V;    Op-amp positive source.
VEE   0   3   DC   10V;    Op-amp negative source.
```

```
* CIRCUIT ELEMENTS
R1      1   2    5 kohms
RF      2   5    25 kohms ;       The feedback resistance.
RL      5   0    1 kohms ;        The load resistance.

* CIRCUIT DEVICE, the op-amp uA741.
Xop-amp          0   2   4   3   5    uA741
* Connections    |   |   |   |   |    X calls the subcircuit named uA741.
*                |   |   |   |   Output (load).
*                |   |   |   Negative power supply (10V).
*                |   |   Positive power supply (10V).
*                |   Inverting input (-).
*                Noninverting input (+), and ground connection.
* If the subcircuit named uA741 in not found in the circuit file,
* PSpice expects to find it in the file EVAL.LIB

. END
```

»»

STEP 3 The PSpice simulation can now be run with the circuit file EX1_4_2.CIR. No analysis is specified, so PSpice performs a dc operation (OP) analysis and records the node voltages and source currents in the output file EX1_4_2.OUT.

STEP 4 An inspection of the file EX1_4_2.OUT leads to the following. **(a)** $V_l = -4.9977$V. Voltage gain $V_l/V_1 = 4.9977$. This is slightly lower than the ideal value. **(b)** $I_1 = 0.2$mA. This is the same as that found in EX1.4.1. **(c)** $I_l = V_l/R_l = 4.9977/10^3 = 4.9977$mA.

A conclusion can be drawn from these results. The simple model of the op-amp in EX1.4.1 appears adequate to use in power electronics control circuits.

END OF EXAMPLE 1.4.2

Drill Exercises D1.4.3 and D1.4.4
Repeat Drill Exercises D1.4.1 and D1.4.2 using the PSpice op-amp model uA741.

Note that there are extra Drill Exercises on the WEB, related to the uA741 op-amp.

1.4.3 A COMPARATOR

The comparator compares the magnitude of two input voltages. The value of the output voltage of the comparator indicates which input is the greater. A comparator is a modified op-amp that operates in the open-loop mode. Figure 1.20 illustrates the operation of the basic op-amp comparator. As long as the op-amp is not in saturation the output voltage is $v_l = A(v_2 - v_1)$. Because A is large, a small difference in the input voltages $v_2 - v_1$ puts the op-amp into saturation, with the output limited to VCC or $-VEE$.

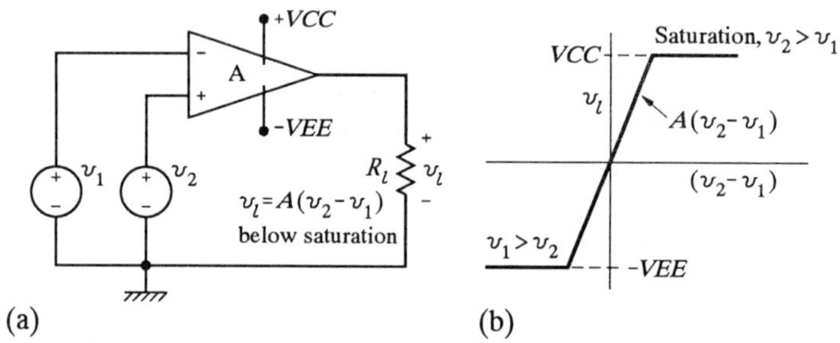

Fig. 1.20 A comparator.
(a) The op-amp circuit, (b) output versus input (compared).

It is the polarity of the saturation levels that is important in the operation of the device. A positive output saturation voltage (approximately VCC) indicates that $v_2 > v_1$. A negative output saturation voltage (approximately VEE) indicates that $v_1 > v_2$. Over a small range of input-voltage differences the output voltage is a linear function of the open-loop gain.

EXAMPLE 1.4.3

Figure 1.21 shows the circuit diagram and ideal waveforms of a zero-crossing detector using an op-amp. From a PSpice simulation, plot traces of the input and output voltages.

Solution

The four steps of the solution are as follows.

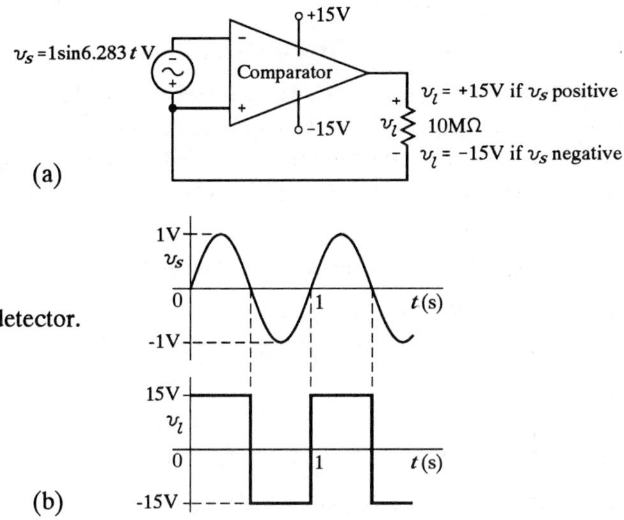

Fig. 1.21
Zero-crossing detector.

STEP 1 A PSpice configuration of nodes and element names can be drawn from an inspection of the circuit diagram in Fig. 1.21a. See Fig. 1.22. A PSpice dependent voltage (VCVS) models the comparator. This requires a special E statement.

STEP 2 From the PSpice configuration in Fig. 1.22, we can write the circuit file named EX1_4_3.CIR. See next page.

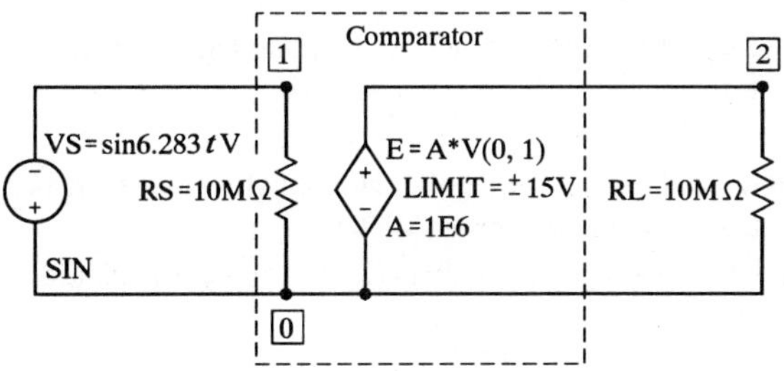

Fig. 1.22 PSpice configuration.

```
»»»»» EX1_4_3.CIR »»»»»»»»»»»»»»»»»»»»»»»»»»»»»»»»»»»»»»»»
COMPARATOR FOR ZERO-CROSSING DETECTION
* To determine the zero crossings of a sinewave voltage. Detection by a
* comparator (at the points of switching from one saturation level to the other).
* See Fig. 1.21b. The comparator is modelled using a dependent source, the
* PSpice option VALUE for E-type sources and LIMIT, a PSpice function.

* SOURCE, of voltage whose zero crossings are to be detected.
VS 1 0 SIN(0 1V 1Hz); Sinewave, amplitude 1V, f=1Hz. (Polarity reversed.)

* SOURCE, VCVS E, the comparator.
E 2 0 VALUE={LIMIT(V(0, 1)*1E6, -15V, 15V)}; .............. (1)

* CIRCUIT ELEMENTS
RS 1 0 10megohms;          Closes a voltage loop (optional).
RL 2 0 10megohms;          Closes a voltage loop.

* ANALYSIS, over an interval covering 1½ cycles.
.TRAN    3ms    1.6s
.PROBE   v(1),  v(2);      Variables of interest to be recorded.
.END
»»»»»»»»»»»»»»»»»»»»»»»»»»»»»»»»»»»»»»»»»»»»»»»»»»»»»»»»»»
```

EXPLANATION OF THE FILE EX1_4_3.CIR

The statement that is new in this circuit file is the model of the comparator.

. . . .

Line (1) of circuit file EX1_4_3.CIR

```
E 2 0 VALUE={LIMIT(V(0, 1)*1E6, -15V, 15V)};                    (1)
```

This statement describes the value of a dependent voltage in terms of an expression. See Fig. 1.20.

There are three fields in this statement.

- The first field contains E. Since the name begins with E, it is classified as a voltage-controlled voltage source (VCVS).
- The second field contains the numbers (2, 0) that are associated with the nodes to which the source E is connected.
- The third field contains VALUE and an expression {...} which includes a PSpice LIMIT function (see Table 2.5 page 61). VALUE= tells PSpice that an expression {...} follows. In this particular case the expression contains a

LIMIT function. Inside the LIMIT function there is the control voltage $v_c \equiv$ v(0, 1) of the VCVS and numbers, divided into three subfields and separated by commas.

- The first subfield contains the product of the control voltage $v_c \equiv$ v(0, 1) of the VCVS and $A = 10^6$, the gain of the source.
- The second subfield is -15V. If the control voltage $v_c < -15$V, then the VCVS is given the value E = -15V.
- The third subfield is +15V. If the control voltage $v_c > 15$V, the VCVS is given the value E = +15V.
  Otherwise, E = v(0, 1)*1E6.
  This statement is described in Fig. 1.4.3b.
  VCC = +15V, VEE = -15V and $v_c = v_2 - v_1$.

- The general form of this particular statement is

  E*xxx* N+ N- VALUE = {LIMIT(y, min, max)}

  y is clipped to the range between *min* and *max*.
  That is, the value of E*xxx* is min if y < min, the value of E*xxx* is max if y > max. Otherwise, E*xxx* = y.
  There is a similar form for the voltage-controlled current source G. See Appendix C.

This ends the explanation of file EX1_4_3.CIR.

. . . .

STEP 3 The PSpice simulation is run with EX1_4_3.CIR.

STEP 4 The results of the simulation can be studied using PROBE. We created traces of the input voltage $v_s \equiv$ v(0, 1) and the output voltage $v_l \equiv$ v(2). Figure 1.23, depicted on the next page, shows the results. The output voltage v_l switches level each time the input voltage crosses 0V. While v_s is positive (V(1) negative), $v_l = +15$V. While v_s is negative, $v_l = -15$V. The output voltage v_l has a rectangular waveform of the same frequency as the input voltage v_s.

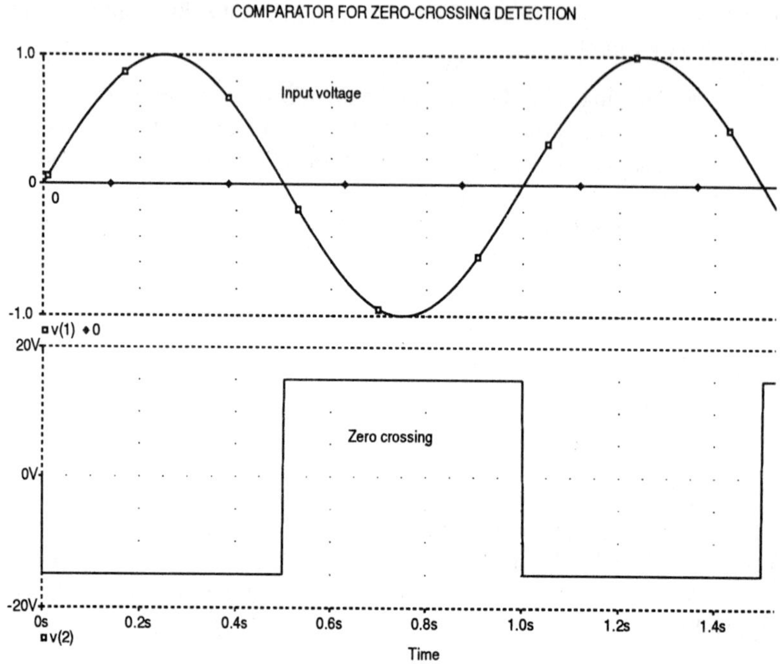

Fig. 1.23

END OF EXAMPLE 1.4.3

Drill Exercise D1.4.5
The circuit diagram in Fig. 1.24a shows a means to detect levels of voltage. The comparator is meant to give a positive 15V output if $v_{in} = V_{ref}$. Otherwise, the output must be -15V. Do a PSpice simulation to check this, using a voltage-controlled voltage source to model the comparator as in EXAMPLE 1.4.3. Let the input voltage v_{in} be a periodic pulse as shown in Fig. 1.24b.

Drill Exercise D1.4.6
Consider the comparator circuit diagram in Fig. 1.25. From a PSpice simulation plot the output voltage v_l and the input voltage $v_{in} - V_{ref}$. Use the comparator model in EXAMPLE 1.4.3.

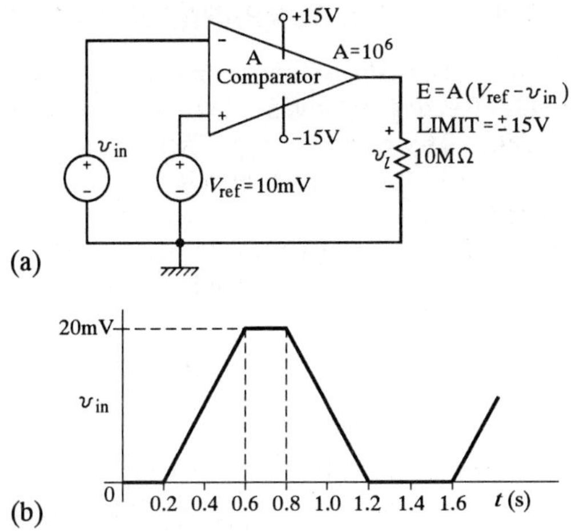

Fig. 1.24

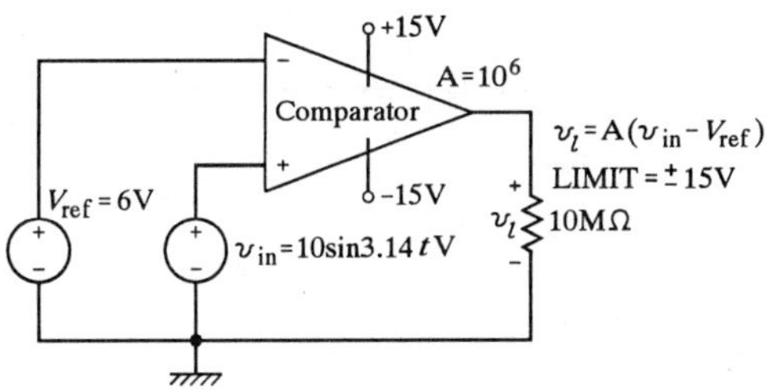

Fig. 1.25

Drill Exercise D1.4.7

Consider the circuit diagram in Fig. 1.26. It incorporates a comparator and a time delay. From a PSpice simulation plot traces of $(v_2 - V_{ref})$ and v_l over an interval 2s. Use the comparator model in EXAMPLE 1.4.3.

Use initial conditions UIC in the .TRAN statement.

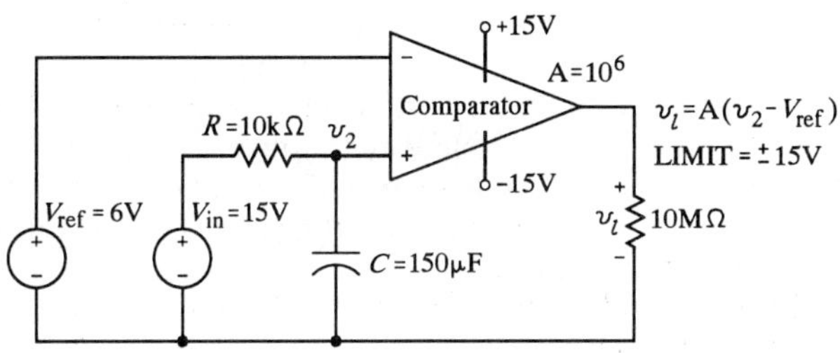

Fig. 1.26

1.4.4 BUILT-IN COMPARATOR MACRO MODEL

The use of an analogue behavioural model (VCVS E) has proved to give near-ideal results of a comparator in Section 1.4.3. Job times were short and few nodes were needed. The comparator has a macromodel built into PSpice's library file EVAL.LIB. Inspect it; it models the comparator LM111, a silicon monolithic integrated circuit suitable for driving lamps, relays and solenoids. It is found in EVAL.LIB in a subcircuit with the name LM111. See Fig. 1.27. Associated with the LM111 there is an example in Section 1.4.4 on the WEBsite.

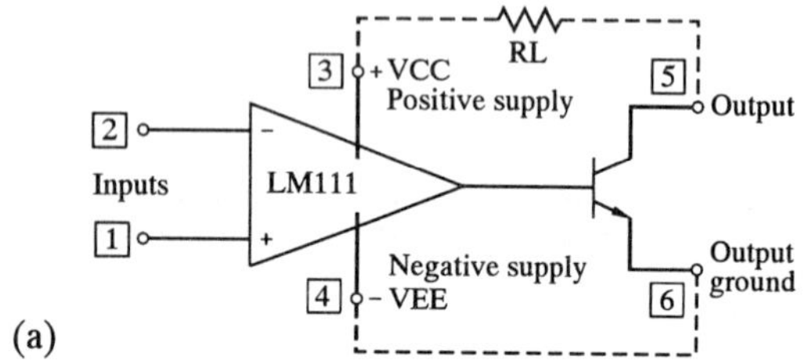

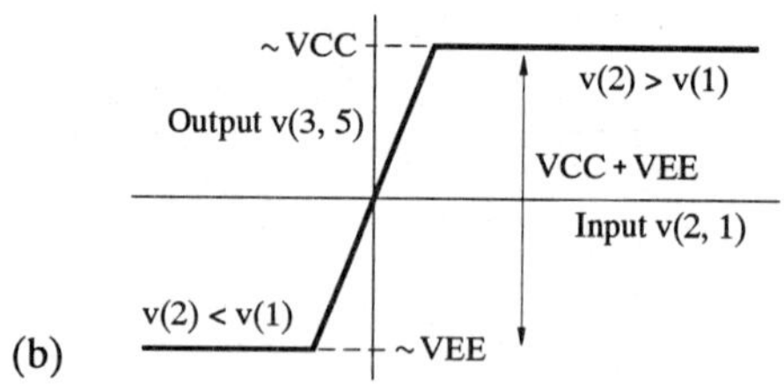

Fig. 1.27 The LM111 comparator. (a) PSpice configuration, (b) characteristics.

```
* Refer PSpice to the appropriate library file.
.LIB     EVAL.LIB

* The next statement calls the comparator model into the circuit file.
Xcomparator  1  2  3  4  5  6  LM111;      X signifies a subcircuit.
*    |       |  |  |  |  |  |    |
*    |       |  |  |  |  |  |    Subcircuit name in library.
*    |       |  |  |  |  |  Output ground.
*    |       |  |  |  |  Node for output.
*    |       |  |  |  Node for negative supply, VEE.
*    |       |  |  Node for positive supply, VCC.
*    |       |  Node for inverting input connection (-).
*    |       Node for noninverting input connection (+).
* Device name in main circuit.
* The node numbers can be arbitrary. The order of the comparator
* connections is important.
```

Chapter 2

PSpice Switches and Power Modulation

2.1 INTRODUCTION

Power to an electric load can be controlled by the use of power electronics switches. The action of switches and the simulated response of voltages and currents in electric circuits that have switches is described in this chapter. To do this we make use of a special PSpice switch that is almost ideal.

2.2 DC-DC MODULATION SIMULATION

Figure 2.1, shown on the next page, illustrates a dc-dc converter which is called a chopper. The action of a chopper is to convert a dc voltage from one level at the input to another level at the output. The source voltage V_s is an unvarying dc voltage. The average voltage $V_{l\,av}$ across the load can be adjusted from zero to V_s by turning the switch *Sw* on and off repeatedly in a defined manner. The most common technique for the switching sequence of a chopper is called pulse-width modulation (PWM). This technique involves switching the fixed dc source on and off so that the output voltage is a train of pulses for which the switching period T is fixed and the on-time t_{ON} is adjustable. The average voltage $V_{l\,av}$ across the load using a PWM chopper is given by

$$V_{l\,av} = V_s\,(t_{ON}/T) = m\,V_s \tag{2.2.1}$$

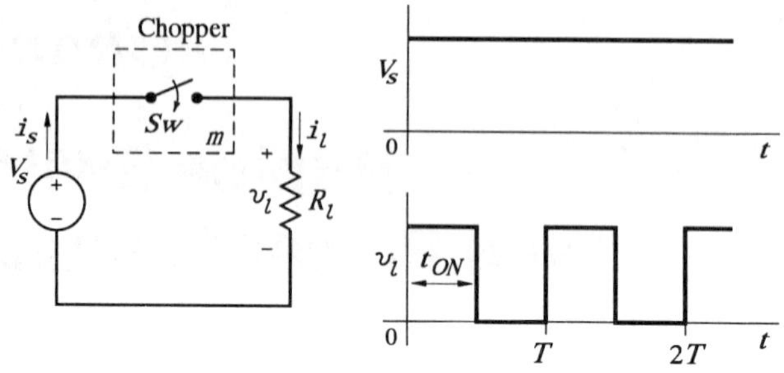

Fig. 2.1 DC chopper with input and output waveforms.

where $m = t_{ON}/T$ and is referred to as the duty cycle. The adjustable voltage across the load provides the means to modulate the average power $V_{l\,\mathrm{rms}}^2/R_l$.

The dc chopper uses a switch *Sw* which is considered to be near ideal. An ideal switch takes no time to switch on or off. While the switch is on it offers no impedance to current, and while it is off it offers an infinite impedance to current. There is no need for protection since the ideal switch has no current or voltage limitations.

2.2.1 CHOPPER SWITCH

Before proceeding with a PSpice representation of the circuit in Fig. 2.1, we must introduce one of the PSpice models of a switch. The switch available in PSpice is a special kind of voltage-controlled resistor that is a four-terminal element. The resistance *R* between the two main nodes N+, N− of the switch depends on the controlling voltage v_c between two input nodes, CN+, CN− as illustrated in Fig. 2.2. The resistance of the switch has a smooth transition between values RON, if the switch is fully on, and ROFF, if the switch is fully off. This switch is not ideal, since RON cannot be zero and ROFF must be finite. In addition, the transition time of the switch between on and off cannot be zero. However, this model can closely approximate an ideal switch by setting RON very small and ROFF very large compared with other resistances in the circuit. To minimize the possibility of numerical problems during the simulation process of analysis, it is recommended that the ratio of ROFF to RON be less than 1E+12. In PSpice and in practice, it must take a finite time for a switch to open, otherwise the voltage induced across a series inductor in the circuit would be infinite.

We will use an independent voltage source v_c in the form of a pulse to control the state of the switch. Besides the dc-voltage, the sinusoidal-voltage and the PWL-voltage sources that were introduced in Section 1.3, PSpice has a number of additional independent voltage sources including a PULSE source. The pulse source produces a periodic waveform with adjustable parameters that include amplitude, delay time, rise and fall time, pulse width and period. The pulse waveform used to control the state of the switch *Sw* is illustrated in Fig. 2.3.

Fig. 2.2
Voltage-controlled switch.

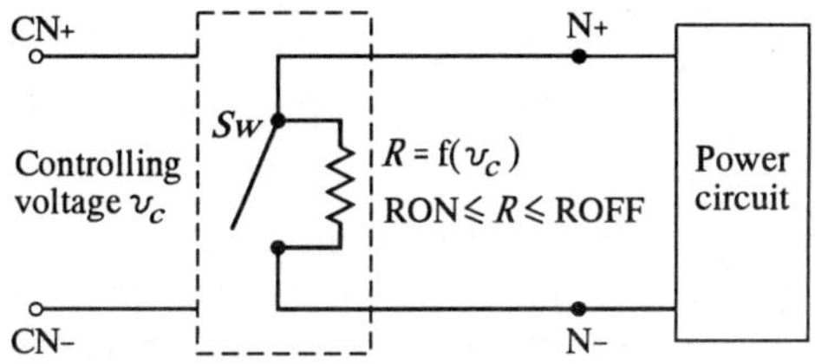

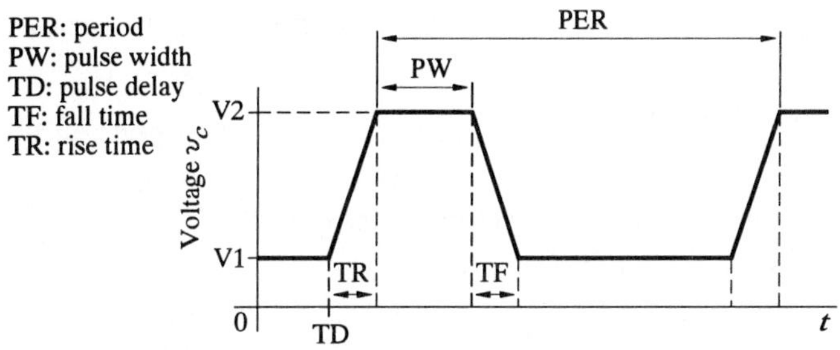

Fig. 2.3 Controlling voltage for the switch (periodic PULSE).

EXAMPLE 2.2.1

A chopper is connected to a direct-voltage source $V_s = 100\text{V}$ in order to modulate power to a load whose resistance is $R_l = 10\Omega$. If the switching frequency of the chopper is $f = 10\,\text{kHz}$ and if the duty cycle is $m = 0.6$, view the load-voltage waveform from a PSpice simulation using PROBE.

Solution

For a PSpice simulation the input *circuit file* must be constructed, the program must be run and the results must be viewed. The solution is described in four steps.

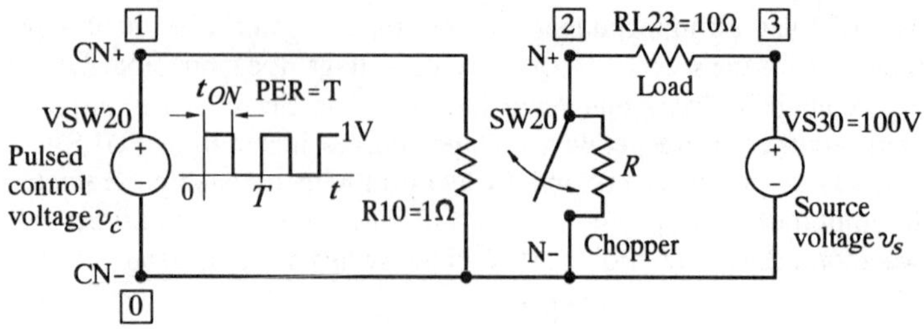

Fig. 2.4 PSpice circuit configuration for a chopper.

STEP 1 Configure the circuit diagram of Fig. 2.1 by assigning unique numbers to each node and unique names to each element in the circuit. The resulting circuit configuration which was prepared for entry into PSpice is shown in Fig. 2.4.

STEP 2 The commented circuit file, named EX2_2_1.CIR, is listed below.

```
»»»»» EX2_2_1.CIR »»»»»»»»»»»»»»»»»»»»»»»»»»»»»»»»»»»»»»»»»»
DC-DC CONVERSION     THE CHOPPER;            Title.
* Determine the load-voltage waveform of the circuit.

* SOURCES
VS30     3  0  DC  100V ;     A 100-V dc source at nodes (3, 0).
VSW20    1  0  PULSE(0V 1V 0 1E-9s 1E-9s 60us 100us); ....... (1)

* CIRCUIT ELEMENTS
RL23  2  3  10;               10-ohm load resistor in the power circuit.
R10   1  0  1;  Nominal 1-ohm resistor across the control-voltage source.
SW20  2  0  1  0  SWITCH;     The power-circuit switch. ......... (2)

* MODEL STATEMENT, in association with the voltage switch SW20.
.MODEL SWITCH VSWITCH(RON=1E-3 ROFF=1E6 VON=1 VOFF=0); . (3)

* ANALYSIS, to view the time-varying load voltage.
.TRAN 1us   300us;            Prints every microsecond (TSTEP).
* Simulation stops at 300μs (TSTOP).

* OUTPUT
.PROBE   v(2, 3);             Views the load voltage V(2, 3) graphically.
.END;                         The end of the circuit file.
»»»»»»»»»»»»»»»»»»»»»»»»»»»»»»»»»»»»»»»»»»»»»»»»»»»»»»»»»»»»
```

EXPLANATION OF THE CIRCUIT FILE EX2_2_1.CIR

There are three new statements in this circuit file. The first is a data statement for a pulsed-voltage source on line (1). The second is a data statement describing the switch element on line (2), and, associated with this is the control statement on line (3) that allows a declaration of the switch's parameter values.

. . . .

Line (1) of circuit file EX2_2_1.CIR

```
VSW20  1  0  PULSE(0V 1V  0 1E-9s 1E-9s 60us 100us);          (1)
```

- This is a PSpice data statement about an independent, periodic, voltage source (starts with V). There are three fields in this statement.
- The first field contains the name VSW20 of a voltage source. This source provides a voltage that controls the state of the switch SW20.
- The second field contains the numbers given to the nodes to which the control voltage source is connected. In this case N+ is 1 and N− is 0.
- The third field has two parts, the name of the source type (see Table 1.2) and the values of the parameters of the pulse-source waveform (see Fig. 2.3). The PSpice name given to this kind of voltage source is PULSE. There are seven parameters, five of which are optional. Table 2.1 lists seven parameters in the order that their values must appear in the data statement. TSTEP is the print interval and TSTOP is the time interval of the transient analysis.
  The values given here are V1 = 0, V2 = 1V, TD = 0, TR = 1ns, TF = 1ns, PW = 60μs and PER = 100μs. The value PER = 100μs satisfies the stipulation that the chopper switching frequency is f = 10 kHz. The value PW = 60μs provides a duty cycle m = 0.6.

Table 2.1 PSpice parameters of a pulse source, PULSE

Parameter	Meaning	Units	Default value
V1	Initial voltage	volts	0 (aborts)
V2	Peak voltage	volts	0 (aborts)
TD	Initial delay time	seconds	0
TR	Rise time	seconds	TSTEP
TF	Fall time	seconds	TSTEP
PW	Pulse width	seconds	TSTOP
PER	Pulse period	seconds	TSTOP

- The general data statement for a periodic pulsed voltage source is

```
Vxxx N+  N-  PULSE (V1 V2 <TD <TR <TF <PW <PER>>>>>)
```

The options for omission in the angle brackets are nested.

. . . .

Line (2) of circuit file EX2_2_1.CIR

```
SW20 2 0 1 0   SWITCH; The power-circuit switch.                (2)
```

- This is a data statement with four fields and is also known as a device statement. This describes the switch partially.
- The first field contains the name of an element SW20 whose first letter S signifies that it is a voltage-controlled switch†.
- The second field contains the numbers of the nodes (N+ is 2 and N− is 0) to which the switch element is connected.
- The third field contains the numbers of the nodes (CN+ is 1 and CN− is 0) to which the controlling voltage VSW20 is connected.
- The fourth field contains the model name of the voltage-controlled switch. This model name, which is SWITCH here, must appear in a control statement. This links the data statement for SW20 with a model statement that gives the parameter values of the same switch. The model name is your choice.
- The general data statement for a voltage-controlled switch is

```
Sxxx  N+  N-  CN+  CN-  Model_name
```

. . . .

Line (3) of circuit file EX2_2_1.CIR

```
.MODEL SWITCH VSWITCH(RON=1E-3 ROFF=1E6 VON=1 VOFF=0);   (3)
```

- This is a control statement that is also called a model statement. Its purpose is to define the parameter values of a device. PSpice uses these parameter values in the equations to be solved. The model statement is a companion statement for a data statement that requires device definition. This model statement has three fields and is associated with the switch.
- The first field contains the dot command .MODEL to signify that device or element parameters are to be declared in this statement.
- The second field contains SWITCH which is the model name of the device given in the last field of the device's data statement. The link between the model statement and the relevant data statement has been made.

---

† A PSpice name with a first letter W signifies a current-controlled switch.

Table 2.2 Names given to PSpice model types in .MODEL

Model-type name	Device or element	Element name
CAP	Capacitor	Cxxx
IND	Inductor	Lxxx
RES	Resistor	Rxxx
D	Diode	Dxxx
NPN	NPN BJT	Qxxx
PNP	PNP BJT	Qxxx
NMOS	N-channel MOSFET	Mxxx
PMOS	P-channel MOSFET	Mxxx
CORE	Transformer	Kxxx
ISWITCH	Current-controlled switch	Wxxx
VSWITCH	Voltage-controlled switch	Sxxx

- The third field contains two parts, the model type and the model parameter values. In this case VSWITCH indicates that the model type is a voltage-controlled switch. Some other model types of devices are listed in Table 2.2 .

  The parameter values of the switch are given in parentheses after the model type, each parameter being separated by a space and/or a comma.

  RON = 1E−3 is the value of the switch's resistance R in the on-state.

  ROFF = 1E6 is the value of the switch's resistance R in the off-state.

  VON = 1 is the control-voltage value at which the switch turns on if the control voltage v_c is rising. For $v_c \geq$ VON, R = RON. In this example, the threshold value of the control voltage for the on-state is 1 volt.

  VOFF = 0 is the control-voltage value at which the switch turns off if the control voltage is falling. For $v_c \leq$ VOFF, R = ROFF. Here, the threshold value of the control voltage for the off-state is 0 volt.

  Table 2.3, shown on the next page, lists the default values of the switch's parameters.

  The functional relation between the resistance R and the control voltage v_c is illustrated in Fig. 2.5, shown on the next page. For some protection against simulation failure it is advised that the conductance of any branch (including the switch) has a minimum value that is called GMIN. The value of GMIN can be set in an OPTIONS statement. Otherwise, the default value is 1.0pS. This suggests that the maximum value of ROFF must not be greater than 1/GMIN.

Table 2.3 Model parameters of the voltage-controlled switch (Model type VSWITCH)

Model	Description	Units	Default value
RON	Resistance of the closed switch	ohms	1.0Ω
ROFF	Resistance of the open switch	ohms	1.0MΩ
VON	Minimum control voltage to close the switch	volts	1.0V
VOFF	Maximum control voltage to open the switch	volts	0V

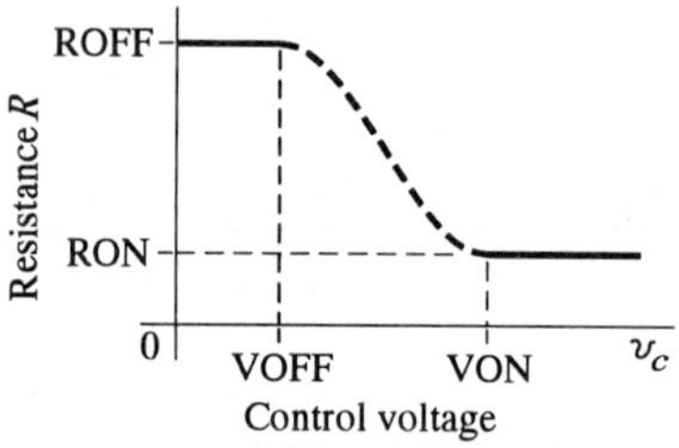

Fig. 2.5
Characteristics of a voltage-controlled switch.

GMIN can be used by PSpice without our knowledge. If we had not put a nominal resistance of 1.0Ω across the control voltage VSW20, the voltage source would have been left hanging. This is against the rules. However, PSpice gives protection against an incomplete voltage loop. A conductance GMIN is automatically connected across the source for us.
The rate of change of the control voltage v_c must be limited otherwise the simulation iteration step will be very short and the time for solution will be long. Capacitors are used to slow down the change of voltage. For this example with a control voltage rise time of 1ns the iteration time step reduces to 10^{-12} s.

- The general format of the .MODEL statement is

```
. MODEL  Switch_name  VSWITCH < (parameter = value ...) >
```

This ends the explanation of circuit file EX2_2_1.CIR

. . . .

STEP 3 The next step is to run the program. There will be two complete output files, EX2_2_1.OUT and EX2_2_1.DAT.

STEP 4 The results of the simulation of the chopper can be investigated using PROBE. The example calls for the waveform of the load voltage V(3, 2). This is shown in Fig. 2.6. The waveform is as we would expect it to be. The chopped waveform has a frequency of 10 kHz, a duty cycle of 0.6 and a magnitude of 100V. Also, Fig. 2.6 shows the variation of the switch resistance *R* by using the expression V(2)/I(SW20).

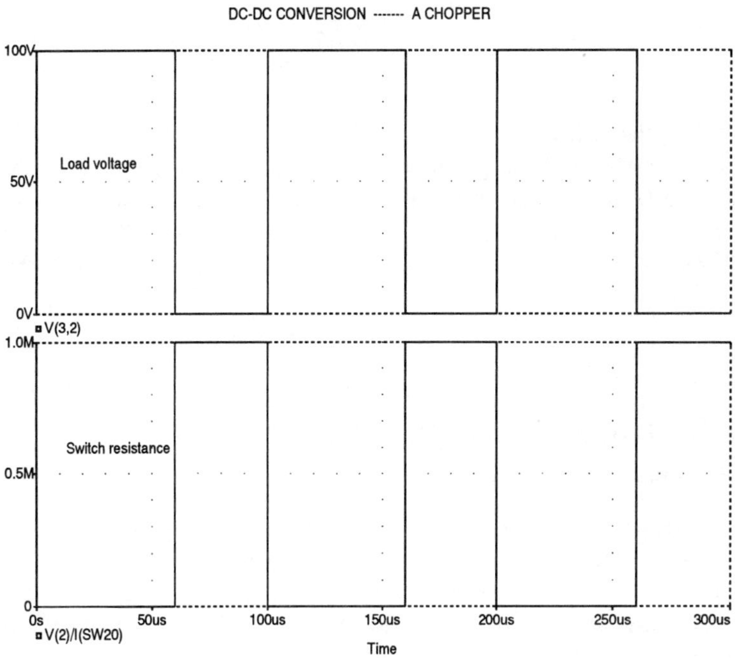

Fig. 2.6 Load voltage and switch resistance waveforms.

END OF EXAMPLE 2.2.1

Drill Exercise D2.2.1

The circuit diagram in Fig. 2.7 represents a chopper circuit. Switch $Sw1$ is the chopper. Switch $Sw2$ is switched on to provide a path for load current while switch $Sw1$ is in the off-state. The chopper switching frequency is $f = 1\text{kHz}$ and the duty cycle is $m = 0.5$ (for switch $Sw1$, time on t_{ON} = time off t_{OFF} in each period $T = 1\text{ms}$). Use PSpice to simulate the circuit. Plot the load current i_l as a function of time over an interval of 3 ms. What is the peak value of the load current?

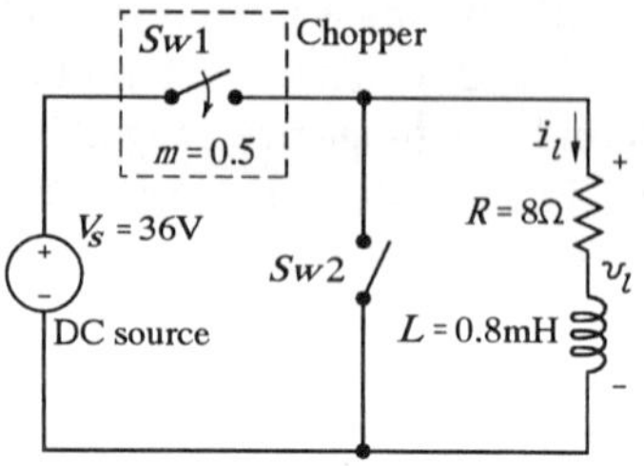

Fig. 2.7

Drill Exercise D2.2.2

Consider the PSpice chopper configuration in Fig. 2.4. Change the source voltage to $V_s = 150\text{V}$, the load resistance to $R_l = 20\Omega$, the switching frequency to $f = 400\text{Hz}$ and the duty cycle to $m = 0.5$. From a PSpice simulation, use PROBE to view the instantaneous power absorbed by the load. Estimate the average power dissipated in the load. (Ans: 562.5W).

2.3 DC-AC MODULATION SIMULATION

The configuration of switches that converts power from a dc supply to provide ac power at the load is called an inverter. An elementary single-phase inverter, comprising a bridge arrangement of four switches, is illustrated in Fig. 2.8a. With this configuration we have the opportunity of controlling the frequency and the magnitude of the voltage at the output of the inverter.

Figure 2.8b shows waveforms of the direct voltage source V_s and the ac voltage output v_l of the inverter. In order to obtain the rectangular waveform of the voltage v_l, the switches Swa and Swb are on for an interval T' seconds, switches Swc and Swd are on for an interval T' seconds and the frequency of the switching is $1/T$ hertz. If one pair of switches is on the other pair must be off. Adjustment of the variable T' controls the rms value of the output voltage. Adjustment of the variable T controls the value of the ac frequency.

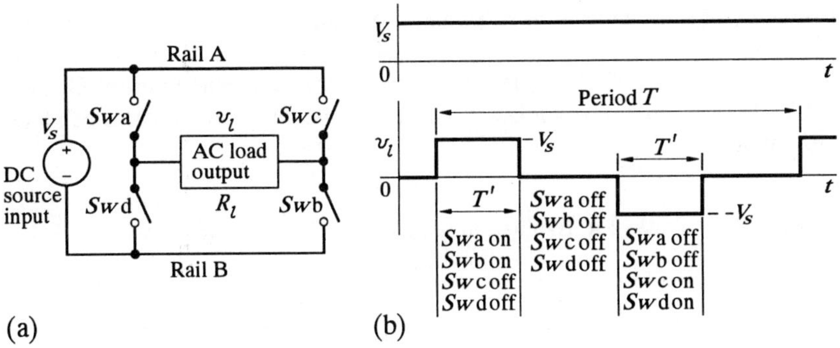

Fig. 2.8 DC-AC power modulation.
(a) Circuit diagram of a single-phase bridge inverter,
(b) input and output waveforms.

2.3.1 INVERTER SWITCHES and SUBCIRCUITS

We have enough background of PSpice from this chapter and Chapter 1 to configure the circuit diagram of the inverter in Fig. 2.8a. We know how to represent a dc voltage source, a voltage-controlled switch, a load resistor and a controlling voltage source for a switch. PSpice dot commands .LIB, .SUBCKT, .PARAM and .STEP will be described in this section.

.LIB is a dot command that calls up a library† in PSpice. In the student version of PSpice there is a library file named EVAL.LIB. In that file there are some model names of devices with their characteristics in terms of parameter values. You can create your own library†† files with extensions .LIB. We have created a library file POWER.LIB, in which we have listed subcircuits that are used repeatedly throughout the text. This saves time and space when creating a new circuit file that includes a repeated subcircuit. The devices that are characterized in the file EVAL.LIB are listed in Table 2.4 on the next page. You can use an editor to browse in your version of EVAL.LIB.

.SUBCKT is the dot command that begins a PSpice subcircuit description. A subcircuit is any *subnetwork* of connected elements made available so that repeated networks need not be tediously entered multiple times. For example, the

† A library is like a *toolbox* of common parts that can be called and used in any circuit file.

†† The master library defaults to NOM.LIB that can be used to reference individual library files.

same type of switch that is used a number of times can be referenced once as a subcircuit and called whenever needed by using an abbreviation. If the subcircuit is used in a number of circuit files, it is usual to write it in some library file (such as EVAL.LIB). In this way it can be called into the main circuit file much like a subroutine.

Table 2.4 Evaluation library (EVAL.LIB) parts list
(A selection from Version 6.1.)

Model name	Description	Model name	Description
SCR	Thyristor subcircuit	A4N25	Opto-isolator subcircuit
2N1595	Thyristor subcircuit	MBD101	Not-carrier diode
TRIAC	Triac subcircuit	MV2201	Varactor diode
2N5444	Triac subcircuit	Q2N2222	NPN BJT model
D1N4148	Diode (switching)	Q2N2907A	PNP BJT model
D1N750	Zener diode	Q2N3904	NPN BJT model
LF411	Op-amp subcircuit	Q2N3906	PNP BJT model
LM111	Comparator subcircuit	IRF150	N power MOSFET model
LM324	Op-amp subcircuit	IRF9140	P power MOSFET model
uA741	Op-amp subcircuit		

Plus models of four nonlinear magnetic cores.

.PARAM is the dot command that is associated with a statement that defines a variable or an element as a parameter.

.STEP is the dot command that is used in a statement to specify different stepped values of an element named in the .PARAM statement. The range and increment of the element value is called the parametric *sweep*. PSpice will run an ac or transient analysis for each value of the parameter given by the sweep.

We have created a library file that we have named POWER.LIB. Into this file we have written a model statement that defines the parameters of a switch. We have also written a subcircuit description of driver circuits that control the on and off states of switches in a bridge. The format is the same for a library file as it is for a circuit file. The limitations are that a library file may contain only .MODEL statements, subcircuits and additional .LIB statements. Browse through the actual file, POWER.LIB. It can be downloaded from the WEBsite. A section from POWER.LIB follows.

```
****************** SWITCH ******************
* A GENERIC MODEL OF A VOLTAGE-CONTROLLED SWITCH.
* On-resistance is 0.001ohm and the off-resistance is 1.0Megaohm.
. MODEL SWITCH
+ VSWITCH(RON=1E-3 ROFF=1E6 VON=1.0 VOFF=0.0); .......... (1)
*********************************************

********* SUBCIRCUIT NAMED SW_PWM *********
* BRIDGE AND DRIVER FOR VOLTAGE-CONTROLLED SWITCHES.
* The following subcircuit description shows the generation of the gate control
* (on and off) for generic voltage-controlled switches (model type VSWITCH) * in
a full-bridge configuration.
* Nodes 101 and 102 are the gate-drive output terminals for the switch pairs * and
node 100 is the reference node.
. SUBCKT SW_PWM 101 102 100; .......................... (2)
* This is a description of a switch driver circuit that changes the on/off states.
* The parameters must be defined in the main circuit file in this case.
* The global parameters are PERIOD and PULSE_WIDTH
VSWab 101 100 PULSE(0 1 {0.25*PERIOD-PULSE_WIDTH/2}
+ 1ns 1ns {PULSE_WIDTH} {PERIOD}); ...................... (3)
VSWcd 102 100 PULSE(0 1 {0.75*PERIOD-PULSE_WIDTH/2}
+ 1ns 1ns {PULSE_WIDTH} {PERIOD}); ...................... (4)
Rab 101 100 1ohm
Rcd 102 100 1ohm
* The call for the subcircuit from the circuit file has the form
* XSW 1 2 0 SW_PWM

.ENDS SW_PWM; ........................................ (5)
*********************************************
```

EXPLANATION OF THE LIBRARY FILE POWER.LIB

There are five lines in this library file that are numbered. Line (1) is a model statement. Line (2) is the first line of a subcircuit description of the inverter gate drives. Lines (3) and (4) are statements about the voltage sources of the gate drives. Unlike other sources encountered earlier in the text, these sources include general expressions in terms of parameters in place of fixed numerical values. This gives the latitude to adjust values in the expressions as required. Line (5) indicates to PSpice the end of this particular subcircuit description.

. . . .

Line (1) of library file POWER.LIB

```
. MODEL SWITCH
+ VSWITCH(RON=1E-3 ROFF=1E6 VON=1.0 VOFF=0.0);
```
(1)

- This is a control statement and a model statement. It provides the model name SWITCH of the device. This same model name can also appear in the last field of the device's data statement in a circuit file. If PSpice comes across SWITCH while reading a data statement in the circuit file, it can be directed to a library file to read the parameter values of the device. We will see how this is done in the following example.
- The device type is named VSWITCH in the model statement. It is a voltage-controlled switch that is modelled in Fig. 2.9.
- Whenever this switch is used in a circuit file, the switch will offer an on-state resistance $R = 1.0\text{m}\Omega$ if the control voltage has a value $v_c \geq 1.0\text{V}$. The switch will offer an off-state resistance $R = 1.0\text{M}\Omega$ if the control voltage has a value $v_c \leq 0.0\text{V}$.

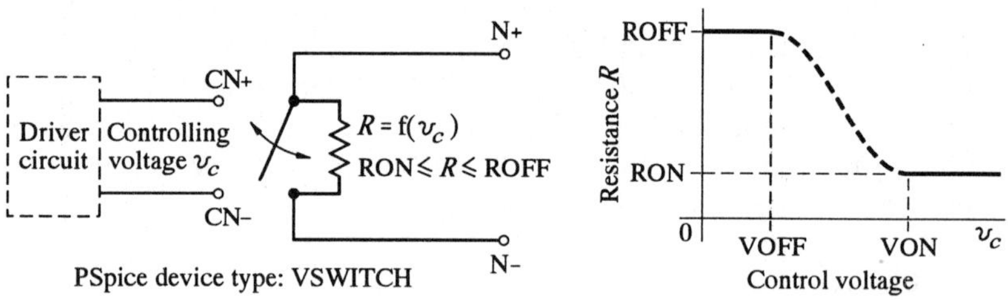

Fig. 2.9 A switch model.

. . . .

Line (2) of library file POWER.LIB

```
. SUBCKT SW_PWM 101 102 100;
```
(2)

- .SUBCKT is the dot command of a control statement that precedes a description of a part of a circuit called a subcircuit. This description saves repetition if this part of a circuit is used more than once.
- This control statement has three fields.
- The first field .SUBCKT gives the information that a subcircuit description in PSpice terms is to follow. The PSpice configuration of this subcircuit is drawn in Fig. 2.10.

- The second field SW_PWM is the name given to this particular subcircuit. This name will appear in the circuit file in a statement that calls for this subcircuit. This particular name is associated with SWitches that give rise to Pulse-Width Modulation of the output voltage of the circuit. The name is the link between the circuit file and the library-file subcircuit.
- The third field contains all the external node names. These nodes are the connections to the main circuit described in the circuit file. Here the three nodes are named 101, 102 and 100, and they are shown in Fig. 2.10. These node numbers only relate to the subcircuit and have nothing to do with any similar numbers on the main circuit diagram. The global node 0 is an exception, but should be avoided.

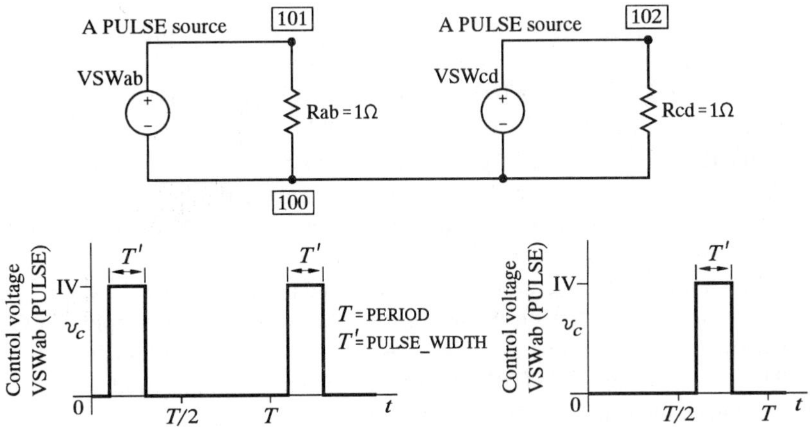

Fig. 2.10 Gate-driver subcircuit for the inverter switches.

- The general form of this control statement is

```
.SUBCKT Subcircuit_name N1 N2 N3 ...
```

. . . .

Line (3) of library file POWER.LIB

```
VSWab 101 100 PULSE(0 1 {0.25*PERIOD-PULSE_WIDTH/2}
+ 1ns 1ns {PULSE_WIDTH} {PERIOD});
```
(3)

- This is a data statement about an independent voltage source (it begins with V). The source produces periodic pulses (it contains PULSE). We have already used a similar source in EXAMPLE 2.2.1. The circuit diagram containing the source element VSWab is shown in Fig. 2.10. The intention is to use this voltage source as the control voltage v_c for the switches *Sw*a and *Sw*b of the inverter circuit diagram in Fig. 2.8a.
- There are three fields in this data statement.
- Field one contains the name VSWab of the voltage source.
- Field two contains the numbers (101, 100) allocated to the source nodes.
- Field three names the type of source (PULSE) and describes the ordered parameters of the source waveform (Table 2.1 and Fig. 2.3). Here, we use expressions and parameters in order to generalize.

  V1 = 0 V2 = 1V
  TD = {0.25*PERIOD-PULSE_WIDTH/2} ≡ $T/4 - T'/2$
  TR = 1ns TF = 1ns
  PW = {PULSE_WIDTH} ≡ T' PER = {PERIOD} ≡ T

  Note: PSpice parameters are enclosed within braces { } .
- In the third field we made the use of general expressions.
  An expression is recognized by PSpice if it is enclosed in braces { }.
  An expression can contain mathematical functions (see Table 2.5), circuit variables (currents and voltages), parameters (defined in the circuit or subcircuit file), the operators +, −, * and /, and parentheses ( ).
  Each expression must be limited to one line. This does not apply to the statement. PSpice recognizes TIME as the time variable in seconds.
  The dot command .FUNCTION can help to keep within this limit.
- We repeat the general format of a periodic, pulse voltage

```
Vxxx N+ N- PULSE (V1 V2 < TD < TR < TF < PW < PER >>>>> )
```

. . . .

Line (4) of library file POWER.LIB

```
VSWcd 102 100 PULSE(0 1 {0.75*PERIOD-PULSE_WIDTH/2}
+ 1ns 1ns {PULSE_WIDTH} {PERIOD} );                                    (4)
```

- This data statement is similar to that written on line (3). This statement pertains to the voltage source named VSWcd in Fig. 2.10 and represents the control voltage v_c that triggers the switches *Sw*c and *Sw*d (Fig. 2.8a). The parameter expressions give rise to pulses that coincide with the negative voltage output v_l of the inverter (Fig. 2.8b).

. . . .

Table 2.5 Mathematical functions and operations in PSpice

Function	Meaning
ABS (y)	$\lvert y \rvert$ (absolute value)
SQRT (y)	$\sqrt{\ } \equiv (y)^{1/2}$ (square root)
EXP (y)	e^y
LOG (y)	$\ln y$ (log base e)
LOG10 (y)	$\log y$ (log base 10)
PWR (y, x)	$\lvert y \rvert^x$
PWRS (y, x)	PWR (y, x) if $y>0$, -PWR (y, x) if $y<0$
SIN (y)	sin (y) (y in radians)
COS (y)	cos (y) (y in radians)
TAN (y)	tan (y) (y in radians)
ATAN (y)	$\tan^{-1}(y)$, result in radians
ARCTAN (y)	$\tan^{-1}(y)$, result in radians
TABLE ($y, y1, x1, y2, x2, ...$)	Piecewise linear function of y
LIMIT (y, min, max)	y clipped to range [min, max]

. . . .

Line (5) of library file POWER.LIB

```
.ENDS  SW_PWM;                                                  (5)
```

- This is a control statement with two fields.
- The first field contains the dot command .ENDS. The dot command signifies to PSpice that this is the last line of the subcircuit description.
- The second field contains the name of the subcircuit. This field is optional. Its use helps the reader if there are nested subcircuits; that is, if one or more subcircuits are described or called within the main subcircuit. The name in this control statement tells which subcircuit is being ended.

This is the end of the subcircuit description.

. . . .

EXAMPLE 2.3.1

A single-phase inverter is depicted in the circuit diagram of Fig. 2.8. Consider the dc source voltage to have a value $V_s = 100$V and let the load resistance R_l be an adjustable parameter whose value may be 2Ω, 5Ω or 10Ω. The inverter operates at a frequency $f = 100$Hz with an output whose pulse width is $T' = 3$ms. View the load-current waveforms from a PSpice simulation using PROBE.

Solution

The circuit diagram in Fig. 2.8 includes four switches. If we put one .MODEL statement of a switch into a library file, this statement need be written only once instead of in every subsequent circuit file where the switch appears. Every switch must have a driver circuit to turn the switch on and off. Figure 2.4 depicts a PSpice circuit configuration as a voltage-controlled switch and its driver circuit. Both are described in the circuit file EX2_2_1.CIR. The driver circuits for this example can be written as a subcircuit description in a library file and can be called by any circuit file to be included in the PSpice simulation. This cuts down on space and repetition.

The solution of this example is described in four steps.

STEP 1

The first step is to transform the conventional circuit diagram and the switches' driver circuits (Fig. 2.8a and Fig. 2.9) to a PSpice configuration. This result is shown in Fig. 2.11. The driver circuit is shown as a block diagram of a subcircuit with external nodes that are numbered 4, 5 and 0. The subcircuit is called XDRVR and it is to produce control voltages, as shown, to turn the switches on and off. We have already defined the subcircuit according to the PSpice configuration in Fig. 2.10; the description is in the file POWER.LIB.

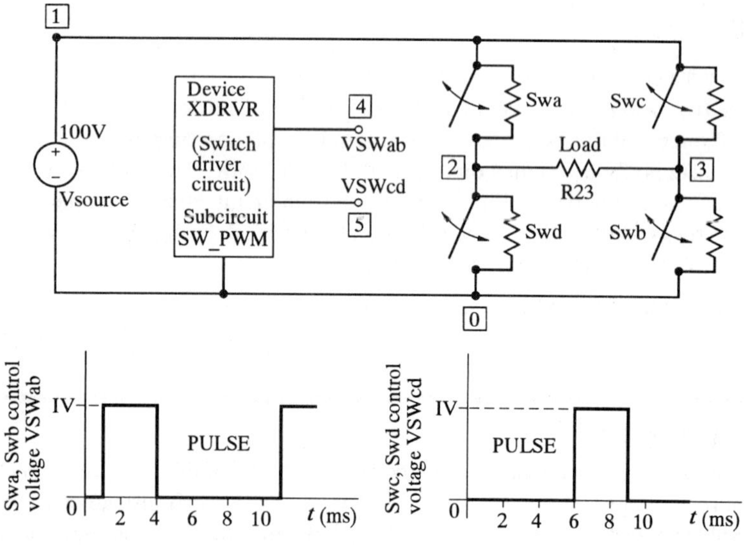

Fig. 2.11 PSpice inverter circuit and gate pulses for switches.

STEP 2

We will compose a circuit file from the diagram in Fig. 2.8a and link it with the subcircuit that we have named SW_PWM. We will call the circuit file EX2_3_1.CIR.

```
»»»»» EX2_3_1.CIR »»»»»»»»»»»»»»»»»»»»»»»»»»»»»»»»»»»»»»»»
DC-AC CONVERSION  IDEAL-SWITCH INVERTER;                          Title.
* Determine the load-current waveforms for resistive loads 2, 5 and 10ohms.

* Include the power-electronics library and define parameters.
. LIB       POWER.LIB;      A library call. (Illegal comment.) ..... (1)
. PARAM PERIOD=10ms, PULSE_WIDTH=3ms; ................. (2)
* Make the load resistance a parameter for multiple-run simulation.
. PARAM RLOAD=1.0; ........................................ (3)

* SOURCE, define a dc independent voltage source of 100V.
Vsource  1    0    DC    100V

* Call a subcircuit that is defined in the file POWER.LIB.
XDRVR  4    5    0       SW_PWM; ......................... (4)

* CIRCUIT ELEMENTS and DEVICES
* Define the inverter bridge using four voltage-controlled switches.
Swa 1    2    4    0       SWITCH;   Similar to line (2) of EX2_2_1.CIR.
Swb 3    0    4    0       SWITCH;   Swb is an element name.
* S, is for voltage-controlled switch.
Swc 1    3    5    0       SWITCH;   Nodes 1, 3 are for the switch.
* Nodes 5, 0 are for the control voltage.
Swd 2    0    5    0       SWITCH;   SWITCH is the model name.
* A model statement for SWITCH is not necessary.    PSpice will find it in the
* library file POWER.LIB because there is a reference link in .LIB.
R23 2    3    {RLOAD};  A value of R23 is not given here. .......... (5)

* ANALYSIS, to view the time-varying load current.
. TRAN     50us    40ms;          Prints every 50 microseconds (TSTEP).
* Simulation stops at 40ms (TSTOP).

* Perform a multiple run for different values of load resistance.
. STEP     PARAM  RLOAD  LIST  2    5    10; ............... (6)

* OUTPUT
. PROBE;                     View the load current i(R23) graphically over 40ms.
. END
»»»»»»»»»»»»»»»»»»»»»»»»»»»»»»»»»»»»»»»»»»»»»»»»»»»»»»»»»»
```

EXPLANATION OF THE CIRCUIT FILE EX2_3_1.CIR

This file refers to Fig. 2.11. There are six lines in the file with new forms of PSpice statements. Line (1) has a dot command .LIB. Lines (2) and (3) both have the dot command .PARAM. Line (4) refers to a subcircuit because it begins with X. In line (5) a data statement has a parameter name in place of a value. Line (6) has a dot command .STEP.

. . . .

Line (1) of circuit file EX2_3_1.CIR

```
.LIB     POWER.LIB;          A library call.                    (1)
```

- This is a PSpice control statement that indicates there is a need to refer to the library of devices and subcircuits in a particular file.
  There are two fields in this statement.
- In the first field there is the dot command .LIB. This prepares PSpice that information regarding a device or a subcircuit is to be read from a library file and included in the simulation.
- In the second field there is a filename POWER.LIB. This tells PSpice that data about the circuit-file devices and subcircuits can be found in the library file named POWER.LIB. The data will be retrieved. There is no need to repeat the data in the circuit file.
  Users can create their own library file and write device and subcircuit data in it. Your file will be named by you, but you must give it the extension .LIB. It is convenient to have the library files in the PSPICE subdirectory. However, the library file can be placed in any accessible directory, but the full address must be included in the second field if it is called, or the path must be included in AUTOEXEC.BAT.
  If the dot command .LIB is written in the circuit file without a filename in the second field, the default library file NOM.LIB is called. In the student version of PSpice the library file that would be referred to would be EVAL.LIB. Browse through this file to see what your version has to offer. You can add to it at any time. We prefer to add model statements and subcircuit descriptions to the file POWER.LIB for this guide to PSpice.
  The .LIB statement is not to have a comment in the circuit file.
  The general statement for reference to a device or subcircuit library is

  ```
  .LIB <file.LIB>
  ```

  If the optional file name is omitted, PSpice defaults to NOM.LIB.

. . . .

. . . .

Line (2) of circuit file EX2_3_1.CIR

```
.PARAM    PERIOD=10ms,    PULSE_WIDTH=3ms;                    (2)
```

- This line is a form of data statement about parameters of the circuit. It allows parameter names instead of numerical values to be used. The parameter names can be associated with elements (as in line (5)) or with model descriptions. They can also be used in expressions.
  There are two fields in this statement.
- The first field contains the dot command .PARAM. This is a keyword command that indicates that this data statement is to contain parameter names together with their values or expressions.
- The second field contains the parameter names and their values. The values can be changed by using a statement that begins with the dot command .STEP. In this case there are two parameters, separated by a comma.
  One parameter is named PERIOD. Its value is given as 10 ms. This represents the inverter switching period T, shown in Fig. 2.8b, and the driver-circuit control-voltage period T, shown in Fig. 2.10.
  The second parameter is named PULSE_WIDTH. Its value is given as 3 ms. This represents the pulse width T' of both the inverter output voltage and the switch control voltages, shown in the same figures. These parameters appear in the subcircuit named SW_PWM and have their numerical values expressed in the main circuit file.

. . . .

Line (3) of circuit file EX2_3_1.CIR

```
.PARAM    RLOAD=1.0;                                          (3)
```

- This is a data statement similar to line (2). In this case it relates the parameter name RLOAD of the resistance element R23, which is described in line (5). The parameter name RLOAD also appears in line (6). The value of RLOAD in line (3) is given as 1.0Ω.

. . . .

Line (4) of circuit file EX2_3_1.CIR

```
XDRVR    4    5    0    SW_PWM;                               (4)
```

- This statement is a call to PSpice to bring a defined subcircuit into the program that is described by the circuit file. There are three fields involving the subcircuit call, the appropriate nodes of the circuit and the name of the subcircuit. The description of the subcircuit can reside in the circuit file or in a library file.

- The first field contains the call XDRVR. This call is to introduce a subcircuit into the program. Every subcircuit begins with an X which is followed by an appropriate group of characters. This can be thought to be a new device defined by the user. In the PSpice circuit configuration we call the device DRVR (driver circuit). (See Fig. 2.11.)
- The second field contains an optional list describing all the circuit nodes to which the new device XDRVR is to be connected. Here, there are three nodes, numbered 4, 5 and 0. (See Fig. 2.11.) The ordering of the nodes is important.
- The third field contains the name of the subcircuit that we want to have connected to the circuit at nodes 4, 5 and 0.

  The name of the subcircuit is SW_PWM. PSpice searches the circuit file EX2_3_1.CIR for a SUBCKT statement which is named SW_PWM. Finding none, PSpice searches the circuit file for a reference to a library file. In this case it finds the library file POWER.LIB in the .LIB statement. PSpice searches† the file POWER.LIB and finds the subcircuit named SW_PWM. This subcircuit is now included in the program, with the external subcircuit nodes 101, 102 and 100 corresponding to the circuit configuration nodes 4, 5 and 0 in the same order.

  The general call for a subcircuit is

  ```
  Xxxx   NODE_LIST   SUB_NAME
  ```

  If there is no node list, the device *Xxxx* in the circuit will have the same external-node numbering as the subcircuit. If there is a node list, there must be as many nodes in the list as there are external nodes in the subcircuit description, but the actual node numbers do not have to correspond; only the order of the nodes is important.

. . . .

Line (5) of circuit file EX2_3_1.CIR

```
R23    2    3    {RLOAD};
```
A value of R23 is not given here. (5)

- This is a data statement concerning the resistor named R23 in the circuit configuration of Fig. 2.8a. There are three fields in this statement.
- The first field contains R23, the PSpice name of the load resistor.
- The second field contains the numbers given to the two nodes, to which the load resistor R23 is connected.
- The third field usually contains the value of the resistor. In this case it contains the parameter name RLOAD within braces. The parameter name must be

† In order to speed up searches in library files PSpice creates index files (.IND) for each library file.

defined. It is found in the .PARAM statement of line (3) of the circuit file. The .PARAM statement must also give a numerical value or an expression for RLOAD. Here, it is given a nominal value 1.0Ω, which is used by PSpice to calculate the circuit bias point (.OP analysis). The nominal value is also used in the program if no other values of RLOAD are given. In this example other values are given.

. . . .

Line (6) of circuit file EX2_3_1.CIR

.STEP PARAM RLOAD LIST 2 5 10; (6)

- This is a control statement for the parametric analysis. This particular STEP statement has three fields, one for the dot command, one for the parameter and one for a list of the parameter values.
- The first field contains the dot command .STEP. It indicates that PSpice is to perform multiple analyses.
- The second field contains PARAM RLOAD. PARAM is the keyword that is to be followed by the parameter name whose value is to be changed for each analysis. Here the parameter is known as the sweep variable name.
- The third field contains LIST 2 5 10. LIST is a keyword. What follows is the list of values that the sweep variable RLOAD is to assume in this example. In the first transient analysis RLOAD is assigned the value 2Ω, expressions are evaluated and the program results are recorded. The analysis and recording are redone for each of the load resistor's values 5Ω and 10Ω in the list. The value RLOAD = 1.0 in the .PARAM statement is ignored in the analysis but the parameter statement must be included in the file.
- One general form of the .STEP statement is

  .STEP PARAM name LIST values ...

  There are other forms of the .STEP statement described, in the Appendix C.

This ends the explanation of the circuit file EX 2_3_1.CIR

. . . .

STEP 3 The third step is to run the program. The program will run three times to cover the sweep of RLOAD equal to 2Ω, 5Ω and 10Ω.

STEP 4 We assume the PSpice program has been run using a DOS version and the results have been written in a file with the address C:\PSPICE\EX2_3_1.DAT. To look at the load current as a function of time, at the prompt C:\PSPICE> we type

PROBE EX2_3_1.DAT {The extension .DAT is optional}

The screen message CHECKING DATA FILE appears briefly, followed by a request for all or one of the transient analyses to be chosen.
Hit <↵> for all analyses to be displayed at once.
The screen message LOADING DATA FILE ... appears, followed by SETTING UP DATA STRUCTURE, then an empty graph with a time base and a menu with **Add_trace** highlighted.
Hit <↵> for a request for variables.
Hit F4 for a choice of variables. Choose I(R23) for the load current to be the variable on the *y*-axis of the graph. This can be done by using the arrow keys and <↵> or by clicking the mouse with the cursor arrow on the variable. I(R23) appears on the bottom of the screen. You can experiment with these methods of choosing one or more variables.
Hit <↵> once more. The three traces appear for the currents associated with the three load resistor values. A new menu appears .
From the menu invoke **Hard_copy**. Try hitting the **H** key. You will be requested to enter the file name. We entered EX2_3_1.PS <↵> <↵> to obtain a PostScript file (1_page_long) to disk.
To exit from PROBE to DOS hit **ESC** twice.
A hard copy of the results of the transient analysis appears in Fig. 2.12. The waveforms are similar to that in Fig. 2.8b. With a load resistance of 2Ω and switches *Swa* and *Swb* on, the load voltage is 100V and the load current is equal to V(2, 3)/R23 which is 50A. The trace shows this to be so.

END OF EXAMPLE 2.3.1

Drill Exercise D2.3.1
Consider the inverter circuit whose PSpice configuration is shown in Fig. 2.11. Change the dc supply to 60V, the load frequency to 200Hz and the output pulse width T' to 2.5ms. Using PSpice and PROBE, create time plots of the currents in switch Swa and the load R23 for load-resistance values 6Ω and 12Ω. Incorporate the subcircuits from the library file POWER.LIB. Comment on the switch current value and make the necessary circuit corrections.

Drill Exercise D2.3.2
The circuit diagram in Fig. 2.13a represents an ideal centre-tapped inverter. Write a PSpice circuit file that incorporates the model SWITCH and driver subcircuits already incorporated in the library file POWER.LIB (EXAMPLE 2.3.1). Design the inverter to have a periodic output waveform like that shown in Fig. 2.13b. Use PROBE to check the simulation results.

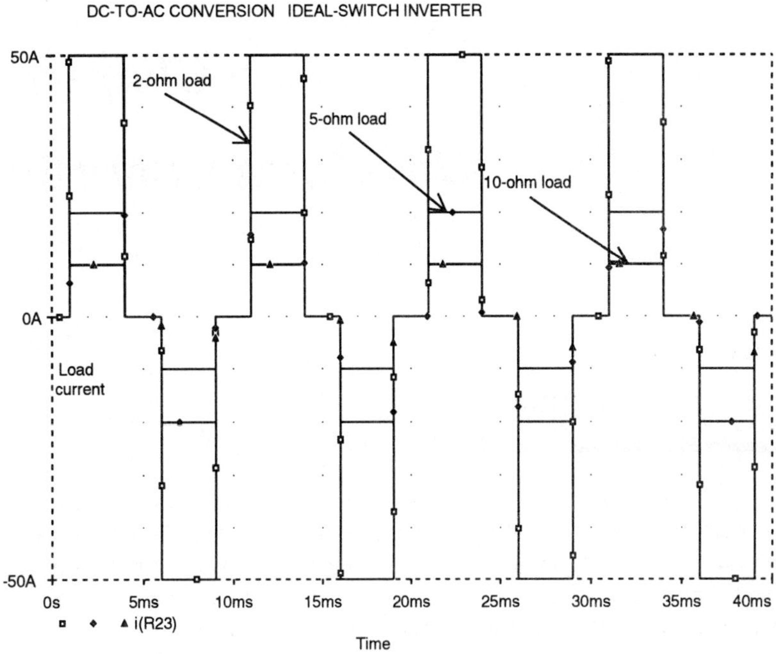

Fig. 2.12

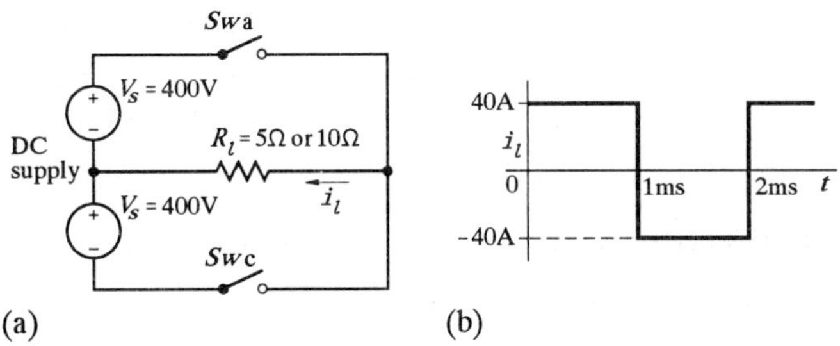

Fig. 2.13 A centre-tapped inverter.
(a) Inverter circuit diagram, (b) output waveform for $R_l = 10\Omega$.

Section 2.4 on AC-AC Modulation Simulation is available with an EXAMPLE and Drill Exercises on our WEBsite.

Chapter 3

Waveform Distortion

3.1 INTRODUCTION

In general, the output voltage of a power electronics circuit is a series of switched pulses. This is not ideal. We would like a constant voltage for a dc output or a sinewave for an ac output. In most cases where we have waveform distortion, we need to know the average or rms value of a variable that is being controlled. The variables are usually current, voltage and power. However, there are other considerations also. Examples to take into account are the effects of the circuit on the power factor and also the harmonics of ac waveforms. We will see how PSpice and PROBE can be used to provide us with all this information.

In the PROBE environment there are functions and operators that can be used to create plots of traces in terms of the circuit variables. Some of the functions are different from the PSpice functions. A list of the PROBE functions is given in Table 3.1 and a list of PSpice functions is given in Table 2.5.

3.2 AVERAGE VALUES

PROBE can calculate and plot a running average of any variable or function with respect to the range of the x-axis variable. A running average is

$$f_{avg} = \frac{1}{t}\int_0^t f(t)\,dt\,. \qquad (3.2.1)$$

Normally, we want the periodic average $f_{av} = \frac{1}{T}\int_0^T f(t)\,dt$ (3.2.2)

where T is the period. In the steady state, the running average is equal to the periodic average at $t=nT$ where n is an integer.

For a running average of a particular variable $v=\mathrm{V}(1)$, PROBE recognizes AVG(V(1)) and will trace this function versus the x-axis variable†.

Table 3.1 PROBE functions

Function	Meaning
ABS (y)	$\lvert y \rvert$ (absolute value)
EXP (y)	e^y
LOG (y)	ln y (log base e)
LOG1O (y)	log y (log base 10)
SGN (y)	+1 for $y>0$, 0 for $y=0$, −1 for $y<0$
SQRT (y)	$\sqrt{y}$ (square root)
M (y)	Magnitude of y, for complex y
R (y)	Real part of y, for complex y
IMG (y)	Imaginary part of y, for complex y
P (y)	Phase of y (degrees), for complex y
MIN (y)	Minimum of the real part of y
MAX (y)	Maximum of the real part of y
G (y)	Group delay of y
PWR (y, x)	$\lvert y \rvert^x$
SIN (y)	Sine of y, y in radians
COS (y)	Cosine of y, y in radians
TAN (y)	Tangent of y, y in radians
ATAN (y)	$\tan^{-1} y$, result in radians
ARCTAN (y)	$\tan^{-1} y$, result in radians
AVG (y)	Running average of y over x-axis range
AVG y (y, d)	Running average of y from y-d over x-axis range
RMS (y)	Running rms average of y over x-axis range
D (y)	Differentiation with respect to x-axis variable
S (y)	Integration with respect to x-axis variable
+ - * /	Operators

† This variable is usually time. Radians is a common alternative x-axis variable.

EXAMPLE 3.2.1

The circuit diagram in Fig. 3.1a is that of a battery charger, which comprises a stiff voltage source V_s and a chopper. A battery that is to be charged is modelled by an ideal voltage source V_b in series with a source resistance R_b (Fig. 3.2). For a state of charge such that V_b=12.2V and R_b=0.11Ω, use a PSpice simulation to find the average charging current if the chopper duty cycle is m=0.7.

Solution

The on/off stage of the switch will be governed by a driver circuit that has a control-voltage output, whose waveform is depicted in Fig. 3.1b.
There are four steps in the solution of this example.

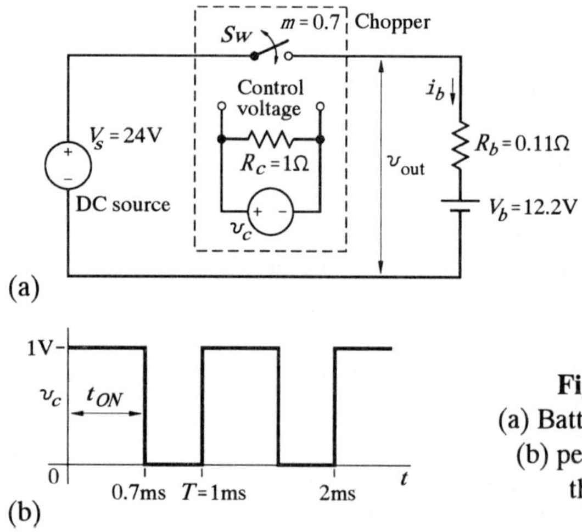

Fig. 3.1 Chopper application. (a) Battery-charger circuit diagram, (b) periodic voltage pulses to turn the switch *Sw* on and off.

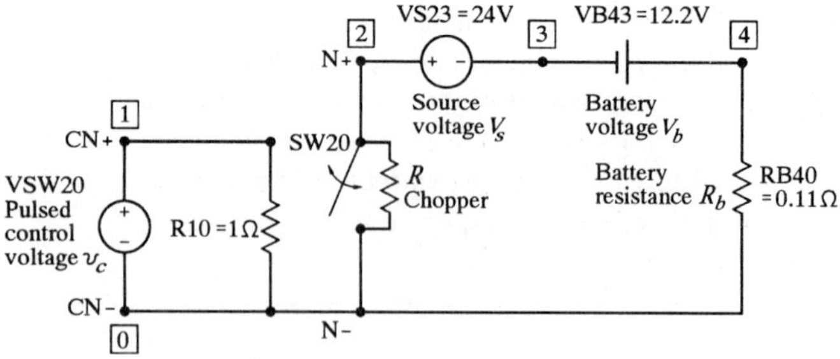

Fig. 3.2 PSpice circuit configuration.

STEP 1 Figure 3.2 depicts a PSpice configuration with node numbers and element or device names. The voltage-controlled switch SW20 simulates the chopper and the pulsed voltage source VSW20 provides the voltage signal v_c to turn the switch on for 0.7ms and off for 0.3ms. The switching frequency has been arbitrarily chosen to be $f = 1$kHz.

STEP 2 The second step is to use the circuit configuration in order to write the circuit file, which we call EX3_2_1.CIR.

```
»»»»» EX3_2_1.CIR »»»»»»»»»»»»»»»»»»»»»»»»»»»»»»»»»»»»
BATTERY CHARGER    CHOPPER    AVERAGE CURRENT
* Determine the average charging current in the battery.

* SOURCES
VS23    2 3   DC 24V;      A 24-V dc source at nodes (2, 3).
VB43    4 3   DC 12.2V;    The ideal 12.2-V battery source at nodes (4, 3).
VSW20  1 0   PULSE(0V 1V 0 1ns 1ns 0.7ms 1ms)
* VSW20 is the control voltage whose waveform is in Fig. 3.1b.
* This control voltage has an on-time 0.7ms and a period 1ms.

* CIRCUIT ELEMENTS
RB40  4   0  0.11ohms;  The source resistance of the battery.
R10   1   0  1;         A nominal resistance across the PULSE voltage.

* CIRCUIT DEVICE
SW20 2 0 1 0 SWITCH;      Chopper, a voltage-controlled switch.
* MODEL STATEMENT in association with the switch device.
* The name SWITCH is the link between the device and the model statement.
.MODEL SWITCH  VSWITCH(RON=1m ROFF=1meg VON=1 VOFF=0)
* VSWITCH is the model type, and the parameters give the switch resistances
* for the control-voltage values.
* This switch is in the POWER.LIB library file and has the name SWITCH.

* ANALYSIS, to view the time-varying current.
.TRAN    10us   4.7ms;        TSTEP=10μs (print), TSTOP=4.7ms.

* OUTPUT
.PROBE  i(RB40);              Limits data storage to this current.
.END;                         The end of the circuit file.

»»»»»»»»»»»»»»»»»»»»»»»»»»»»»»»»»»»»»»»»»»»»»»»»»»»»»»
```

STEP 3 The third step is to run the PSpice simulation.

STEP 4 The fourth step is to use PROBE. See Fig. 3.3.
In response to **Add_trace** in the menu we called in turn for

(1) the battery current -i(RB40) and obtained a chopped waveform with a peak-current value of 107.2A.
(2) the average battery current avg(-i(RB40)) and obtained the running average with a decreasing envelope of rises and falls. The times of the periodic average are $t = nT$ where $n = 1, 2, 3$ ... and T is the period.
(3) the value of the average current 74.41A from the standard calculation over one period ($I_{b\,av} = m(V_s - V_b)/R_b$).

The answer to the problem using PROBE is that the average charging current is approximately 74.4A. This steady-state average value is read from the trace

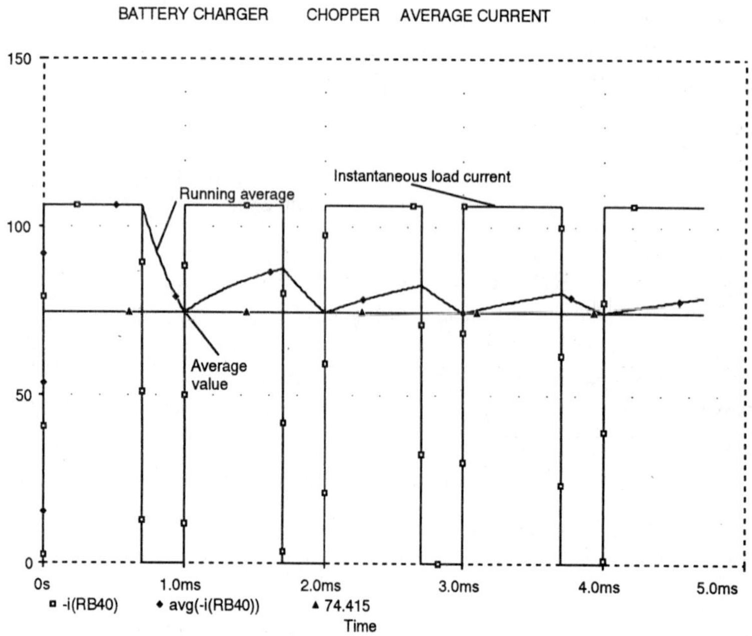

Fig. 3.3

AVG(−I(RB40)) at any time that is a multiple of the period $T = 1.0$ms. A more accurate value for AVG(−i(RB40)) can be obtained by using the ZOOM facility within PROBE or using the CURSOR facility.

END OF EXAMPLE 3.2.1

Drill Exercise D3.2.1
A battery charger comprises an ideal dc voltage source $V_s = 36$V in series with an ideal chopper that can be modelled by a voltage-controlled switch. The circuit diagram is similar to that in Fig. 3.1. A battery under charge is represented by an ideal voltage source of 24V and a series resistance whose value is 0.14Ω. Determine from a PSpice simulation and the use of PROBE the time it takes to charge the battery 1.0Ah if the chopper switching frequency is 1kHz and the duty cycle is **(a)** $m = 0.7$, **(b)** $m = 0.8$ and **(c)** $m = 0.9$. Use a sweep statement provided by the dot step command and use a parameter statement for the duty cycle. (Ans: **(a)** 1.0min, **(b)** 0.88min, **(c)** 0.78min.)

Drill Exercise D3.2.2
Use an ac source and a voltage-controlled switch, as in Fig. 3.4, to act as a controlled half-wave rectifier. The switch is turned on only for positive half cycles of the supply waveform. There is no conduction during the negative half cycles. Determine the average load current from a PSpice simulation and PROBE for firing angles **(a)** $\alpha = 0$ and **(b)** $\alpha = \pi/2$ rad. (Ans: **(a)** 10.8A, **(b)** 5.4A.)

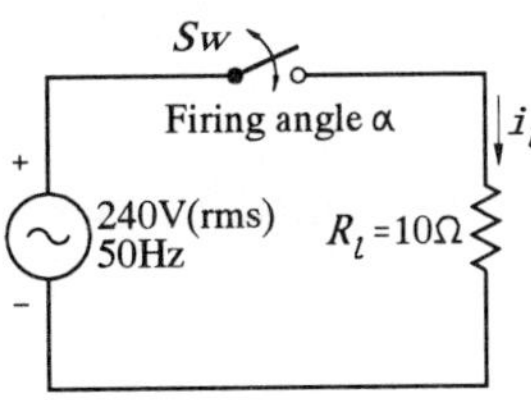

Fig. 3.4

Extra Drill Exercises for this Section can be viewed or downloaded from the WEBsite.

3.3 RMS VALUES

In the PROBE environment you can take any of the circuit variables, or any allowed function (see Table 3.1) or expression and plot a running rms value with respect to the x-axis variable. If the PSpice current is i(RL), the running rms value is traced on the PROBE graph by typing in RMS(i(RL)).

The running rms value is defined by $$I_{\text{rms run}} = \left[\frac{1}{t} \int_0^t i^2(t)\,dt \right]^{1/2} . \qquad (3.3.1)$$

For steady-state conditions the true rms value is $$I_{\text{rms}} = \left[\frac{1}{T} \int_0^T i^2(t)\,dt \right]^{1/2} . \qquad (3.3.2)$$

The running rms value is equal to the periodic rms at a time t equal to an integer multiple of the period T, because PSpice carries out its transient analysis from time $t=0$.

EXAMPLE 3.3.1

A voltage-controlled voltage source v_s with a resistive load R_l is shown in the circuit diagram of Fig. 3.5. Carry out a PSpice simulation and, using PROBE, determine **(a)** the amplitude of the load current, **(b)** the steady rms value of the load current and **(c)** the average power ($I^2_{l\,\text{rms}} R_l$) absorbed by the load resistor.

Solution

There are four steps in this solution.

STEP 1 — The first step is to draw a PSpice circuit configuration. This is shown in Fig. 3.6.

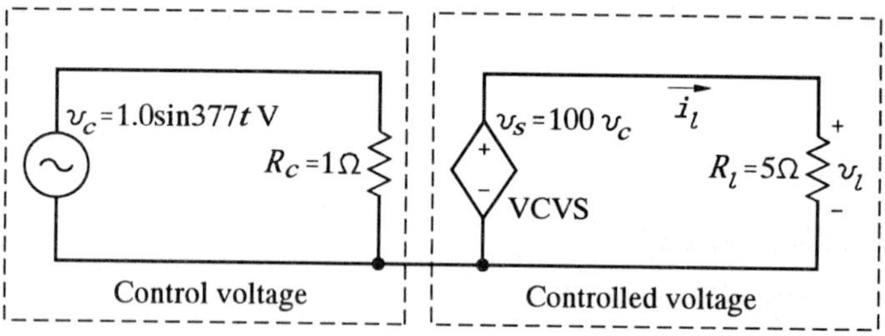

Fig. 3.5 A voltage-controlled voltage source (VCVS).

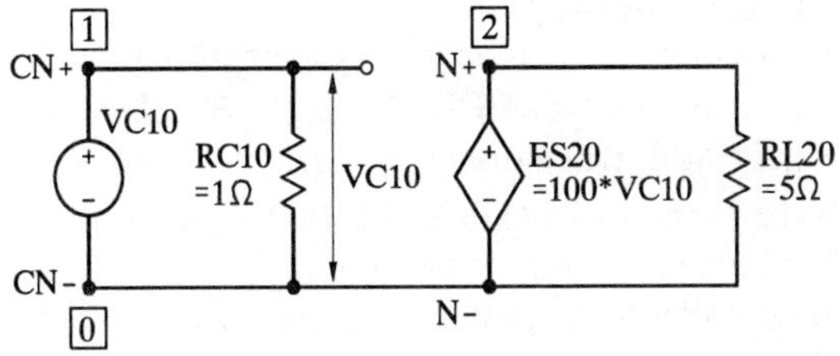

Fig. 3.6
PSpice circuit configuration.

STEP 2 The second step is to compose the PSpice circuit file from the PSpice circuit configuration. This circuit file follows.

```
»»»»» EX3_3_1.CIR »»»»»»»»»»»»»»»»»»»»»»»»»»»»»»»»»»
VOLTAGE-CONTROLLED VOLTAGE SOURCE, PEAK & RMS VALUES
* Determine the steady-state rms value of the load current.

* SOURCE, an independent sinusoidal voltage source.
VC10  1    0  SIN  (0 1 60Hz);          This is the control voltage.
* SOURCE, a dependent source (VCVS).    See file EX1_4_1.CIR
ES20  2    0  1    0    100;            100 is the gain.  See Table 1.1.

* CIRCUIT ELEMENTS
RC10  1    0  1ohm;             An arbitrary value for the control voltage.
RL20  2    0  5ohms;            The load resistor.

* ANALYSIS, transient.
.TRAN    20us  17ms;            TSTEP=20μs, TSTOP=17ms.

* OUTPUT
.PROBE  i(RL20);                Load current data stored.
.END
»»»»»»»»»»»»»»»»»»»»»»»»»»»»»»»»»»»»»»»»»»»»»»»»»»»
```

STEP 3 The third step is to run the PSpice program for EX3_3_1.CIR. A PROBE data file will be created and called EX3_3_1.DAT.

STEP 4 Use PROBE to view the results of the PSpice analysis. This will allow us to find the answers to the three problems.

Part (a) of Solution.

Let us plot load current i(RL20) versus time.

At the prompt type PROBE EX3_3_1.DAT

Select **Add_trace** from the menu.

Type in the variable i(RL20) <↵>. This is trace #1.

Find the peak value of the current.

By calculation $I_{max}=V_{max}/R=100/5=20$A.

Type in 20. This is a tangent to trace #1 i(RL20) and verifies $I_{max}=20$A.

Read the value of I_{max} from trace #1.

You can guess the value or you can use **Zoom** in the menu. Another way to determine values from the graph is to use the cursors.

Select **Cursor**. (Hit **C** or use the mouse.)

Cursors C1 and C2 appear on the graph.

There is a box with the values of C1 and C2 coordinates.

There is a submenu as follows.

Exit Hard_copy Peak Trough slOpe Min maX Search_commands Label_point

Clicking the mouse arrow near the trace gives time and magnitude of the current.

Selecting **maX**, the box shows the values of C1 to be 4.1659ms and $I_{max}=20$A. See Fig. 3.7.

Explore with the **Cursor** submenu items.

Part (b) of Solution.

Plot trace #2, the running rms value of the load current.

Select **Add_trace**

Type in RMS(i(RL20)) <↵> {or RMS(#1) <↵>}.

Select **Cursor**

Move cursor C1 using the mouse arrow and arrow keys → ← to the trace #1 at $t=T/4=4.1659$ms. This is cursor C1 on trace #1 and it gives the value 20A.

Hit CTRL SHIFT and → to set cursor C2 on trace #2.

Move the cursor with SHIFT and → to $t=T=16.667$ms. It gives the rms value of i(RL20)=14.142A.

The calculated value of $I_{l\,rms}=I_{max}/\sqrt{2}=20/\sqrt{2}=14.14$A. See Fig. 3.7 for i(RL20) and RMS(i(RL20)).

Note the cursor box, the legend for the cursors C_1 and C_2 and a label point.

Part (c) of Solution.

Remove all traces except #1=i(RL20) by selecting and deleting.

Select **Add_trace** for trace #2.

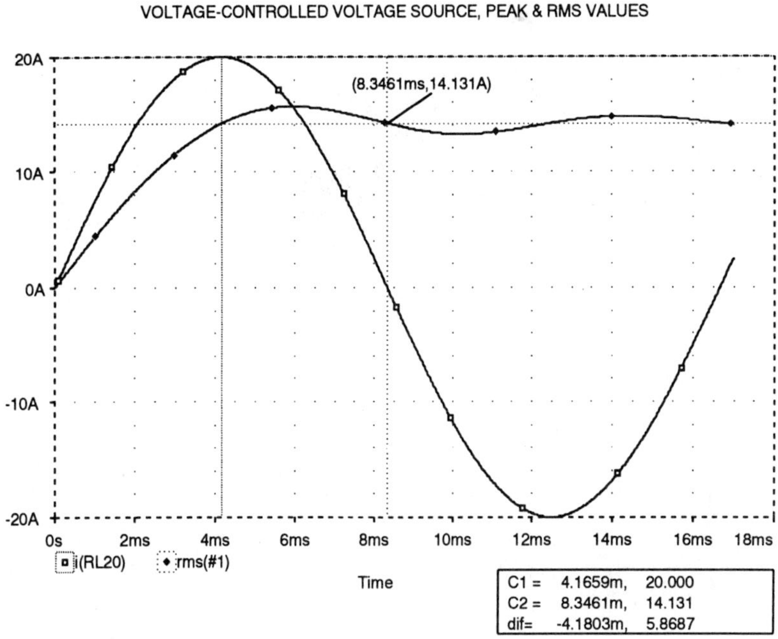

Fig. 3.7

Type in RMS(#1)*RMS(#1)*5. This is the running average power.
The y-axis scales do not match the two traces.
We will provide separate y-axis scales.
Remove trace #2.
Select **Y_axis**
From the new submenu select **add_aXis**, then **Exit**
Select **Add_trace**
Type in RMS(i(RL20))*RMS(i(RL20))*5 <↵> or
PWR(RMS(i(RL20),2)*5 <↵>.
Trace #2 is the running average load power.
Select **Cursor**
Click mouse arrow close to $t=T=16.667$ms.
Repeatedly hit arrow ➔ until the cursor C1 is on $t=16.667$ms. The average power for this point is 1.0kW. This is recorded on the plot by selecting **Label_point**
See Fig. 3.8 illustrated on the next page.

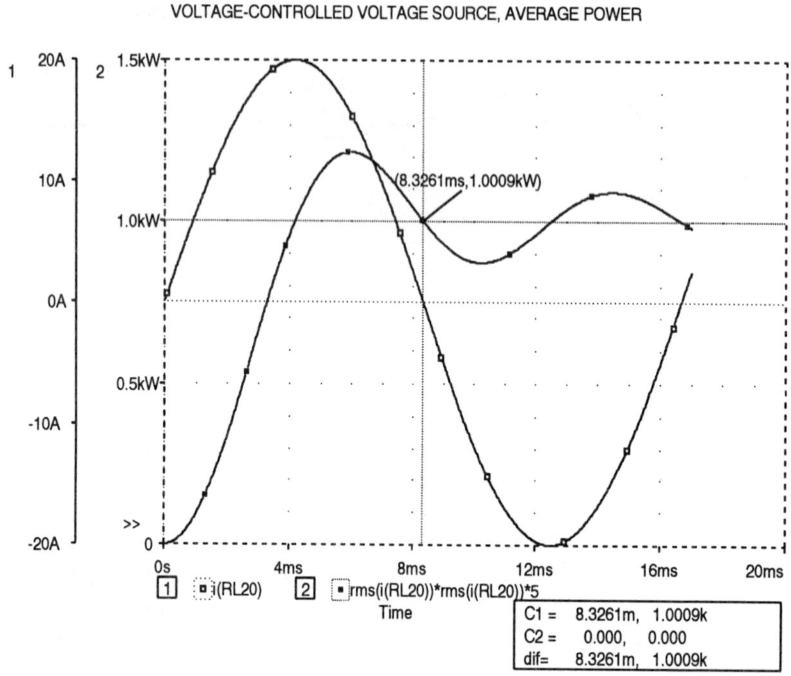

Fig. 3.8

END OF EXAMPLE 3.3.1

Drill Exercise D3.3.1

The power from a source of 240V at 60Hz is modulated in a resistive load of 20Ω by means of a bidirectional switch. The switch can be turned on by a voltage signal and remains on as long as the signal is applied. However, it is required that the switch turns off if the current falls to zero. The circuit diagram is shown in Fig. 3.9. Run a PSpice simulation for a sweep of the switch trigger angle α from 0 to 3.0radians. Use PROBE to obtain from the traces **(a)** $I_{l\text{rms}}$ and **(b)** $I^2_{l\text{rms}}\ R_l$ for $\alpha=2.0$radians. (Ans: **(a)** 5.91A, **(b)** 699.57W.)

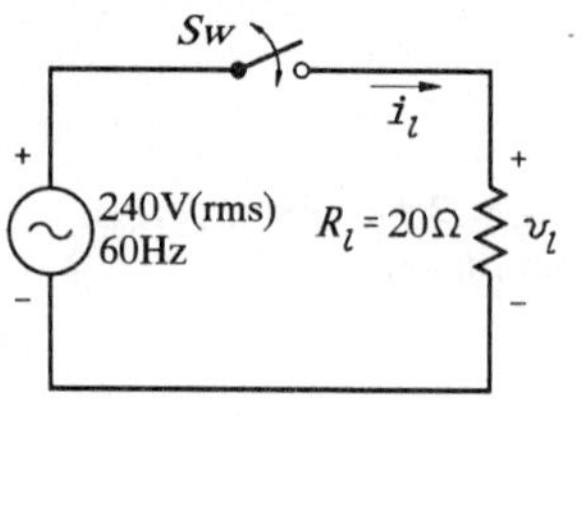

Fig. 3.9

Drill Exercise D3.3.2

Consider the circuit diagram in Fig. 3.10. The switch is closed during the positive half cycles of the supply voltage and open otherwise. Using a PSpice simulation and a voltage-controlled switch obtain PROBE traces of (i) the load voltage v_l over three half cycles, (ii) the running rms value of the load current i_l and (iii) the running value of the expression V^2_{lrms}/R_l. From these traces determine (a) the steady rms value I_l of the load current and (b) the average power absorbed by the load.
(Ans: (a) 4.159A, (b) 147.0W.)

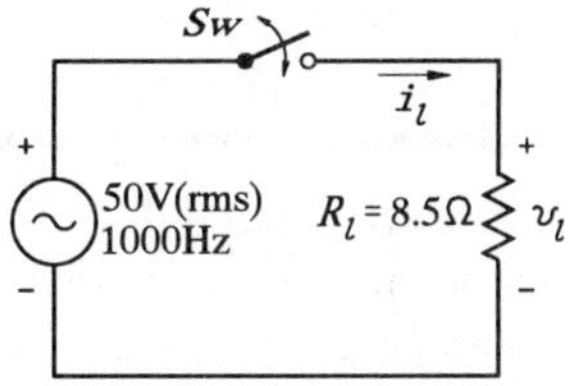

Fig. 3.10

An Extra Drill Exercise for this Section can be viewed or downloaded from the WEBsite.

3.4 POWER AND ENERGY

It is important to be able to determine the value of the power that is being modulated in power electronics circuits. The waveforms of current and voltage variables in power electronics circuits are rarely ideal sinewaves or constants. They are more often partial sinewaves or pulses. However, as long as the time variations of currents and voltages are available from a PSpice analysis, we can determine the power absorbed in any element. The instantaneous value of power $p(t)$ is

$$p(t)=v(t)i(t). \tag{3.4.1}$$

If the waveforms of the voltage and current are periodic, the average power P that is dissipated in a resistor, for example, is

$$P = V^2_{\text{rms}}/R = I^2_{\text{rms}}R. \tag{3.4.2}$$

PROBE offers a number of facilities to plot traces of power and energy as functions of time. Thus, we can create graphs of instantaneous power, or of a running average of power.

If power versus time is known, energy can be determined, because PROBE can integrate (S) a signal. That is, energy W can be expressed by

$$W = W_{\text{initial}} + \int_0^t p\,dt. \tag{3.4.3}$$

Conversely, if energy versus time is known, then power can be determined because PROBE can differentiate (D) functions. That is, power is expressed by

$$p = \frac{dW}{dt}. \tag{3.4.4}$$

EXAMPLE 3.4.1

An ideal voltage source is applied directly across the coil terminals of a superconducting electromagnet. See Fig. 3.11. Simulate the circuit by PSpice and use PROBE to plot over an interval of 15ms (three half cycles) **(a)** the instantaneous power $p=vi$ absorbed by the load, **(b)** the average power $P=P_{av}$ absorbed by the load, **(c)** the energy $W=\int p\,dt$ stored in the inductor, **(d)** the energy $W=(½)Li^2$ stored in the inductor, **(e)** the instantaneous power $p=dW/dt$ absorbed by the load and **(f)** the average power $P=(dW/dt)_{av}$ absorbed by the load.

Solution

There are four steps in this solution. Note that numerical analysis does not allow a voltage loop with only inductance. Add a small resistor.

STEP 1 Transform the regular circuit diagram in Fig. 3.11 to a PSpice configuration. This is shown in Fig. 3.12.

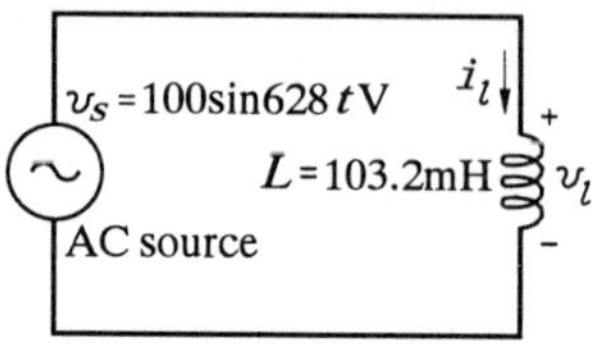

Fig. 3.11 Circuit diagram.

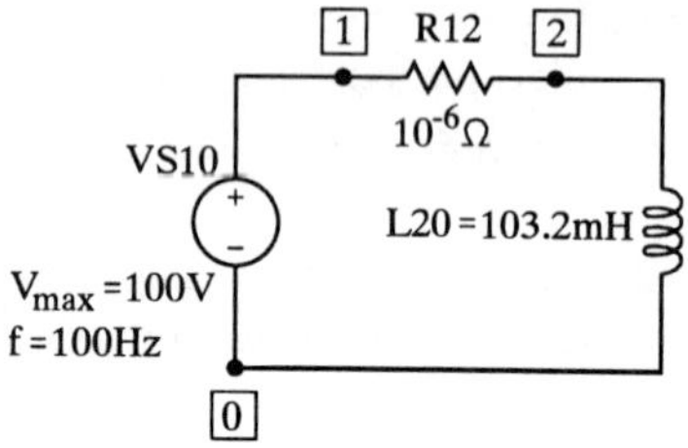

Fig. 3.12 PSpice configuration.

STEP 2 Use the PSpice configuration in order to write a description in the circuit file, which we will call EX3_4_1.CIR.

```
»»»»» EX3_4_1.CIR »»»»»»»»»»»»»»»»»»»»»»»»»»»»»»»»»»»»»»»»

POWER AND ENERGY IN AN AC CIRCUIT
* To determine power and energy in an inductive load.

* SOURCE
VS10  1  0   SIN(0  100V  100Hz); An ac voltage, 100-V amplitude.

* CIRCUIT ELEMENTS
L20   2  0   103.2mH;          The magnet inductance.
R12   1  2   1.0E-6ohm;        Adding some resistance to a voltage loop.

* ANALYSIS
.TRAN     20us    15ms;     TSTEP=20us, TSTOP after 3 half cycles.

* OUTPUT
.PROBE    v(2),    i(L20);    Load variables stored in data file.
.END

»»»»»»»»»»»»»»»»»»»»»»»»»»»»»»»»»»»»»»»»»»»»»»»»»»»»»»»
```

STEP 3

The third step is to run the program by keying in

C:\PSPICE>PSPICE EX3_4_1

STEP 4

Invoke PROBE for the data file EX3_4_1.DAT. This is done automatically in version 5.0 after STEP 3 in the DOS environment. If you exit at this time, you can return by keying in

C:\PSPICE>PROBE EX3_4_1

In the WINDOWS environment you double click the mouse with the arrow on the PROBE icon and from the FILE menu choose EX3_4_1.DAT.
We are in the PROBE environment and the data has been loaded.
The results of the plotting are shown in Fig. 3.13 on the next page.

Part (a) of Solution. The instantaneous power p=v(2)*i(L20) is shown in the lower plot, y-axis scale 1. The power p is cyclic and has a peak value of 200W. The power is delivered to the inductor from the source for the first half cycle and the power is delivered to the source from the inductor for the second half cycle.

Part (b) of Solution. The average load power
P=AVG(v(2)*i(L20))=AVG(#1) (measured at $t=T=10$ms) in the lower plot, y-axis scale 1 trace #1, is zero in the steady state.

Part (c) of Solution. The energy W=S(v(2)*i(L20)) is shown in the lower plot, y-axis scale2. All energy stored in the inductor is returned to the supply at $t=T$. The peak energy stored is about 500mJ at $t=T/2=$5ms. S(y) is the integral of the function with respect to the x-axis variable.

Part (d) of Solution. An expression for energy W=i(L20)*i(L20)*0.1032/2 is shown in the upper plot, y-axis scale1. It is the same as in part (c).

Part (e) of Solution. Instantaneous power p=D(i(L20)*i(L20)*0.1032/2 is shown in the upper plot, y-axis scale2, trace #1. It is the same as in part (a). D(y) is the differential of the function.

Part (f) of Solution. Running average power P=AVG(#1) is shown in the upper plot, y-axis, scale2. The steady-state average value of P is zero, measured at $t=T=$10ms, and is the same as in part (b).

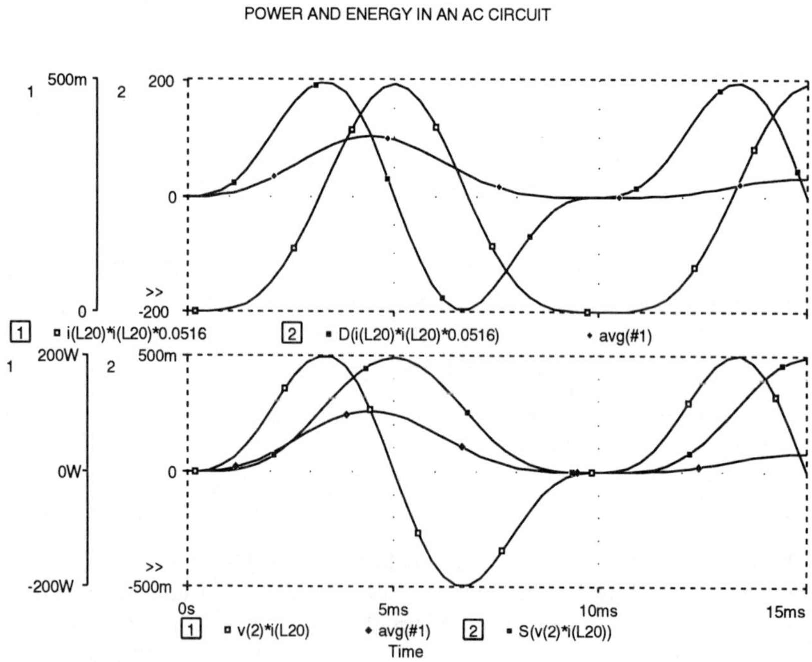

Fig. 3.13

END OF EXAMPLE 3.4.1

Drill Exercise D3.4.1

Consider the circuit diagram in Fig. 3.14. Use PROBE to estimate **(a)** the steady average power dissipated in the load resistance R_l and **(b)** the energy dissipated in the load over an interval of 20ms. (Ans: **(a)** 4.0475kW, **(b)** 81J.)

Fig. 3.14

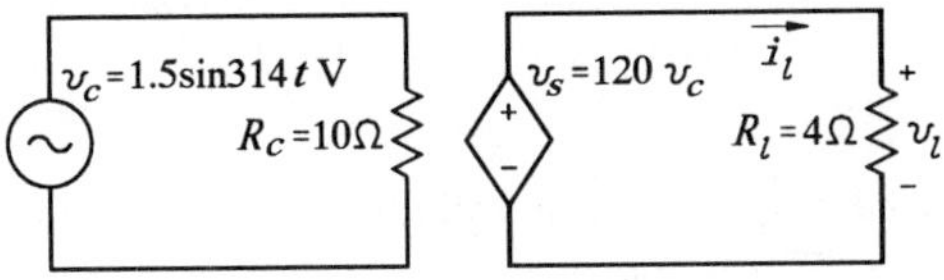

Drill Exercise D3.4.2

Power is modulated from a 100-V dc source to a resistive load R_l=4.5Ω by means of a chopper whose switching frequency is 1kHz and whose duty cycle is m=0.6. See the circuit diagram in Fig. 3.15. Model the chopper with a PSpice voltage-controlled switch and use PROBE to estimate **(a)** the average power absorbed by the load resistor and **(b)** the energy delivered by the source over an interval of 1ms. (Ans: **(a)** 1.333kW, **(b)** 1.333J.)

Fig. 3.15

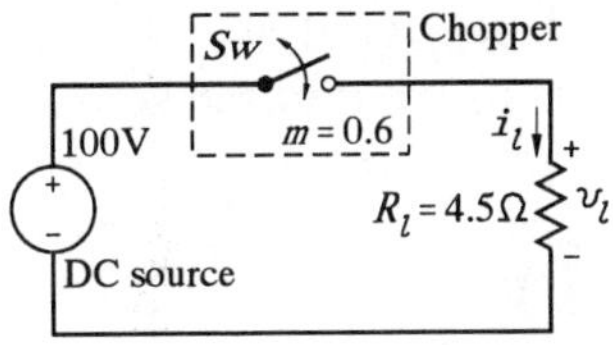

3.5 HARMONICS

The on/off operation of power-electronics switches introduces distortion into a waveform. The ideal ac waveform of a voltage or current is usually a pure sinewave. A Fourier analysis of a periodic waveform gives information about the deviation from the ideal, that is, the amplitudes and phases of the harmonics that are introduced by the switching action of power modulation.

The Fourier series of a periodic waveform can be written in the form

$$v(t) = V_{dc} + V_{1\max}\sin(\omega t + \theta_1) + V_{2\max}\sin(2\omega t + \theta_2) + V_{3\max}\sin(3\omega t + \theta_3) + \dots \qquad (3.5.1)$$

By means of a dot command .FOUR, PSpice can provide the amplitudes ($V_{1\,max}$, $V_{2\,max}$, $V_{3\,max}$ ...) and the phases (θ_1, θ_2, θ_3 ...) of the harmonics. In addition, the average value (dc level V_{dc}) and the total harmonic distortion (THD) are calculated. The total harmonic distortion is given by the expression

$$\text{Per unit } (\text{THD})^2 = \frac{V_{rms}^2}{V_{1\,max}^2/2} - 1 \tag{3.5.2}$$

where V_{rms} is the rms value of the periodic waveform.

The harmonic analysis is a useful tool to use in the design process of power electronics circuits. We can quickly determine how good, or bad, the design is and whether modifications need be made to the switching strategy.

Another dot command .INC will be introduced in this section so that we can include an existing file into a new circuit file.

EXAMPLE 3.5.1

A bidirectional switch modulates power from an ac source of 120V at 100Hz to a resistive load whose value is $R_l = 6.0\Omega$. The switch is to be turned on at 90 degrees every half cycle ($\alpha = \pi/2$rad) and turned off at 180 degrees every half cycle of the voltage waveform of the source. For these conditions, determine **(a)** the form factor of the load-voltage waveform and **(b)** the Fourier series for the load-voltage waveform. **(c)** From the information gathered in part (b) construct a voltage waveform from the ten harmonic values and compare the result with that waveform obtained from the transient analysis.

Solution

The circuit diagram for the ac voltage regulator is shown in Fig. 3.16a. A PSpice voltage-controlled switch is used for the switch *Sw*.

There are eight steps in the solution procedure.

STEP 1 The first step is to transform the circuit diagram into the PSpice configuration. This is shown in Fig. 3.16b.

STEP 2 Create a PSpice circuit file from the configuration depicted in Fig. 3.16b. The switch has to be turned on at twice the frequency of the supply voltage. There is one new statement to be found in this file. It concerns the Fourier analysis.

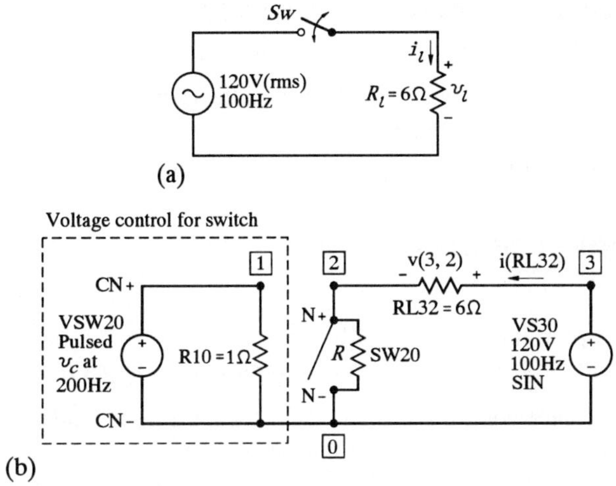

Fig. 3.16
(a) Voltage-regulator circuit diagram, (b) PSpice configuration.

»»»»» EX3_5_1.CIR »»

```
FOURIER ANALYSIS OF A VOLTAGE-REGULATOR AC WAVEFORM
* Determine the form factor and harmonics of the load-voltage waveform.

* SOURCE, for power circuit.
VS30    3  0   SIN(0 169.7V 100Hz);                         RMS value 120V.

* SOURCE, to control on/off action of switch SW20.
VSW20   1  0   PULSE(0 1 2.5ms 1ns 1ns 2.5ms 5ms); Voltage 0 or 1V.

* DEVICE, the switch.
SW20    2  0  1  0  SWITCH; S signifies a voltage-controlled switch and
* SWITCH is the model name. The control voltage is across nodes (1, 0).
* MODEL STATEMENT, for a voltage-controlled switch, type VSWITCH.
.MODEL SWITCH  VSWITCH(RON=1E-3  ROFF=1E6  VON=1  VOFF=0)

* CIRCUIT ELEMENTS
RL32    3  2   6ohms;                       Load resistor, value 6 ohms.
R10     1  0   1;            Nominal resistance for the control-voltage loop.

* ANALYSIS
.FOUR   100Hz   v(3, 2);   Harmonic values to the output file. . . . . . . (1)
```

```
. TRAN    20us  35ms 20ms;  Print TSTEP=20μs.  Analysis TSTOP=35ms.
* Print TSTART=20ms.

. PROBE;                    To determine the form factor.
. END
```

» »

EXPLANATION OF THE FILE EX3_5_1.CIR

. . . .

Line (1) of circuit file EX3_5_1.CIR

. FOUR 100Hz v(3, 2); Harmonic values to the output file. (1)

- This is a PSpice control statement. Here there are three fields.
- The first field contains the dot command .FOUR.
  This tells PSpice to perform a Fourier analysis on the last complete cycle (TSTOP - PERIOD to TSTOP) of the transient analysis.
  The transient analysis must cover enough cycles for a steady state.
  A Fourier analysis will be performed only if there is a .TRAN statement.
  The output of the Fourier analysis is automatically sent to the output file.
- The second field contains 100Hz, the fundamental frequency.
- The third field contains v(3, 2), the voltage whose waveform is to be analyzed. The third field can contain a number of variables.
- The general form of this Fourier statement is

  .FOUR FREQUENCY VARIABLES ...

- In PSpice versions 5.1 and later there are four fields in this statement.
  The four fields are

  .FOUR, FREQUENCY, NUMBER_OF_HARMONICS (optional) and VARIABLES.

. . . .

STEP 3 The third step is to run PSpice for the circuit file EX3_5_1.CIR.

STEP 4 The fourth step is to examine the results. Fig. 3.17 shows the plots of the output voltage $v_l(t)$ = v(3, 2), the running rms value RMS(v(3, 2)) and the running average value AVG(v(3, 2)). At t = 25ms the half-wave average is shown to be 54.008V. At t = 30ms the rms value over one cycle is shown to be 84.836V.

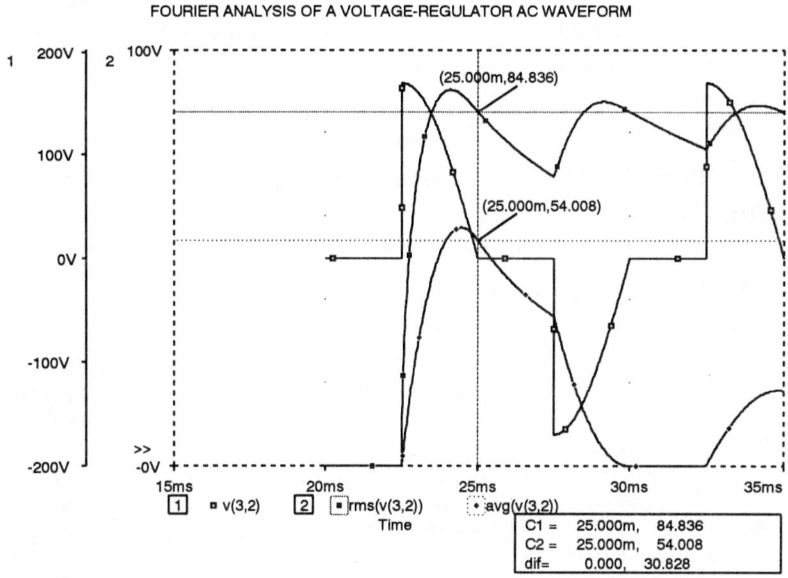

Fig. 3.17

Part (a) of Solution. The form factor FF is $\dfrac{\text{rms value}}{\text{average value}} = \dfrac{84.836}{54.008} \approx 1.57$.

Part (b) of Solution. In Table 3.2, shown on the next page, the harmonic values are listed. The dc component and the values of the even components are negligible. The total harmonic distortion is about 60%.

Part (c) of Solution.

STEP 5 The fifth step is to plot the results of the harmonic data that are found in the file EX3_5_1.OUT. The data are

$$
\begin{aligned}
v_l(t) \approx 1.4\text{E}-7 \quad & + 101.1\sin(628t + 2.578) + 3.9\text{E}-7\sin(2 \times 628t - 2.409) \\
& + 54.01\sin(3 \times 628t - 1.56) + 7.3\text{E}-7\sin(4 \times 628t - 1.479) \\
& + 18.0\sin(5 \times 628t + 1.60) + 1.7\text{E}-6\sin(6 \times 628t - 0.73) \\
& + 18.0\sin(7 \times 628t - 1.537) + 2.13\text{E}-6\sin(8 \times 628t + 0.03) \\
& + 10.8\sin(9 \times 628t + 1.627).
\end{aligned}
$$

The dc value and even harmonics can be neglected. The rest of this expression can be incorporated into a PSpice dependent voltage source E of the form

```
EOUT  N+  N-  VALUE = {expression}
```

We will add this voltage expression into the circuit file that we have already used, that is EX3_5_1.CIR. Let us write a new circuit file called EX3_5_1A.CIR. The PSpice configuration in Fig. 3.18, is used as the reference.

Table 3.2

```
* * * 11/30/95 20:24:01 * * * * * Evaluation PSpice (September 1991) * * * *
FOURIER ANALYSIS OF A VOLTAGE REGULATOR AC WAVEFORM
* * * * * * * * * * * * * * * * * * * * * * * * * * * * * * * * * * * * * * * * * * *

FOURIER COMPONENTS OF TRANSIENT RESPONSE V(3,2)
DC COMPONENT = 1.432484E-07
```

HARMONIC NO	FREQUENCY (HZ)	FOURIER COMPONENT	NORMALIZED COMPONENT	PHASE (DEG)	NORMALIZED PHASE (DEG)
1	1.000E+02	1.011E+02	1.000E+00	1.477E+02	0.000E+00
2	2.000E+02	3.901E-07	3.859E-09	-1.380E+02	-2.857E+02
3	3.000E+02	5.401E+01	5.343E-01	-8.936E+01	-2.371E+02
4	4.000E+02	7.325E-07	7.247E-09	-8.476E+01	-2.325E+02
5	5.000E+02	1.800E+01	1.781E-01	9.193E+01	-5.578E+01
6	6.000E+02	1.701E-06	1.683E-08	-4.189E+01	-1.896E+02
7	7.000E+02	1.800E+01	1.781E-01	-8.807E+01	-2.358E+02
8	8.000E+02	2.130E-06	2.108E-08	1.945E+00	-1.458E+02
9	9.000E+02	1.080E+01	1.069E-01	9.321E+01	-5.449E+01

TOTAL HARMONIC DISTORTION = 6.003169E+01 PERCENT

STEP 6 Create the circuit file EX3_5_1A.CIR associated with the configuration in Fig. 3.18. This circuit file follows.

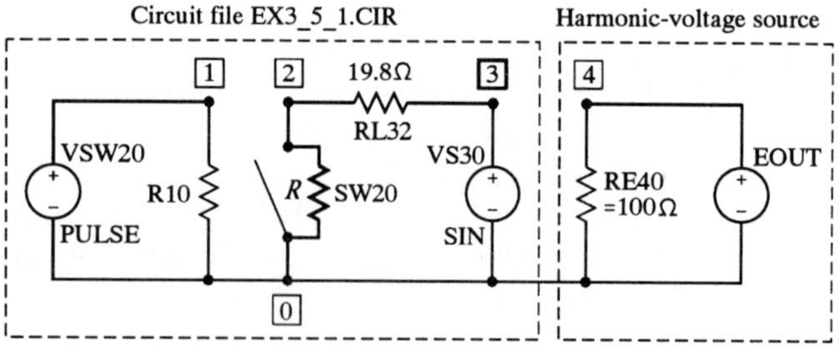

Fig. 3.18

»»»»» EX3_5_1A.CIR »»

```
HARMONIC ANALYSIS VERSUS GENERATION
* A voltage source to generate the harmonics that were calculated in the
* PSpice analysis of EX3_5_1.CIR and written in EX3_5_1.OUT.

* The following is a description of individual harmonics in the form of
* function statements.                    The following is ............. (1)
.FUNC HARM1()    101.1*SIN(628*time+2.578)
.FUNC HARM3()    54.01*SIN(3*628*time-1.56)
.FUNC HARM5()    18.0*SIN(5*628*time+1.60)
.FUNC HARM7()    18.0*SIN(7*628*time-1.537)
.FUNC HARM9()    10.8*SIN(9*628*time+1.627)
.FUNC HARM1to9() HARM1()+HARM3()+HARM5()+HARM7()+HARM9()

* The following is a voltage source incorporating these functions.
* The following line is ...................................... (2)
EOUT 4 0  VALUE = {HARM1to9()}
RE40 4 0  100Ω; A nominal value to close the voltage loop for EOUT.

.INC     a:\EX3_5_1; ................................. (3)

.PROBE  v(4)  v(3,2);      To view the recorded Fourier series.
.END;                      The end of the complete file with the included file.
```

»»

EXPLANATION OF THE FILE EX3_5_1A.CIR

. . . .

Line (1) of circuit file EX3_5_1A.CIR
.FUNC HARM1() 101.1 * SIN(628 * time + 2.578)

- This is a control statement for PSpice. It must appear in the file before the expression in which it is used.
- There are three fields in this control statement. The fields are the dot command, the name given to the function and the function description.
- The first field contains .FUNC. This means that a name of a function and its description are to follow in this statement. The aim is to use the name rather than the actual function in an expression in order to limit the length of the expression. An expression is limited to one line in PSpice.
- The second field contains HARM1( ). HARM1 is the arbitrary NAME given to the fundamental harmonic part of the Fourier series for the output voltage

waveform that was obtained from the analysis. The parentheses () can contain from zero to 10 arguments. Here, there are zero arguments. On the following file lines there are functions and function names for all the harmonic components. The function NAME in the .FUNC statement must not be the name of a function that is already in PSpice.

- The third field has a description of the function that is to be used by the name HARM1(). Any of the PSpice functions and operators, that are listed in Table 2.5, can be used. Here, the function is $V_{1\,max}\sin(\omega t+\theta_1)$ in general, and 101.1sin(628t+2.578) in particular. The length of this function description must not exceed one line.
- The general form of the .FUNC statement is

  .FUNC NAME (arguments) Function_description

- There cannot be any comments on the same line as the .FUNC statement.

. . . .

Line (2) of circuit file EX3_5_1A.CIR
EOUT 4 0 VALUE = {HARM1to9()}

- This is a data statement for PSpice. It must not appear before the .FUNC statements that describe the functions within the expression {...} of line (2).
- There are three fields in this data statement. The fields describe a voltage source, node names, and an expression of the source voltage.
- The first field contains the name of the source EOUT. The name must begin with E since it represents an independent voltage source.
- The second field contains 4, 0, the numbered nodes of the source
- The third field contains VALUE = {HARM1to9()}. That part VALUE = tells PSpice that the source voltage EOUT is described by an expression. The term within the braces {...} is the expression. PSpice expands the expression HARM1to9() from the .FUNC statements to be the Fourier series of the load voltage.
- The general form for source voltage that can be described by a time-varying expression is

  E*xxx* N+ N- VALUE = {expression}

  The expression that specifies the VALUE as a function of time must fit on one line. There must be no comments on this statement line. In the jargon of PSpice this statement represents an *Analogue Behavioural Model.*
  If the expression had specified a current, the source name would have been G*xxx* instead of E*xxx*.

. . . .

. . . .

Line (3) of circuit file EX3_5_1A.CIR

. INC a:\EX3_5_1 ; (3)

- There are two fields in this control statement, a dot command and a filename.
- The first field contains the dot command .INC. This tells PSpice to INClude or insert a file into the current circuit file.
- The second field contains a:\EX3_5_1. This is the filename whose contents are to be inserted in the current circuit file. Here, EX3_5_1 is the circuit file EX3_5_1.CIR without the introductory title or the extension .CIR.
  The purpose of the .INC statement is simply to save the time rewriting in a circuit file what already exists in another file.

. . . .

STEP 7 Run the circuit file EX3_5_1A.CIR that has the statements of EX3_5_1.CIR plus the Fourier series harmonic generator.

STEP 8 Use PROBE to compare the traces of the load voltage $v_l = v(3,2)$ with the Fourier series $v_l = EOUT = v(4)$. The results are shown in Fig. 3.19. The conclusion is that a knowledge of the harmonic voltages up to the ninth is not enough to describe the load voltage.

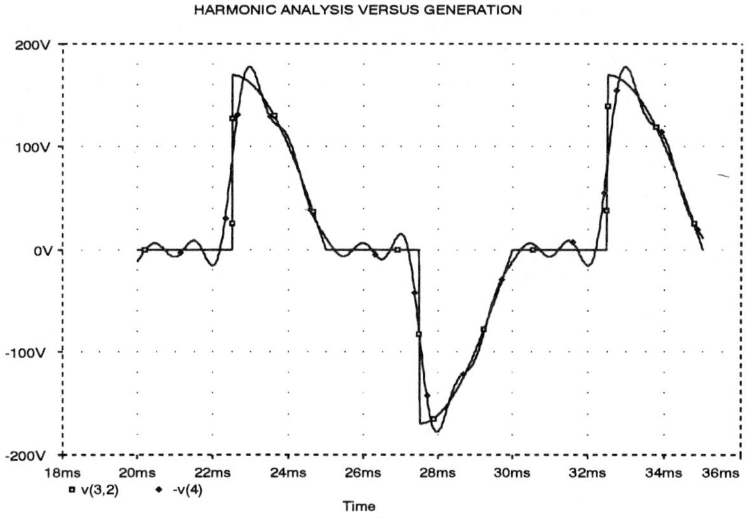

Fig. 3.19

END OF EXAMPLE 3.5.1

Drill Exercise D3.5.1

Consider the ac voltage-regulator circuit depicted in Fig. 3.16. Change the ac source voltage to an rms value of 250V and the load resistance value to 7.5Ω. The switch is to be turned on at 60 degrees and turned off at 180 degrees every half cycle of the voltage waveform of the source. Determine from a PSpice transient analysis **(a)** the form factor and **(b)** the Fourier series for the load-voltage waveform. **(c)** From the information gathered in part (b) construct a voltage waveform from the harmonic values given in the output file and compare the result with that waveform of the load voltage obtained from the transient analysis.
(Ans: **(a)** 1.328, **(b)** $v_l = 296.3\sin(628t + 2.852) + 84.59\sin(3\times628t - 0.528)$ $+ 48.72(5\times628t - 2.103) + 24.36\sin(7\times628t + 2.077) + 22.41(9\times628t - 0.208)$ volts.)

Drill Exercise D3.5.2

Figure 3.20a shows the primitive circuit diagram of a centre-tapped inverter. The ac voltage at the load has a waveform as shown in Fig. 3.20b. From a PSpice transient analysis, determine **(a)** the form factor for the load-voltage waveform and **(b)** the Fourier series for the load-voltage waveform. **(c)** From the information gathered in part (b) construct a voltage waveform from the harmonic values given in the output file and compare the result with that waveform of the load voltage obtained from the transient analysis. You can adapt the circuit file used in the drill exercise D2.3.2 or you can write a new circuit file.
(Ans: **(a)** 1.414, **(b)** $v_l = 90.03\sin(785.4t + 3.138) + 30.01\sin(3\times785.4t - 0.0118)$
$+ 18.01\sin(5\times785.4t - 0.02) + 12.86\sin(7\times785.4t + 3.114)$
$+ 10.01\sin(9\times785.4t + 3.107)$ volts.)

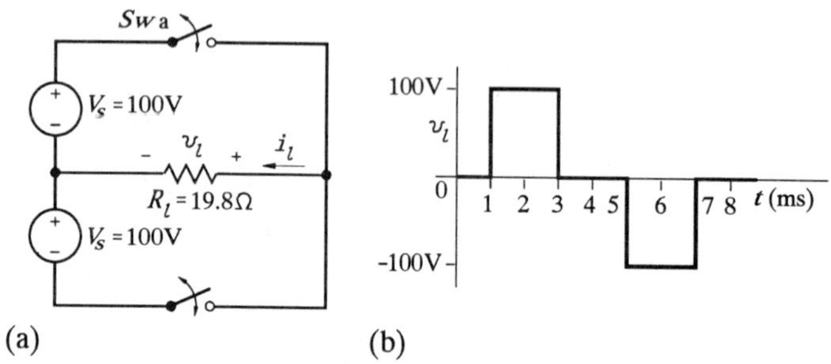

Fig. 3.20

Chapter 4

Ideal Semiconductor Switches

4.1 INTRODUCTION

Semiconductor switches that are used in power electronics circuits are fast acting compared with most circuit elements. The switches have on-state voltage drops that are small compared with most elements of the circuit. Further, the off-state switch currents are negligible compared with most branch currents. With such characteristics, semiconductor switches can be considered to be virtually ideal, if the circuit responses are to be determined from a theoretical analysis.

In system analysis in which we look at load characteristics or supply quality, we allow the semiconductor switches to have characteristics close to the following.

- Turn-on switching time and turn-off switching time :- negligible.
- On-state switch impedance :- negligible (a virtual short circuit).
- Off-state switch impedance :- very large value (a virtual open circuit).
- In the on-state there is no current limit.
- In the off-state there is no limit to the forward blocking voltage.

PSpice provides two switches that can be used to model closely the characteristics of ideal semiconductor switches. The two PSpice switches are the voltage-controlled switch and the current-controlled switch. The voltage-controlled PSpice switch was described in Chapter 2.

The current-controlled switch has two PSpice statements of the form

```
WSWITCH N+ N- VX IMOD
.MODEL IMOD ISWITCH(RON=value ROFF=value
+                   ION=value IOFF=value)
```

where

WSWITCH is the statement name of the switch. The name must begin with W.
N+ N- are the nodes to which the switch is connected.

VX	is an ideal, independent voltage source. It is used to sense the control current that causes the switch to change states. VX is usually a dummy source of zero volts.
IMOD	is a model name that links the data and model statements.
.MODEL	introduces the statement with the switch parameters.
ISWITCH	is the PSpice model type of a current-controlled switch.
RON, ROFF	are the switch's on/off resistances.
ION, IOFF	are the control currents that cause the switch to change states. If $I_{control} \geq ION$, $R_{switch} = RON$. If $I_{control} \leq IOFF$, $R_{switch} = ROFF$. If $ION > I_{control} > IOFF$, $RON < R_{switch} < ROFF$.

In real power electronics circuits we have a choice of semiconductor switches. Some of them are

1. the diode.
2. the BJT. (Bipolar Junction Transistor)
3. the thyristor. (Silicon Controlled Rectifier. SCR)
4. the MOSFET. (Metal-Oxide Semiconductor Field-Effect Transistor)
5. the IGBT. (Insulated-Gate Bipolar Transistor)
6. the triac. (Three-terminal ac switch)
7. the GTO. (Gate Turn-Off thyristor)
8. the MCT. (MOSFET-Controlled Thyristor)

These switches may conduct current unidirectionally or bidirectionally; they may be uncontrolled or controlled; they may block voltage of one polarity or both polarities; they may be turned on by current or by voltage; they may be turned off by current or voltage. Whichever of these characteristics are displayed by a particular type of semiconductor switch, we can model the behaviour by using one or other of the PSpice switches.

The models are described in this chapter. This enables power electronics circuits, that incorporate any kind of semiconductor switch, to be readily analyzed using PSpice.

4.2 THE IDEAL DIODE

The diode is an uncontrolled switch. A simple ideal-diode circuit and the diode's responses and characteristic are shown in Fig. 4.1.

In order to turn the diode on there must be a forward-bias voltage $+v_{AK}$. As soon as the ideal diode turns on the voltage drop v_{AK} across the diode assumes a zero value because the diode has zero resistance. In the on-state, the diode current i_D assumes a value that depends on the values of the source voltage v_s and the load elements R and L.

Once the ideal diode has been turned on, it remains on until the current i_D in the device falls to zero. At this zero value the diode will turn off, assume a high resistance value, and block any reverse bias voltage $-v_{AK}$. In the diode off-state, there is no current, $i_D = 0$, the voltage v_l across the load is zero, so that $v_{AK} = -v_s$.

This diode *I-V* characteristic, illustrated in Fig. 4.1c, can be modelled by a PSpice switch to have similar behaviour. The PSpice switch is an adjustable resistor whose value is controlled between a low (on-state) value and a high (off-state) value by means of a control voltage or a control current. The diode condition is governed by the circuit current and voltage.

We will describe four models of a diode for use in PSpice configurations. If one model gives nonconvergence problems in an application, we can try one of the other models.

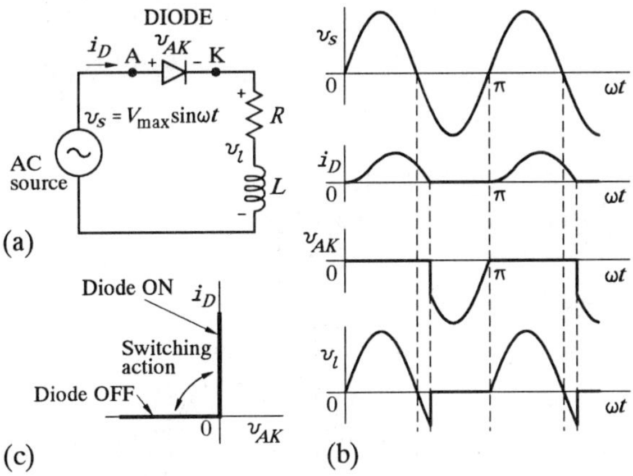

Fig. 4.1 Diode circuit. (a) Circuit diagram with *RL* load, (b) circuit responses, (c) the ideal diode characteristic.

4.2.1 VOLTAGE-CONTROLLED SWITCH AS THE DIODE

The aim of this analysis is to use a PSpice voltage-controlled switch (symbol S, model type VSWITCH) to represent the ideal diode. We will create subcircuits for the diode models and place them in the library file POWER.LIB for future use.

EXAMPLE 4.2.1

A diode is to act as a half-wave rectifier for a circuit with an alternating-voltage source, so that there is direct current in the *RL* load. Values are shown in the circuit diagram of Fig. 4.2a. From a PSpice transient analysis, use PROBE to record the load current and voltage responses. Use a PSpice voltage-controlled switch to model the diode.

Solution

There are four steps to follow in the solution of this problem.

STEP 1 Transform the circuit diagram of Fig. 4.2a to a PSpice configuration. See Fig. 4.2b. The statement name of the voltage-controlled switch is SD10, the diode model name is DIO_V and the control voltage is chosen to be the diode voltage drop $v_c = v_{AK} \equiv$ v(1).

STEP 2 The second step is to use the PSpice configuration to create a circuit file. The circuit file follows.

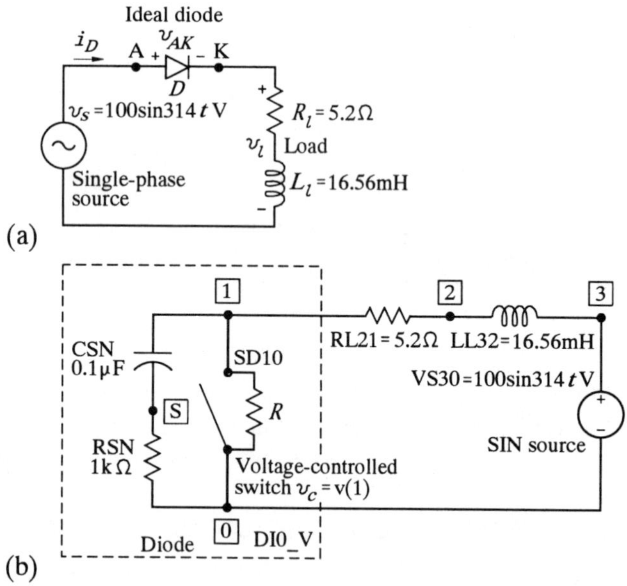

Fig. 4.2 (a) Half-wave rectifier, (b) PSpice configuration.

```
»»»»» EX4_2_1.CIR »»»»»»»»»»»»»»»»»»»»»»»»»»»»»»»»»»»»»»»»»
DIODE TEST CIRCUIT    DIO_V
* To check the operation of a simulated diode (a voltage-controlled PSpice
* switch) that turns on with a forward-bias voltage and turns off when the current
* falls to near-zero value.
* The diode blocks if the current is not positive and the voltage is negative.

* SOURCE
VS30  3   0   SIN(0 100V 50Hz);          Sinusoidal-voltage supply.

* CIRCUIT ELEMENTS
RL21  2   1   5.2ohms;                   Load resistance.
LL32  3   2   16.56mH;                   Load inductance.

* CIRCUIT DEVICE
* This is a voltage-controlled switch that models the diode.
SD10  1   0   1   0   DIO_V;      Switch (1, 0), control voltage (1, 0).
.MODEL   DIO_V  VSWITCH(RON=1E-3  ROFF=1E7
+                       VON=0.1  VOFF=0); ............. (1)
CSN   1   S   0.1uF;          A snubber circuit across the diode to help
RSN   S   0   1kohms;         numerical convergence. An experiment.

* ANALYSIS, always in the transient mode.
.TRAN    20us  30ms  0  40us;  Analysis covers three half cycles.
.PROBE;                  Graphics postprocessor to view all variables.
.END
»»»»»»»»»»»»»»»»»»»»»»»»»»»»»»»»»»»»»»»»»»»»»»»»»»»»»»»»»»»
```

EXPLANATION OF THE FILE EX4_2_1.CIR

The choice of the parameter values RON ROFF VON VOFF of the switch needs some explanation.

. . . .

Line (1) of circuit file EX4_2_1.CIR

```
.MODEL DIO_V  VSWITCH(RON=1E-3  ROFF=1E7
+                     VON=0.1  VOFF=0);                 (1)
```

- The default values of the parameters of this type of switch are given in Table 2.3. Initially the switch is off and $R = \text{ROFF} = 10^7\Omega$. At $v(1) = 0.1V$, $R = \text{RON} = 10^{-3}\Omega$. The resistance value RON has little effect on the actual circuit response. While $R = \text{RON}$, the diode is on and the current is determined by the other circuit elements.

Once the diode turns off $R = \text{ROFF} = 10^7\Omega$, so that the maximum reverse current is $i_D = -V_{s\,max}/\text{ROFF} = -100/10^7 = -0.01\text{mA}$. This is an acceptably small current compared with the peak forward current of about 15A. The value of ROFF is a compromise. As ROFF is increased, the reverse current decreases, but the turn-on time is longer.

This ends the explanation of circuit file EX4_2_1.CIR.

. . . .

STEP 3 The third step is to run the PSpice simulation using the circuit file EX4_2_1.CIR.

STEP 4 The results of the simulation of the rectifier can be studied by using PROBE. The diode current and voltage over three half cycles of the supply waveform are shown in Fig. 4.3.

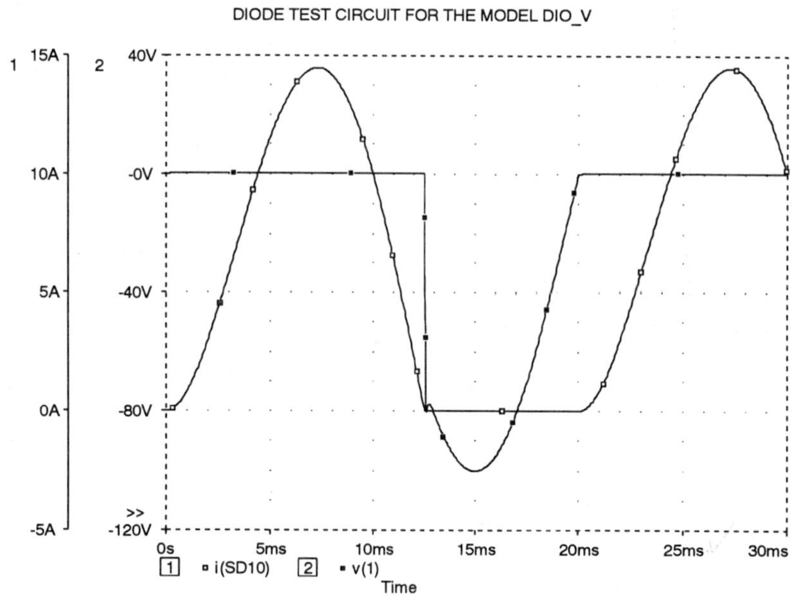

Fig. 4.3

END OF EXAMPLE 4.2.1

SUBCIRCUIT FOR THE DIODE MODEL DIO_V

The voltage-controlled switch with the model name DIO_V models an ideal diode reasonably well. We can create a subcircuit and put it in the library file POWER.LIB. This subcircuit configuration is shown in Fig. 4.4.

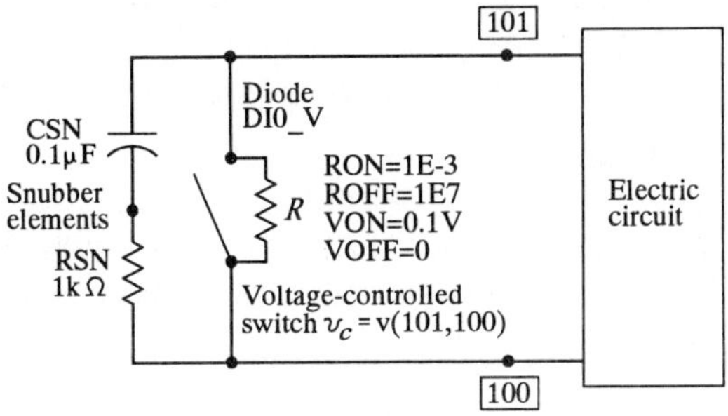

Fig. 4.4
A diode subcircuit configuration named DIO_V.

```
* * * * * * * * * * * * * * * * * * * * * * * * * * * * * * * * * * * * * * * * * * * *

* AN IDEAL DIODE MODELLED BY A VOLTAGE-CONTROLLED SWITCH
* A subcircuit named DIO_V.
.SUBCKT   DIO_V   101   100;      Anode (101), cathode (100).
SWDIODE   101   100  101  100  DIO
.MODEL    DIO   VSWITCH(RON=1E-3 OFF=1E7  VON=0.1  VOFF=0)
CSN       101   103   0.1uF;      A snubber circuit to aid convergence.
RSN       103   100   1kohms
.ENDS     DIO_V

* * * * * * * * * * * * * * * * * * * * * * * * * * * * * * * * * * * * * * * * * * * *
```

This subcircuit is called from the library file POWER.LIB by two statements

```
.LIB   POWER.LIB
XD   N+   N-   DIO_V
```

PARAMS:

In the above subcircuit the snubber circuit elements, CSN and RSN are fixed. The same values will be used in the circuit file that calls the subcircuit with the statement

```
Xname   N1   N2   DIO_V
```

We can alter this situation so that in the circuit file different values can be given to the snubber elements. This is done using parameters in a special form. We will rewrite the diode subcircuit as follows

```
* * * * * * * * * * * * * * * * * * * * * * * * * * * * * * * * * * * * * * * * * * *
* AN IDEAL DIODE MODELLED BY A VOLTAGE-CONTROLLED
* SWITCH WITH A SNUBBER CIRCUIT.

* DIODE MODEL
.SUBCKT DIO_V 101 100 PARAMS: CSNUB=0.1uF, RSNUB=1kohm
SWDIODE 101     100 101 100 DIO; DIO is the model name.
.MODEL  DIO     VSWITCH(RON=1E-3 ROFF=1E7 VON=0.1 VOFF=0)

* SNUBBER CIRCUIT
CSN 101 103 {CSNUB}; The element CSN has a parameter name CSNUB.
RSN 103 100 {RSNUB}; The element RSN has a parameter name RSNUB.
* The default parameter values are given in the .SUBCKT statement.

.ENDS     DIO_V;        Subcircuit descriptions end with .ENDS.

* * * * * * * * * * * * * * * * * * * * * * * * * * * * * * * * * * * * * * * * * * *
```

In a circuit file, a call for a subcircuit, for example

```
Xname 1 0 DIO_V
```

would place the subcircuit DIO_V across the nodes 1, 0 of the PSpice configuration. The default values CSNUB = 0.1μF and RSNUB = 1kΩ would be used in the main circuit.

In a circuit file, a call for the same subcircuit in the form

```
Xname 1 0 DIO_V PARAMS: CSNUB=0.2uF, RSNUB=500ohms
```

would place the subcircuit DIO_V across the nodes 1, 0 in the main circuit. However, the values CSNUB = 0.2μF and RSNUB = 500Ω would replace the default values. This gives added flexibility to the use of subcircuits.

Drill Exercise D4.2.1

A half-wave diode rectifier is in series with a single-phase supply to provide direct current in an *RL* load. The circuit details are given in Fig. 4.5. Use a PSpice voltage-controlled switch to model the diode. From a simulation, use PROBE to determine **(a)** the average current in the load and **(b)** the average power consumed by the load. Use the subcircuit named DIO_V.

(Ans: **(a)** 8.8583A, **(b)** 834W (I_{drms} = 12.915A).)

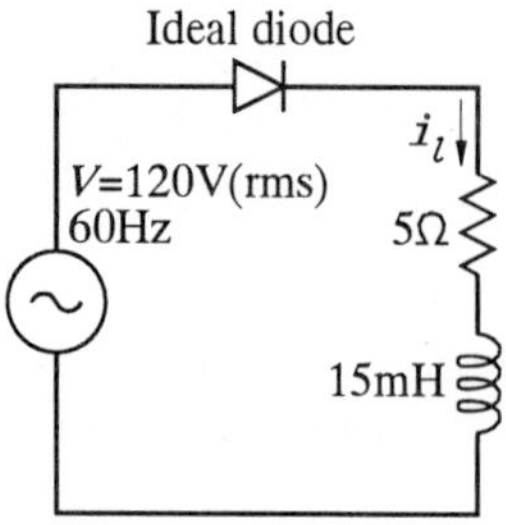

Fig. 4.5

Exercise D4.2.2

Consider the circuit diagram in Fig. 4.6. Simulate this circuit with a PSpice configuration. Model the diode by means of the PSpice voltage-controlled switch. Use PROBE to view the current response and the capacitor voltage over an interval of 50ms.

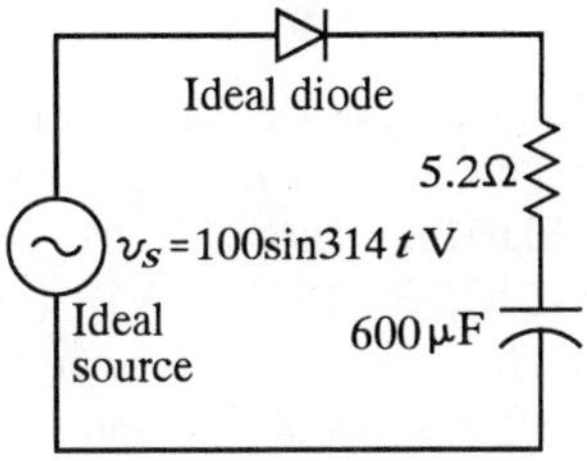

Fig. 4.6

4.2.2 CURRENT-CONTROLLED SWITCH AS THE DIODE

A PSpice current-controlled switch (symbol W, model type ISWITCH) can represent the ideal diode. See Table 4.1. Together with this switch we use a current sensor in the form of a voltage source which has a zero-voltage magnitude. Section 4.2.2 describing this diode model is on the WEBsite.

Table 4.1 Model parameters of the current-controlled switch (Model type ISWITCH)

Model parameter	Description	Units	Default value
RON	Resistance of the closed switch	ohms	1Ω
ROFF	Resistance of the open switch	ohms	1MΩ
ION	Minimum control current to close the switch	amperes	1mA
IOFF	Maximum control current to open the switch	amperes	0

SUBCIRCUIT FOR THE DIODE MODEL DIO_I

The current-controlled switch with the model name DIO_I models an ideal diode reasonably well. We can create a subcircuit file and put it in the library file POWER.LIB, so that we can call it in whenever it is needed. This subcircuit configuration is shown in Fig. 4.7 on the next page. It can be used in place of the voltage controlled switch in EXAMPLE 4.2.1.

```
*******************************************************
* AN IDEAL DIODE MODELLED BY A CURRENT-CONTROLLED SWITCH
* This is a subcircuit named DIO_I.
.SUBCKT DIO_I   101   100
WDIODE  101  Z  VZ   DIO;   VZ gives the control current.
* The diode anode is node 101 and the cathode is node Z.
* DIO is the model name.
.MODEL  DIO  ISWITCH(RON=1E-3  ROFF=1E6  ION=0  IOFF=-1E-4)
VZ   Z    100  0
.ENDS    DIO_I
*******************************************************
```

This subcircuit is called from the circuit file by the two statements

```
.LIB   POWER.LIB
XD   N+   N-   DIO_I
```

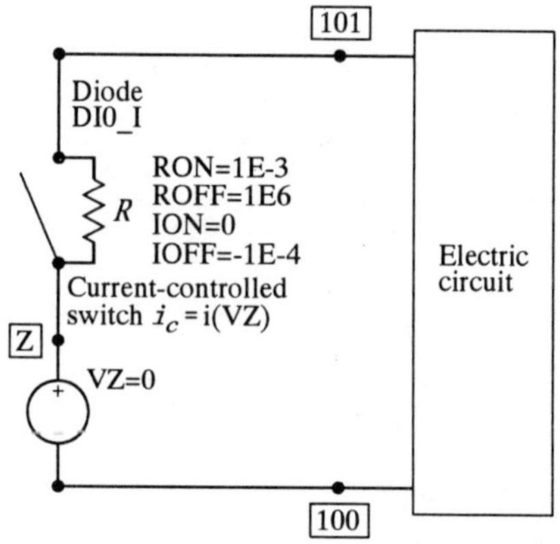

Fig. 4.7
A diode subcircuit named DIO_I.

Drill Exercise D4.2.3
A half-wave diode rectifier and a single-phase supply provide direct current in an *RL* load. The circuit details are given in Fig. 4.5. Use a PSpice current-controlled switch to model the diode. From the simulation use PROBE to plot the circuit current and determine **(a)** the average current in the load and **(b)** the average power consumed by the load. Use the subcircuit named DIO_I. Compare the results with those from Drill Exercise D4.2.1.
(Ans: **(a)** 8.8883A, **(b)** 839.5W (I_{rms} = 12.958A).)

Drill Exercise D4.2.4
Repeat Drill Exercise D4.2.2 using a current-controlled switch (for example, from the subcircuit DIO_I in the file POWER.LIB) for the ideal diode. In the simulation do the transient analysis with a ceiling step size of 20μs and a print step (TSTEP) of 20μs. View the circuit current and the capacitor voltage over an interval of 50ms using PROBE.

4.2.3 VCCS AS THE DIODE

A PSpice voltage-controlled current source (VCCS, symbol G) can represent a near-ideal diode. Arbitrarily, we will design the diode model to have a resistance of 0.1mΩ in the conduction mode and a resistance of 100MΩ in the blocking mode. This model has an *I-V* characteristic as shown in Fig. 4.8. There is negligible resistance to forward conduction and high resistance to reverse bias.

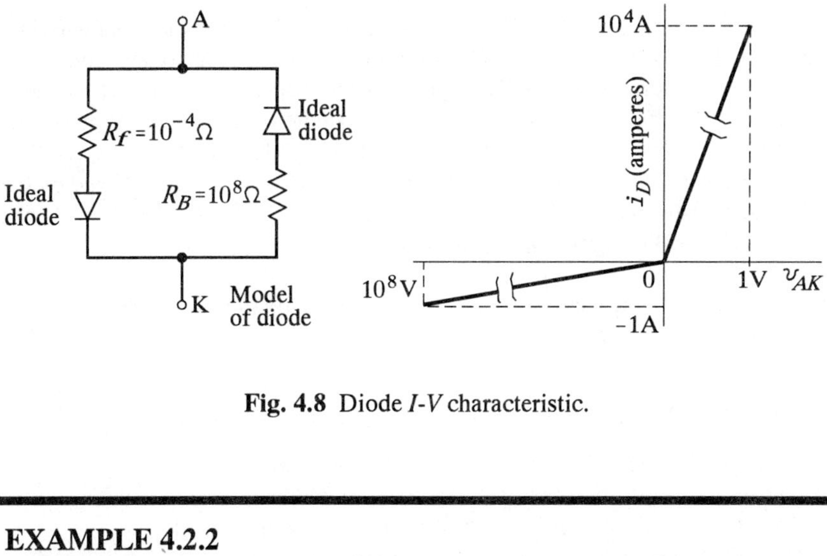

Fig. 4.8 Diode *I-V* characteristic.

EXAMPLE 4.2.2
A diode is to act as a half-wave rectifier for a circuit with an alternating-voltage source. The *RL* load is to conduct direct current. Values are shown in the circuit diagram of Fig. 4.9a on the next page. Create a suitable circuit file.

Solution
There are two steps to follow in the solution of this example.

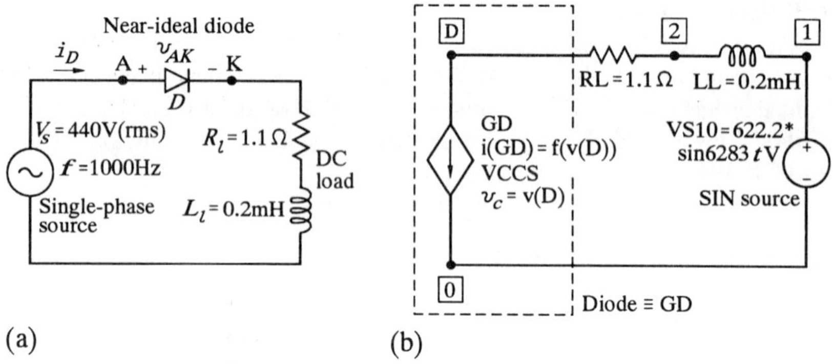

Fig. 4.9 Half-wave rectifier.
(a) Circuit diagram, (b) PSpice configuration.

STEP 1 Transform the circuit diagram of Fig. 4.9a to a PSpice configuration, as shown in Fig. 4.9b. In this configuration the voltage-controlled current source GD represents the near-ideal diode. The control voltage is v(D), the voltage across the VCCS (statement name GD) which models the diode *D*. This voltage v(D) controls the value of the current output i(GD) of this dependent source. The functional form i(GD) = f(v(D)) is determined by the *I-V* characteristic in Fig. 4.8 as a piecewise linear (PWL) relation. This is explained immediately following the circuit file description.

STEP 2 Create a circuit file and run the simulation to view the diode voltage and current.

»»»»» EX4_2_2.CIR »»»»»»»»»»»»»»»»»»»»»»»»»»»»»»»»»»»»»»»

```
DIODE TEST CIRCUIT MODELLED BY A VCCS GD
* To check the operation of a VCCS acting as a diode.
* The VCCS turns on if the dependent source is forward biased.
* The VCCS turns off if the dependent source is reverse biased.

* SOURCE
VS10   1   0   SIN(0 622.16V 1000Hz); Sinusoidal voltage source.

* CIRCUIT ELEMENTS
RL     D   2   1.1 ohms;            The load resistance.
LL     2   1   0.2mH;               The load inductance.
```

```
* THE DIODE MODELLED BY A VCCS (name must begin with G).
GD      D  0   TABLE  {v(D)} = (-1E8, -1)  (0, 0)  (1, 1E4) ;  . . . . . . . . (1)

* ANALYSIS, transient mode.
.TRAN   10us    3.01ms  0    0.5us    UIC
* UIC for convergence, avoids bias calculation.
.PROBE
.END
```

»»

EXPLANATION OF THE FILE EX4_2_2.CIR

. . . .

Line (1) of circuit file EX4_2_2.CIR

GD D   0 TABLE   {v(D)} = (-1E8, -1) (0, 0) (1, 1E4) ;   (1)

- There are four fields in this statement. See the Appendix.
- The first field contains GD, the element name. The first letter G identifies the element as a PSpice dependent current source, voltage controlled.
- The second field contains D, 0. These characters identify the two nodes to which the current source is connected. Current has direction from D to 0.
- The third field has TABLE, that allows nonlinear modelling in the form of a piecewise-linear (PWL) function. There must be a space after TABLE.
- The fourth field contains

  {v(D)} = (-1E8, -1) (0, 0) (1, 1E4)

  In this case the expression is the VCCS control voltage (v(D)) and represents the input of the PWL function. The right-hand side of the equality in this field has a number of pairs of values. The first number in each pair refers to the input voltage v(D) (the expression value) and the second number refers to the output (the value given to the current source GD). The first pair of values (-1E8, -1) means that if the input expression v(D) is equal to 10^{-8} volts, then the source current is -1A. The second pair of values (0, 0) means that for v(D)=0, then i(GD)=0. The third pair of values (1, 1E4) means that if v(D)=1, the current is i(GD)=10^4 A. The piecewise-linear function is obtained by joining straight lines between each of the pairs monotonically. This is shown in Fig. 4.8. PSpice interpolates to find i(GD) for any given value of voltage v(D). In this way the *I-V* characteristic of the diode is modelled by a PSpice dependent voltage source G. The controlled source modelling the diode is called an analogue behavioural model.

This ends the explanation of circuit file EX4_2_2.CIR

. . . .

END OF EXAMPLE 4.2.2

SUBCIRCUIT FOR DIODE DIO_GD

The results of the simulation of the diode circuit using the circuit file EX4_2_2.CIR are normally satisfactory. The near-ideal diode with its PWL *I-V* characteristic can be modelled by the PSpice dependent source GD and the TABLE facility. If we use this model repeatedly we can create a subcircuit. See the library file POWER.LIB under DIO_GD.

Drill Exercise D4.2.5

Consider the circuit diagram in Fig. 4.10. Make a PSpice simulation using a current-controlled voltage source for the diode as described in the subcircuit named DIO_GD in the library file POWER.LIB. **(a)** Plot the diode's *I-V* characteristic over a range of source voltages ±600 volts. **(b)** Determine from the PROBE trace the diode voltage drop for a current of 400A. (Note. To change the *x*-axis variable in PROBE use the X-axis submenu.) (Ans: **(b)** 40mV.)

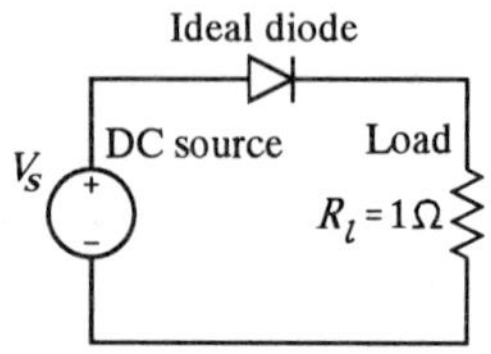

Fig. 4.10

Drill Exercise D4.2.6

Consider the circuit diagram in Fig. 4.11. This half-wave diode rectifier has an *RL* load, and a free-wheeling diode *DF* is used. Do a PSpice simulation. Use a voltage-controlled current source G (subcircuit DIO_GD in the library file POWER.LIB) to model both ideal diodes. Use PROBE to determine **(a)** the load-voltage trace, the load-current trace and the trace of the freewheeling-diode current, **(b)** the average load current, **(c)** the freewheeling diode peak-reverse voltage, **(d)** the freewheeling-diode average-current rating, **(e)** the rectifying-diode peak-reverse voltage and **(f)** the rectifying-diode average-current rating. Do a transient analysis over at least two cycles of steady-state operation. (Ans: **(b)** 12.72A, **(c)** 200V, **(d)** 1.8A, **(e)** 200V, **(f)** 10.92A.)

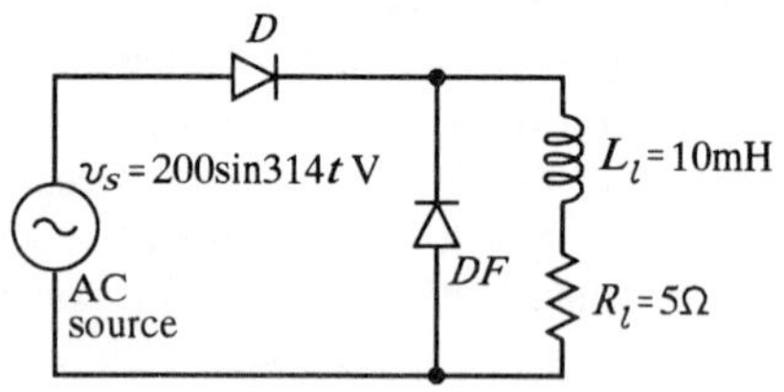

Fig. 4.11

There are Extra Drill Exercises on the WEBsite.

4.2.4 PSPICE RECTIFIER, THE DIODE

This section makes use of a built-in PSpice junction diode model (symbol D, model type D). This model is determined in terms of the device physics. The model has 29 parameters to be described. However, in this chapter we want the diode to be nearly ideal, so we will allow 28 of the parameters to have default values. Accordingly, we do not have to consider them. The only parameter of concern for a virtually ideal diode is the emission coefficient N. This coefficient is given a low value for simulation purposes only. In practice the low value of N used in this section is unrealistic, but it does give a low diode voltage drop.

EXAMPLE 4.2.3

The circuit diagram in Fig. 4.12a represents a half-wave rectifier circuit. Obtain a PSpice configuration and a circuit file using a built-in semiconductor device to model a near-ideal diode.

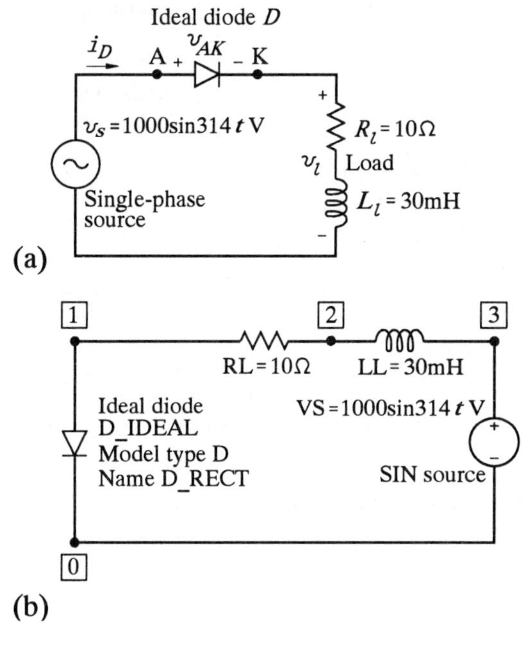

Fig. 4.12 Half-wave rectifier.
(a) Circuit diagram, (b) PSpice configuration.

Solution

There are two steps in this solution.

STEP 1 Transform the circuit diagram to a PSpice configuration as shown in Fig. 4.12b. The PSpice diode is given a statement name D_RECT and a model name D_IDEAL. Its model-type symbol is D. See Table 2.2. Also, see the subcircuit D_IDEAL in POWER.LIB.

STEP 2 Create a circuit file called EX4_2_3.CIR. Run the simulation to view the diode voltage and current.

```
»»»»» EX4_2_3.CIR »»»»»»»»»»»»»»»»»»»»»»»»»»»»»»»»»»»»»»

DIODE TEST CIRCUIT  D_RECT/D_IDEAL  RL RECTIFIER CIRCUIT
* To view the results of rectification using a near-ideal, PSpice, built in diode.

* SOURCE
VS     3  0   SIN(0 1000V 50Hz);          Sinusoidal voltage supply.

* CIRCUIT ELEMENTS
RL     2  1   10ohms;                     Load resistance.
LL     3  2   30mH;                       Load inductance.

* CIRCUIT DEVICE, the near-ideal diode.
D_RECT   1  0   D_IDEAL; ............................... (1)
.MODEL   D_IDEAL   D(N=0.001); ......................... (2)

* ANALYSIS, to cover 3 half cycles (30ms). TSTEP=20μs. TSTART=0.
.TRAN      20us  30ms   0   40us; Time-step ceiling is TMAX=40μs.
.PROBE   v(1),  i(D_RECT);        Records diode voltage and current only.
.END

»»»»»»»»»»»»»»»»»»»»»»»»»»»»»»»»»»»»»»»»»»»»»»»»»»»»»»»»
```

EXPLANATION OF THE FILE EX4_2_3.CIR

. . . .

Line (1) of circuit file EX4_2_3.CIR

D_RECT 1 0 D_IDEAL; (1)

- This is a statement of a device that must have with it a .MODEL statement.
- There are four fields in this statement.

- The first field is for the statement name D_RECT. The name is to indicate a rectifying diode. The name must begin with D.
- The second field contains the names (1, 0) of the nodes to which the diode is connected. The anode A of the diode is connected to the first node. The cathode K is connected to the second node.
- The third field contains D_IDEAL which is the arbitrary model name given here to the diode. This name is repeated in the .MODEL statement to form a link between the two statements.
- The general form of the device statement is

```
Dxxx  N+  N-  model_name
```

. . . .

Line (2) of circuit file EX4_2_3.CIR

```
.MODEL   D_IDEAL   D(N=0.001);                                   (2)
```

- There are four fields in this statement.
- The first field contains .MODEL. This tells PSpice that the device parameter values are to follow in the same statement.
- The second field contains D_IDEAL, which is the name of the device, linking it with the device data statement in line (1).
- The third field contains D, which is the symbol of the model type of the device. In this case the device is the diode.
- The fourth field contains (N = 0.001), one of the 29 possible parameters. This low value of the emission coefficient N makes the on-state voltage drop v_{AK} negligibly small, so that the diode appears almost ideal. The other twenty eight parameters are unimportant to us in this chapter. See Chapter 11.
- The general model statement for the diode is

```
.MODEL  diode_name  D(param1=value1  param2=value2 ... )
```

This ends the explanation of circuit file EX4_2_3.CIR.

. . . .

END OF EXAMPLE 4.2.3

Drill Exercise D4.2.7

A half-wave diode rectifier is in series with a single-phase supply to provide direct current in an *RL* load. The circuit details are given in Fig. 4.13, depicted on the next page. Use the PSpice diode model with parameter values to provide near-ideal characteristics. Use PROBE to plot **(a)** the average current in the load. **(b)** What is the average power consumed in the load? **(c)** From the output file

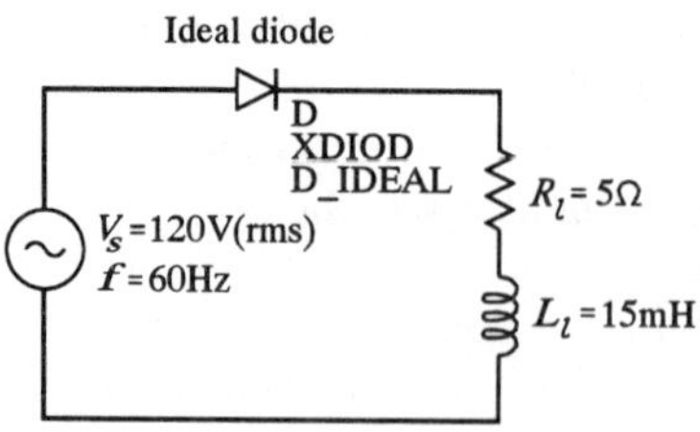

determine the time of the program execution. Use the subcircuit named D_IDEAL from POWER.LIB. Compare the results with those from Drill Exercise D4.2.3.
(Ans: **(a)** 8.8924A, **(b)** 839.8W ($I_{d\,\mathrm{rms}}$ = 12.96A), **(c)** 39.2s.)

Fig. 4.13

Drill Exercise D4.2.8

Use a PSpice simulation of the circuit depicted in Fig. 4.14 to plot the ideal *I-V* characteristic of the diode over a range of source voltages ±600 volts. Use any of the diode models D_IDEAL, DIO_V from Section 4.2.1, DIO_I from Section 4.2.2, or DIO_GD from Section 4.2.3. (Note: To change the *x*-axis variable in PROBE use the X_axis submenu.)

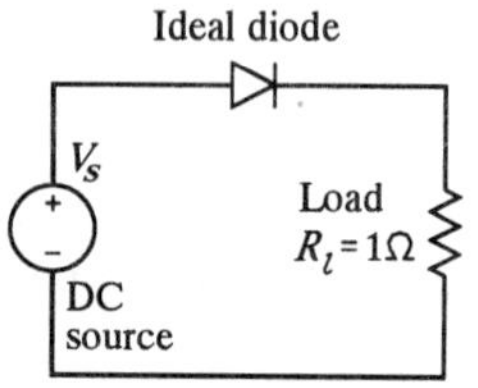

Fig. 4.14

4.3 THE IDEAL BJT TRANSISTOR AS A SWITCH

A simple circuit diagram that contains a bipolar junction transistor (BJT) is shown in Fig. 4.15a. The BJT is to be used as a switch, on and off in a sequence, to modulate power from the source v_s to the load R_l.

The BJT is turned on (virtually zero resistance between the collector C and the emitter E) if a sufficient current is injected into the base region B while the collector terminal is biased positively with respect to the emitter terminal. If the BJT is in the on-state, it is turned off (virtually infinite resistance between the collector and the emitter) by reducing the base current to zero.

Consider the terms in Fig. 4.15a. In essence, if v_{CE} is positive, then a sufficient V_B turns the BJT on. The absence of a source V_B means the BJT is off. The *I-V* characteristic of the ideal BJT switch is shown in Fig. 4.15b. If the BJT is off, no current is conducted between C and E, and the voltage v_{CE} assumes the same value as the source voltage v_s. If the BJT is on, the voltage drop v_{CE} is virtually zero and the value of the current i_C depends solely on the values of the elements, source and load. The current is unidirectional in a BJT.

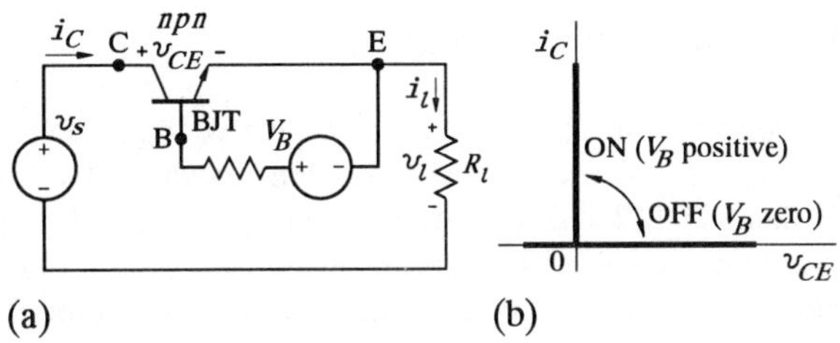

Fig. 4.15 A BJT circuit.
(a) Circuit diagram, (b) ideal *I-V* characteristic of a BJT switch.

The real BJT cannot withstand any significant reverse bias voltage without being destroyed. In any circuit in which a BJT is likely to be subjected to a reverse voltage, protection by a series diode or a reverse-connected parallel diode is usually practical. Because of its reverse-voltage limitations, the BJT can be used in chopper and inverter circuits but not in rectifier or ac voltage regulation applications.

4.3.1 IDEAL BJT CHOPPER

We are familiar with a chopper to modulate power from a dc supply. In this example we wish to represent the BJT as an ideal switch that can conduct unidirectional current for as long as a signal is applied to the base of the transistor. We will use the circuit diagram in Fig. 4.15a as the chopper circuit and employ previously used PSpice elements to do the simulation. A current-controlled switch and an ideal diode will form the BJT model, because the BJT is a current-controlled device.

EXAMPLE 4.3.1

The circuit diagram in Fig. 4.16a, depicted on the next page, gives the component values of the circuit elements of a chopper. Model the *npn* BJT by means of a current-controlled switch and an ideal diode. Create a suitable circuit file. For a switching frequency of 2kHz and a duty cycle of 0.5 do a PSpice simulation to find the average load power (3.99kW). Use PSpice parameter values (.PARAM) for the switching frequency, the duty cycle and the load resistance in order to make the circuit file adaptable to adjustable values.

Solution

There are two steps in the solution.

STEP 1

The first step is to create a PSpice configuration as shown in Fig. 4.16b. The current-controlled switch W_BJT and the ideal diode D_SERIES together constitute the ideal BJT. The BJT base drive is essentially modelled by the independent current source IBASE which is a periodic source of pulses. The existence of a pulse means that the BJT has been turned on. No current pulse indicates that the BJT is in the off-state.

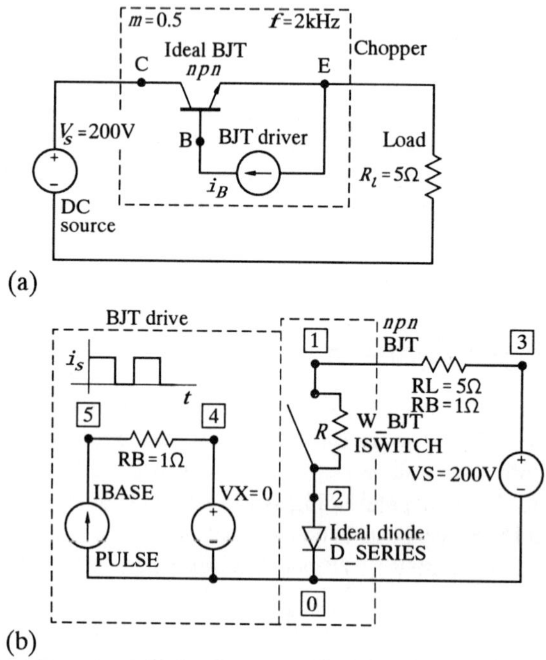

Fig. 4.16 BJT chopper.
(a) Circuit diagram, (b) PSpice configuration.

STEP 2

Create the PSpice circuit file from the configuration of Fig. 4.16b. Now do a simulation. Plot the load current and the running rms value.

```
»»»»» EX4_3_1.CIR »»»»»»»»»»»»»»»»»»»»»»»»»»»»»»»»»»»»

IDEAL BJT CHOPPER CIRCUIT  (BJ_IDEAL)
* To model the ideal BJT as a current-controlled switch.
* The BJT is on if the control current is finite and positive.
* The BJT is off if the control current is zero.

* Parameter definitions and nominal values
.PARAM  FREQ=1kHz;              The chopper switching frequency.
.PARAM  PERIOD={1/FREQ};        The chopper switching period.
.PARAM  DUTY=0.5;               The chopper duty cycle m.
.PARAM  TON={DUTY*PERIOD};      Chopper on-time.
.PARAM  RLOAD=5ohms

* SOURCE
VS  3  0  DC  200V;          This is the dc supply of the power circuit.

* CIRCUIT ELEMENT
RL  3  1  {RLOAD};           Resistance is characterized by a parameter.

* CIRCUIT DEVICE, the ideal BJT as a switch.
W_BJT     1   2  VX  BJ_IDEAL;       The device statement for a switch.
.MODEL  BJ_IDEAL  ISWITCH(RON=1E-2 ROFF=1E6  ION=1  IOFF=0)
D_SERIES      2  0   D_BLOCK;        The reverse-blocking diode statement.
.MODEL  D_BLOCK  D(N=0.001);         Near-ideal PSpice diode.

* BJT DRIVER CIRCUIT
IBASE  0  5    PULSE(0  1.2A  0  10ns  10ns  {TON}  {PERIOD})
*                            The periodic base current pulse of 1.2A.
RB     5  4    1ohm;         An arbitrary loop resistance.
VX     4  0    0;      A zero-voltage source to sense the base current to turn
*                      on the switch W_BJT.

* ANALYSIS, to determine circuit response over 2 cycles of switching.
.TRAN  20us   2ms   0   40us  UIC;      TSTEP=20μs, TSTOP=2ms,
*                                       TSTART=0,  TMAX=40μs.
.PROBE  i(RL);               The problem solution requires this load current.
.END
»»»»»»»»»»»»»»»»»»»»»»»»»»»»»»»»»»»»»»»»»»»»»»»»»»»»»»
```

END OF EXAMPLE 4.3.1

SUBCIRCUIT FOR THE MODEL OF THE IDEAL BJT BJ_IDEAL

The results of EX4_3_1.CIR make it worth while to create a subcircuit for the ideal BJT using a current-controlled switch. There could be some tuning to make the model more stable. The subcircuit description is written in the library file POWER.LIB and named BJ_IDEAL. View it.

Drill Exercise D4.3.1
Follow the pattern of EXAMPLE 4.3.1 to simulate a BJT chopper to operate at 4kHz and duty cycle $m=0.7$. The dc source is 400V and the load resistance is $R_l=10\Omega$. Plot the trace of the load current and determine the average power consumed by the load. (Ans: $16.72^2 \times 10 = 2.8$kW.)

Drill Exercise D4.3.2
A BJT is used as a chopper to modulate power from a 96-V dc source to a load resistance of 4.2 ohms. Use the subcircuit BJ_IDEAL in the library file POWER.LIB. If the switching frequency of the chopper is 500Hz and the duty cycle is $m=0.3$, determine the average load current **(a)** from PROBE and **(b)** from a hand calculation. (Ans: **(a)** 6.841A, **(b)** 6.857A.)

4.4 THE IDEAL THYRISTOR

A simple circuit diagram that contains a thyristor (SCR) is shown in Fig. 4.17a. The thyristor can be used only as a switch. It has applications in high power choppers, controlled rectifiers and inverters at frequencies up to about 1kHz. Figure 4.17a illustrates the use of the thyristor as a controlled rectifier that modulates the power to the load.

The essential functions of the ideal thyristor are as follows.

1. The thyristor is OFF. The thyristor blocks any voltage, forward bias or reverse bias that is applied across its terminals anode A and cathode K if there is no signal applied to the thyristor terminals gate G and cathode K.
2. The thyristor changes state from OFF to ON only if a positive current i_G is applied to the gate terminal and if the anode voltage v_{AK} is positive with respect to the cathode.
3. Once on the thyristor will remain ON, even if the gate signal is removed.

4. Once on the thyristor will not turn off until the gate signal i_G has been removed and the anode current i_A reduces to zero. With both of these conditions satisfied the thyristor turns off and blocks any voltage v_{AK} of any polarity.
5. The thyristor, like the BJT, is a current-controlled device. That is, it is current (or charge) at the gate region that causes turn-on.

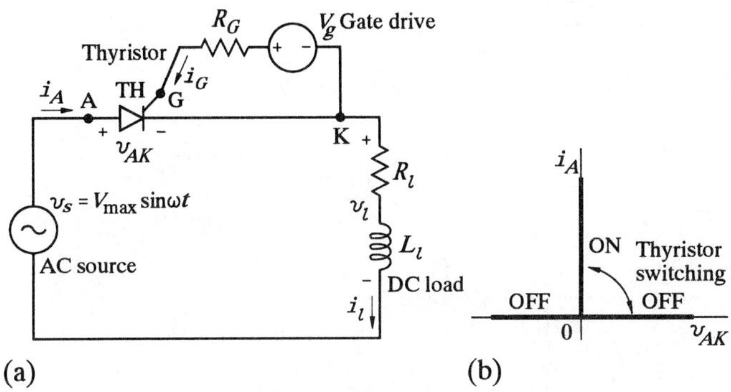

Fig. 4.17 A thyristor rectifier.
(a) Circuit diagram, (b) the ideal thyristor *I-V* characteristics.

4.4.1 IDEAL THYRISTOR RECTIFIER

In this example we wish to adapt the PSpice current-controlled switch so that it behaves like an ideal thyristor. We will introduce a dependent current source with a TABLE of values in this example. It allows us to model the thyristor in the on-state with the gate signal removed. PSpice classifies modelling with dependent sources as analogue behavioural modelling (ABM).

EXAMPLE 4.4.1

The circuit diagram in Fig. 4.18a, shown on the next page, gives the component values of a single-phase, half-wave, controlled-rectifier circuit. By adjusting the firing angle α the power delivered to the dc load can be modulated. Create a PSpice circuit file, and, for a firing angle of $\pi/3$ rad, determine from PROBE the average power delivered by the source. (Ans: 588.9W.)

Solution

There are two steps involved in the simulation and solution of this example.

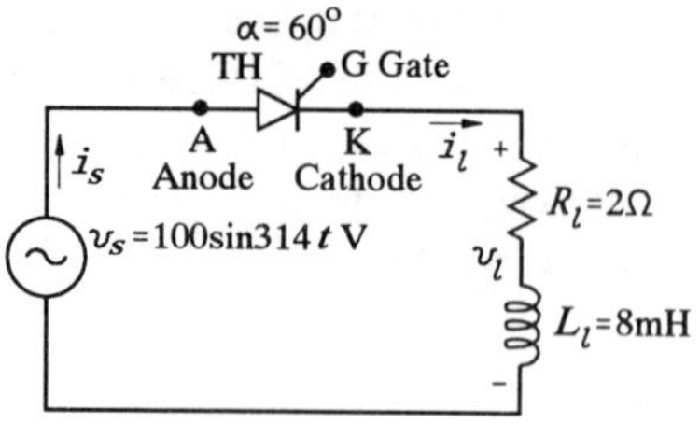

(a)

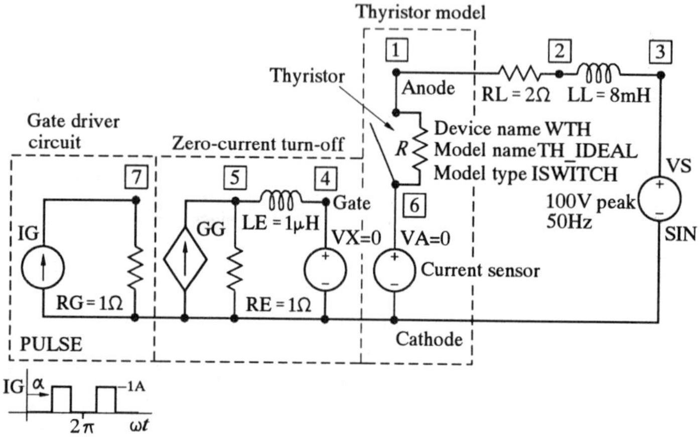

(b)

Fig. 4.18 Half-wave thyristor rectifier. (a) Circuit diagram, (b) PSpice configuration.

STEP 1

Step 1 involves the creation of a PSpice configuration from the circuit diagram in Fig. 4.18a. Once we have chosen the current-controlled switch, type ISWITCH, to model the thyristor, we need a pulse gate drive to represent turn-on firing pulses at the firing angle α every cycle of the source-voltage waveform. The firing pulses must be synchronized with this voltage-source waveform. These pulses are not applied directly to turn on the switch. In order to keep the PSpice switch latched on while the thyristor has no gate signal but conducts current, we must have an equivalent OR circuit. This OR circuit takes the PSpice form of a dependent current source (G) with a TABLE that will allow a source's current to exist in an expression. It is the dependent source that gives rise to the switch turning on or off. This overall current-control circuit is shown in Fig. 4.18b to the left of the power circuit.

STEP 2

Write a circuit file EX4_4_1.CIR with the aid of the PSpice configuration in Fig. 4.18b. A circuit file follows. Now a simulation run can be done and the rms load current viewed.

```
»»»»» EX4_4_1.CIR »»»»»»»»»»»»»»»»»»»»»»»»»»»»»»»»»»»»»»»»

AN IDEAL THYRISTOR RECTIFIER CIRCUIT   (TH_IDEAL)
* To determine the power delivered to the RL load.
* PARAMETERS
.PARAM FREQ=50Hz;                        The voltage-supply frequency.
.PARAM PERIOD={1/FREQ};                  The voltage-supply period.
.PARAM ALFA=60;                          Firing angle in degrees.
.PARAM DELAY={ALFA*PERIOD/360};          Time for firing the thyristor.
.PARAM P_WIDTH=1ms;                      The pulse width of the gate signal.

* SOURCES
VS 3   0   SIN(0 100V {FREQ});  A 100-V peak, 50-Hz source.
VA 6   0   0;                   Current sensor, a zero-voltage source.

* CIRCUIT ELEMENTS
RL 2   1   2ohms;                    Load resistance.
LL 3   2   8mH;                      Load inductance.

* CIRCUIT DEVICE, current-controlled switch modelling the thyristor.
WTH    1   6   VX   TH_IDEAL;            The current switch statement.
* WTH is the name of the device across nodes 1, 6.
* VX is the sensor for the control current. TH_IDEAL is the model name.
.MODEL TH_IDEAL  ISWITCH(RON=1E-3 ROFF=1E6 ION=0.1 IOFF=0)
* This current switch models the thyristor. VX measures ION and IOFF.

* GATE DRIVER
IG  0 7  PULSE(0 0.1A {DELAY} 5ns 5ns {P_WIDTH} {PERIOD})
* A single 0.1A pulse for 1ms at ALFA every cycle.
RG 7 0  1ohm;                 1ohm is used so that i(RG) = v(7) numerically.

* CURRENT CONTROL, thyristor on if gate signal or thyristor current exists.
GG 0 5  TABLE {v(7)+i(VA)}=(0,0) (1E-3,0) (1.5E-3,0.1) (0.1,0.1);.  (1)
* Cannot use i(RG) in expression. Must use i(RG) = v(7)/RG = v(7)/1 = v(7).
RE 5 0  1ohm;                 The current-control circuit loop resistance.
LE 5 4  1uH;   A delay for current to rise in sensor VX.
VX 4 0  0;     A current sensor (zero-voltage source) i(VX) for turn-on.

* ANALYSIS, transient for 3 half cycles (1.5*PERIOD).
.TRAN   20us 30ms 0 40us  UIC;     TSTEP=20μs, TSTOP=30ms,
*                                  TSTART=0,  TMAX=40μs.
.PROBE;        To view all variables.
.END
»»»»»»»»»»»»»»»»»»»»»»»»»»»»»»»»»»»»»»»»»»»»»»»»»»»»»»»»»»»
```

EXPLANATION OF THE FILE EX4_4_1.CIR

. . . .

Line (1) of circuit file EX4_4_1.CIR

```
GG 0 5  TABLE {v(7)+i(VA)}=(0,0) (1E-3,0) (1.5E-3,0.1) (0.1,0.1);     (1)
```

- This is one of the seven possible kinds of statements for a dependent-current source. See the Appendix C.
- There are four fields in this statement.
- The first field contains GG, the element name. The first letter G identifies the element as a PSpice dependent-current source. The second G refers to the gate circuit, but could be anything relevant.
- The second field contains 0, 5. These numbers identify the two nodes to which the current source GG is connected. Current is directed from positive to negative node.
- The third field contains TABLE, a feature that allows nonlinear analogue behavioural modelling. There must be a space after TABLE.
- The fourth field contains

  {v(7) + i(VA)} = (0, 0) (1E−3, 0) (1.5E−3, 0.1) (0.1, 0.1)

  This includes an expression between the braces { }. The expression must be on a single line. In this case the expression is intended to be the sum of the gate current and the thyristor anode current. The right-hand side of this equality has a number of pairs of values. The first number in each pair refers to the input (the expression value) and the second number refers to the output (the value given to the current source GG).
  The first pair of values (0, 0) means that if the expression {v(7) + i(VA)} is equal to zero, then the source current GG is equal to zero. For an expression value less than zero, the source current would be zero in this case.
  The second pair of values (1E−3, 0) indicates that if the expression is equal to 10^{-3}, the output of GG is zero.
  The third pair of values (1.5E−3, 0.1) indicates that if the expression has a value of 1.5×10^{-3} then the source current GG assumes a value of 0.1A. This would turn on the switch (thyristor), because ION = 0.1A. Implicit, is that between the second and third pairs there is linear interpolation to evaluate the source current GG.
  The fourth pair (0.1, 0.1) is to be interpreted to mean that for v(7) + i(VA) = 0.1 then GG = 0.1A. Between the third and fourth pairs linear interpolation is done. In this example the output source current would be 0.1A for any input between 0.0015 and 0.1.

With regard to the last pair, for any actual value of the expression greater that the input value, the output value remains the same, that is 0.1A.
This whole group of pairs is similar to the method of the piece-wise-linear (PWL) source. The relation between the expression (input value) and GG (output value) is shown in Fig. 4.19.

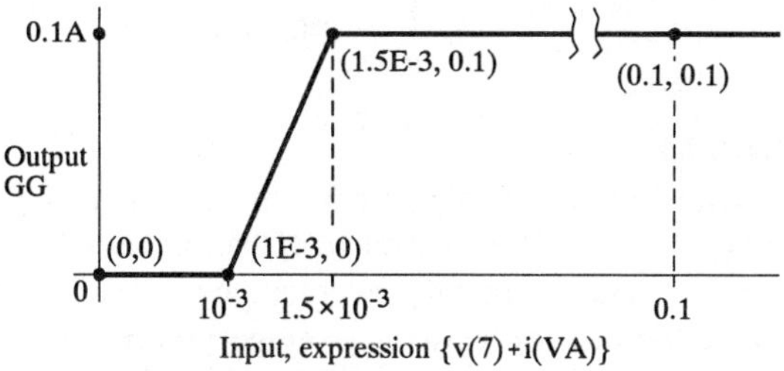

Fig. 4.19
TABLE expression as PWL description.

In the circuit of the dependent source GG, there are *R* and *L* elements to provide a time constant for any change of source current. The zero-voltage source VX measures the current that controls the switch WTH between the on and off states.

This ends the explanation of circuit file EX4_4_1.CIR.

. . . .

END OF EXAMPLE 4.4.1

SUBCIRCUIT FOR THE ANALOGUE MODEL OF THE IDEAL THYRISTOR (TH_IDEAL)

EXAMPLE 4.4.1 uses parameters in the circuit file. We have a choice of using global parameters that are defined in the circuit file and used in the subcircuit. There is also the choice of setting parameters in the subcircuit through PARAMS: and changing their values in the subcircuit call Xname or leaving them with their default values. The subcircuit TH_IDEAL that follows from the circuit file is written in the library file POWER.LIB. View this subcircuit.

Drill Exercise D4.4.1
Consider the circuit diagram in Fig. 4.20 shown on the next page. It represents a half-wave controlled rectifier to control the power transferred from the ac supply to the dc load. Determine **(a)** the average current delivered to the load and **(b)** the fundamental rms value of the supply current for thyristor firing angles 0,

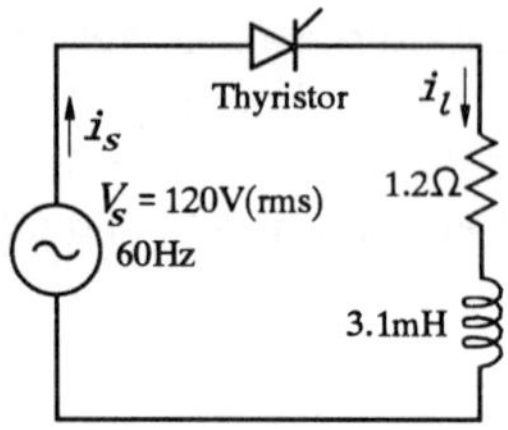

90° and 120°. Use the subcircuit TH_IDEAL and do not use PROBE. (Hint. Use the .FOUR statement.) (Ans: **(a)** 38.41A, 17.162A, 7.4316A, **(b)** 39.46A, 21.34A, 9.83A.)

Fig. 4.20

Drill Exercise D4.4.2

Consider the circuit diagram in Fig. 4.21. For a thyristor firing angle of $\alpha = 60°$ **(a)** plot traces of the load current, the freewheeling diode current over three half cycles, and the running average power supplied by the voltage source, and **(b)** determine the periodic average power delivered by the supply. For the thyristor model use the statements in EXAMPLE 4.4.1 first, then use the subcircuit TH_IDEAL in POWER.LIB. For the freewheeling diode model use the statements in EXAMPLE 4.2.2 or the subcircuit DIO_GD in POWER.LIB. (Ans: **(b)** 780.2W.)

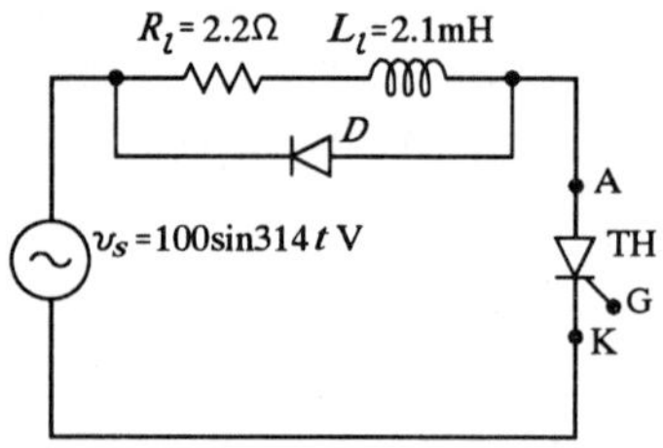

Fig. 4.21

An Extra Drill Exercise is on the WEBsite.

4.5 THE IDEAL TRIAC

The operation of the triac semiconductor switch is similar to the operation of a thyristor except that the triac conducts current bidirectionally. Figure 4.22a shows the symbol of a triac. The switch is used as an ac voltage regulator in this circuit diagram. The triac is an on/off switch. Its *I-V* characteristics are shown in Fig. 4.22b. It is a versatile switch because it can block voltage of either polarity at its terminals when it is off and current can be conducted in either direction when the triac is on.

In its normal state, the triac is off as long as there is no signal applied to the gate terminal. If positive or negative current of sufficient magnitude is applied to the gate region and if one of the main terminals (T_1 or T_2) is forward biased, the triac turns on and conducts current. The magnitude of the current depends on the other circuit elements, since the ideal triac has zero resistance once it is in the on-state. In the on-state, the triac will conduct even if the gate signal is removed. It is not until the main current reduces to zero, and there is no gate signal, that the triac turns off.

Since the triac is a current-controlled device, we like to use a PSpice current-controlled switch (type ISWITCH) for the model.

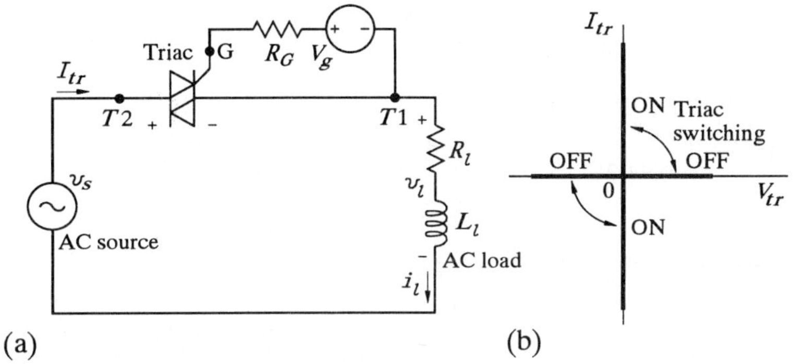

Fig. 4.22 Triac voltage regulator.

4.5.1 IDEAL-TRIAC VOLTAGE REGULATOR

One of the major applications of a triac is the ac-voltage regulator, commonly used as a light dimmer. We will adapt the ideal-thyristor model to allow it to conduct current bidirectionally. This adapted model will represent the ideal triac.

EXAMPLE 4.5.1

The circuit diagram in Fig. 4.23, depicted on the next page, a shows the component values of an ac-voltage-regulator circuit. Simulate the circuit using PSpice for a triac firing angle $\alpha = 90°$. Determine the rms value of the load voltage **(a)** from the Fourier analysis results and **(b)** from a PROBE trace.

Solution

There are four steps involved in the simulation of this ac-voltage regulator.

STEP 1 Create a PSpice configuration from the circuit diagram of Fig. 4.23a. This is shown in Fig. 4.23b. The current-controlled switch models the triac. The gate driver is the in-dependent current source that is to provide a pulse every half cycle of the ac-source waveform. This pulse will indirectly turn on the triac switch to allow conduction in both directions. Similar to the thyristor ideal model, the switch control current must be governed by an OR expression in a TABLE of a voltage-controlled current source GG. The OR expression requires an equivalent gate current (v(7)) and the triac current (i(VA)). The triac switch is to be on if there is a gate signal or if the triac is conducting current in either direction. The elements RE and LE are added to give a delay to the current rise of i(VX), which is the control current of the current-controlled switch.

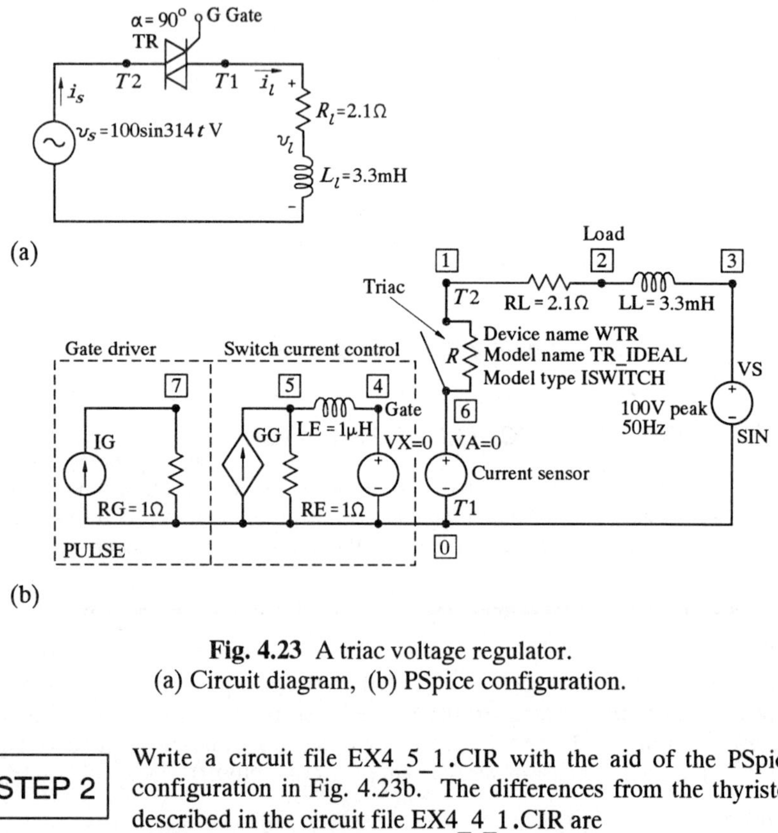

Fig. 4.23 A triac voltage regulator.
(a) Circuit diagram, (b) PSpice configuration.

STEP 2 Write a circuit file EX4_5_1.CIR with the aid of the PSpice configuration in Fig. 4.23b. The differences from the thyristor described in the circuit file EX4_4_1.CIR are

1. the triac is switched every half cycle, whereas the thyristor is triggered every cycle. Thus the current source IG for the triac has a different period.
2. the triac conducts for a part of every half cycle, so the triac current is alternating. This is accounted for by sensing the absolute value of the triac current using ABS(i(VA)) in the control circuit of the VCCS named GG.

A triac subcircuit named TR_IDEAL is written in POWER.LIB

»»»»» EX4_5_1.CIR »»»»»»»»»»»»»»»»»»»»»»»»»»»»»»»»»»»»»»»

```
AN IDEAL TRIAC VOLTAGE REGULATOR  (TR_IDEAL)
* To determine the rms value of the load current for ALFA = 90 degrees.
. PARAM FREQ=50Hz;                          AC source frequency f.
. PARAM PERIOD={1/FREQ};                    AC source period T.
. PARAM ALFA=90;                            Trigger angle in degrees.
. PARAM DELAY={ALFA*PERIOD/360};            Trigger angle in seconds.
. PARAM P_WIDTH=1ms;                        Pulse width of the gate signal.

* SOURCES
VS  3  0   SIN(0 100V {FREQ});          A 100-V peak, 50-Hz source.
VA  6  0   0;                           Zero-voltage source; (current sensor).

* CIRCUIT ELEMENTS
RL  2  1   2.1 ohms;                          Load resistance.
LL  3  2   3.3mH;                             Load inductance.

* CIRCUIT DEVICE, current-controlled switch to model the triac.
WTR   1  6  VX  TR_IDEAL;                The current switch statement.
* WTR is the name (with symbol W) of the device across nodes 1, 6.
* VX is the sensor for the control current that switches WTR on and off.
* TR_IDEAL is the model name given to the ideal triac.
. MODEL   TR_IDEAL  ISWITCH(RON=1E-3 ROFF=1E7 ION=0.1 IOFF=0)
* This current switch models the triac.  ION and IOFF are sensed by VX.
* GATE CIRCUIT, for indirect turn-on.  Current pulse source.
IG  0  7  PULSE(0 0.3A {DELAY} 5ns 5ns {P_WIDTH} {PERIOD/2})
* A single 0.3A pulse for 1ms at ALFA every half cycle.
RG  7  0  1ohm;                  1ohm is used so that v(7) = i(RG) numerically.

* CURRENT CONTROL, of the current-controlled switch.
* This current switches on or maintains switch on if the current is finite.
GG 0  5  TABLE {v(7)+ABS(i(VA))}=(0,0) (1.5E-3,0) (7.5E-2,0.1) (0.1,0.1)
* GG coordinate pairs are sensitive for turn-off to the power elements.
RE  5  0  1ohm;                  The current-control circuit loop resistance.
LE  5  4  1uH;                   A delay for the current i(VX) to rise to 0.1A.
VX  4  0  0;                     A current sensor for the current-controlled switch.

* ANALYSIS, transient for three half cycles (1.5*PERIOD).
. FOUR     50Hz   v(3, 1)
. TRAN 20us   30ms     0     40us;        TSTEP=20μs, TSTART=0,
*                                         TSTOP=30ms, TMAX=40μs.
. PROBE;                                  To view traces of all variables.
. END
```

»»»

STEP 3 Execute EX4_5_1.CIR. Care has to be taken if ABS(i(VA)) is to be used in the control strategy. The numerical method does not bring the load current to zero. The PWL control strategy must provide enough switch resistance at low currents to force the current to a small enough value that the VCCS (GG) output does become zero. For example

GG 0 5 TABLE {v(7)+ABS(i(VA))} = (0, 0) (1E−2, 0) (1.5E−2, 0.1)

does not reduce to zero, so WTR does not switch off.
Fine tuning is necessary for simulation success.

STEP 4 Inspect the Fourier-analysis results in file EX4_5_1.OUT and view the time varying trace of the load voltage v(3, 1) using PROBE.

Part (a) of Solution. From the file EX4_5_1.OUT the important results of the Fourier analysis are total harmonic distortion, THD = 0.6929, and fundamental rms voltage, $V_{l1} = 57.37/\sqrt{2} = 40.83$V.
By definition $\text{THD}^2 = V_{l\,\text{rms}}^2 / V_{l1}^2 - 1$.

So, $V_{l\,\text{rms}}^2 = V_{l1}^2\,\text{THD}^2 + V_{l1}^2 = 40.83^2 \times (0.6929)^2 + 40.83^2 = 2467.47$.

Hence, $V_{l\,\text{rms}} = 49.67\text{V}$.

This value can be in error because PSpice accounted for only 10 harmonics.

Part (b) of Solution. The traces of the load voltage v(3, 1) and the running rms value of the load voltage rms (v(3, 1)) are shown in Fig. 4.24. The actual rms value is the running value over one half cycle from 5ms to 15ms or the running value over one cycle from 5ms to 25ms. This value is 50.86V.

END OF EXAMPLE 4.5.1

Drill Exercise D4.5.1
Consider the circuit diagram in Fig. 4.25. The component values for this ac voltage regulator with a triac are given. Use PROBE to plot a trace of the load current over the interval of time from t=5.555ms to t=25ms. Determine **(a)** the average power delivered to the load using the Fourier analysis results recorded in the output file and **(b)** compare the answer obtained from PROBE. First follow the pattern in EX4_5_1.CIR then use the triac subcircuit TR_IDEAL.
(Ans: **(a)** 378.4W, **(b)** 378.7W.)

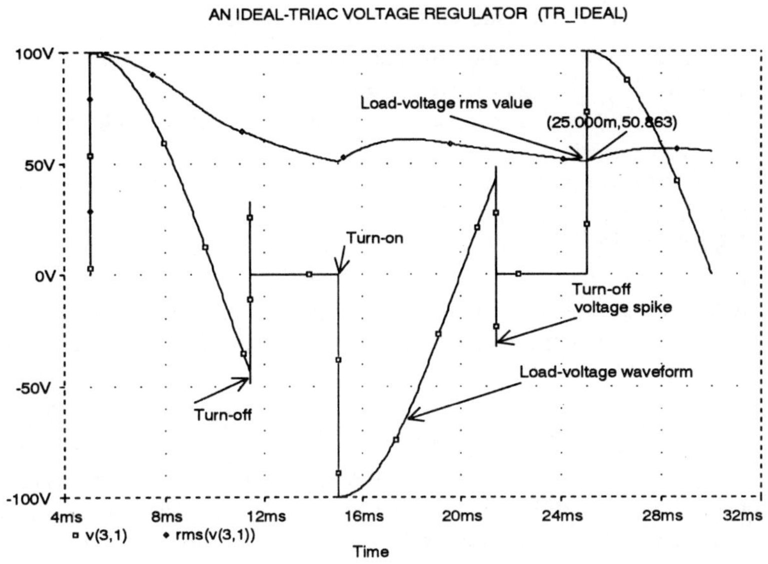

Fig. 4.24

Fig. 4.25

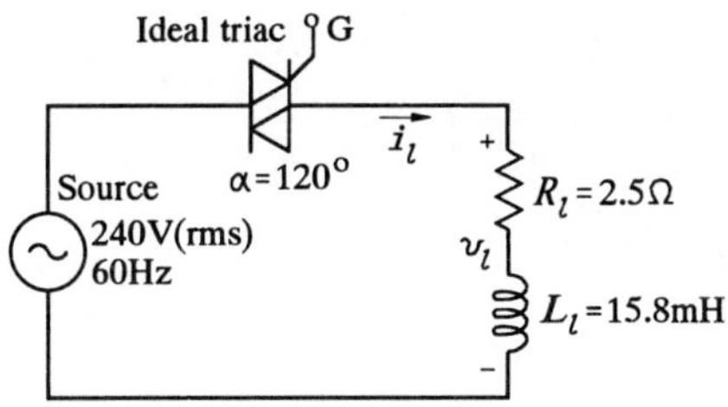

4.6 THE IDEAL GTO

The gate-turn-off thyristor (GTO) is like the thyristor but can be turned off with a negative current pulse at the gate terminal. This provides an advantage in high-power, power electronics circuits. If it is imperative to turn off a semiconductor switch without having to wait for the main current to fall to zero naturally, or not to have special commutation circuits, GTOs can be used instead of thyristors.

The operating actions of the GTO are as follows.

1. The GTO can block a voltage of either polarity at its main terminals, anode A and cathode K, if no gate current is applied.
2. The GTO will turn on if a positive gate signal is applied while the anode is positively biased.
3. Once the GTO is on and conducting the main current, the GTO will remain on even if the gate signal is removed.
4. Once the GTO is on, it will stay on until the anode current reduces to zero. At zero anode current the GTO turns off naturally.
5. If the GTO is on, it can be turned off by applying a reverse current pulse at the gate. The current pulse can typically have a magnitude equal to one third of the value of the anode current.
6. In general, the GTO is current controlled and conducts anode current uni-directionally, just like the thyristor.

A circuit with a GTO is shown in the diagram Fig. 4.26a. It represents a basic chopper circuit. The GTO can be modelled as a controlled switch in series with a diode. The *I-V* characteristics of the GTO are shown in Fig. 4.26b.

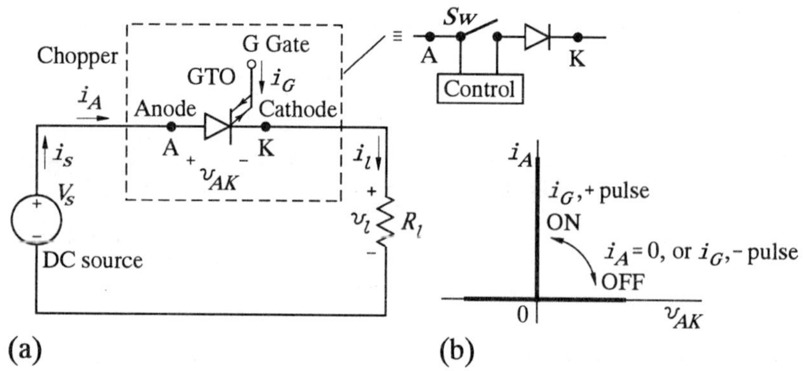

Fig. 4.26 The GTO.
(a) A GTO chopper, (b) GTO *I-V* characteristics.

4.6.1 CHOPPER CIRCUIT IDEAL GTO

It would appear that we can adapt the PSpice model of the ideal thyristor (TH_IDEAL) to create a PSpice model of the ideal GTO (GT_IDEAL). We need to add a gate source to produce a negative current pulse for GTO turn-off. This will be incorporated in the TABLE expression for a piece-wise linear (PWL) relation between input (the value of the expression) and output (the current that controls the PSpice switch).

EXAMPLE 4.6.1

The circuit diagram in Fig. 4.27 shows the component values of an ideal GTO chopper. The chopper is to operate at a switching frequency of 1kHz and a duty cycle of 0.7. Relevant ratings of the GTO are: anode-to-cathode blocking voltage 800V, maximum controllable anode current 50A, maximum turn-on gate current 300mA, maximum turn-off gate current 10A. From a PSpice simulation create a trace of the running average power absorbed by the load. From this determine the average power delivered by the supply. (Ans: 16.83kW.)

Solution

Two of the steps in the solution of this example are described.

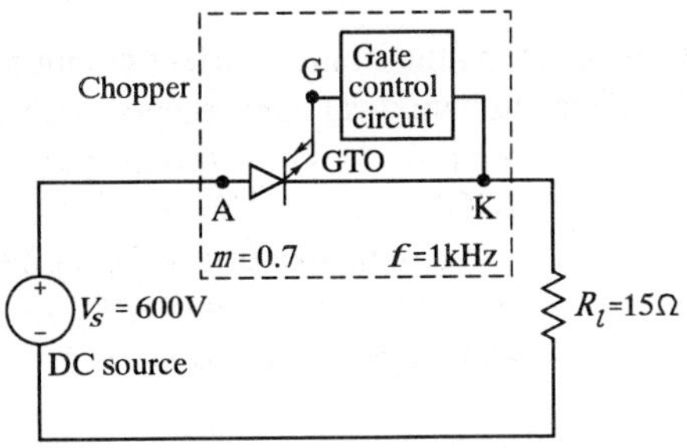

Fig. 4.27
Chopper circuit diagram.

STEP 1 Draw the PSpice configuration as shown in Fig. 4.28 on the next page. The configuration comprises the power circuit, the gate turn-on circuit, the gate turn-off circuit and the switch control circuit. The switch control circuit involves the logic of gate turn-on, gate turn-off and the anode current. The result of this logic determines whether the switch (GTO) is on or off, that is, whether the VCCS (GG) has a finite or zero output.

The GTO acts as a chopper. A PSpice current-controlled switch (device name WGTO, model name GTO, and model type ISWITCH) has been chosen to perform this function. The frequency of switching $f = 1$kHz and the duty cycle $m = 0.7$ is programmed in the gate turn-on and gate turn-off sources. The anode current of the switching device is sensed by the zero-voltage source VA. The value of the current in the zero-voltage source VG is the control current that determines whether the switch WGTO is on or off.

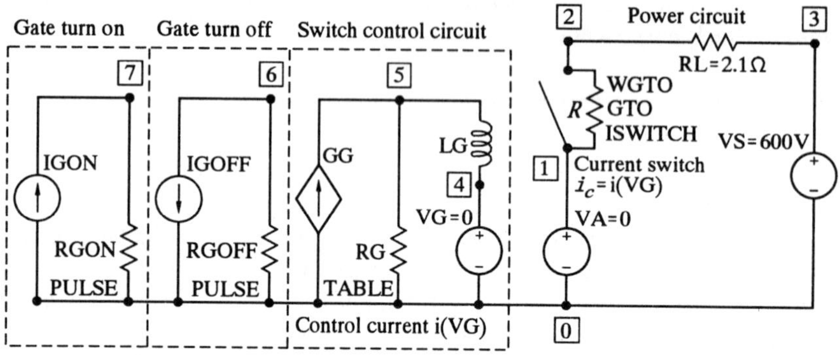

Fig. 4.28 PSpice configuration.

The circuit elements RG and LG are connected in the control-current circuit to provide a short delay for the current i(VG) to aid numerical convergence during analysis. The current pulses IGON and IGOFF have waveforms like that in Fig. 4.29. In this case the period is $T = 1$ms and the duty cycle is $m = 0.7$.

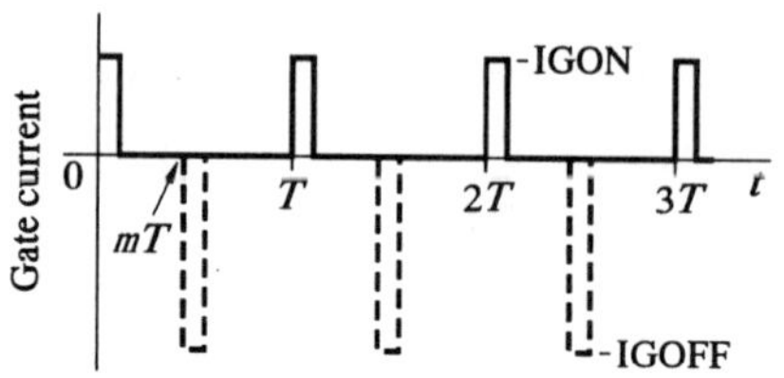

Fig. 4.29 Gate turn-on and turn-off pulses.

STEP 2 Write a circuit file (EX4_6_1.CIR) using the PSpice configuration in Fig. 4.28. The source of the switch's control current i(VG) is the voltage-controlled current source GG whose output depends on the gate pulses IGON, IGOFF and the anode current i(VA). The output of GG is determined by the input expression and the PWL output of TABLE. Now run the PSpice program with the circuit file EX4_6_1.CIR.

» » » » » EX4_6_1.CIR »

```
AN IDEAL GTO CHOPPER        (AVERAGE POWER)
* SOURCES
VS  3  0  600V;                A 600-V dc source at nodes (3, 0).
VA  1  0  0;                   A sensor for the GTO anode current.
* CIRCUIT ELEMENT
RL  3  2  15ohms;              Load resistance.
* CIRCUIT DEVICE, current-controlled switch to model the ideal GTO.
WGTO  2  1  VG  GTO;  Device WGTO is connected to nodes (2, 1).
* VG is the control-current sensor of the switch for ION and IOFF.
* GTO is the model name of the switch. The model statement follows.
.MODEL GTO ISWITCH(RON=1E-3 ROFF=1E8 ION=300mA IOFF=2E-3A)
* ISWITCH is the device type, a current-controlled switch.
* The switch parameters RON, ROFF, ION, IOFF are discussed below.
* GATE TURN-ON CIRCUIT
IGON   0  7  PULSE(0 350mA 0 60ns 60ns 0.1ms 1ms)
* A periodic independent source of current pulses.
* This satisfies m=0.7, frequency f=1kHz and a pulse width of 0.1ms.
RGON  7  0  1ohm;              This gives v(7)=350mV if IGON is on.
* GATE TURN-OFF CIRCUIT
IGOFF 6 0  PULSE(0 41A 0.7ms 150ns 150ns 0.1ms 1ms)
* A periodic independent source of current pulses.
* This satisfies a 0.1ms turn-off pulse every 1ms.
RGOFF  6  0  1ohms;            This gives v(6)=-41V when IGOFF is on.
* CURRENT CONTROL FOR SWITCH WGTO
VG  4 0 0;                     Control-current sensor.
RG  5 0 1ohm
LG  5 4 10nH;                  Time constant 10ns to delay change of i(VG).
GG  0 5 TABLE {v(7)+v(6)+i(VA)}
+                  =(0,1E-3) (1E-3,1E-3) (5E-2,0.35) (0.4,0.35)
* If {v(7)+v(6)+i(VA)}<1E-3,  GG=1E-3,  i(VG)<IOFF,  so WGTO is off.
* If {v(7)+v(6)+i(VA)}>0.05,  GG=350mA,  i(VG)>ION,  so WGTO is on.
* Otherwise, 10^-3<R<10^8. Convergence problems can have their roots in
* the TABLE, or the GG circuit, or the WGTO parameter values.
* ANALYSIS
.TRAN   20us 2ms 0  50us;      A transient analysis over 2 periods.
.PROBE;                        To view all variables.
.END
```

» »

Consider the GG statement. The current source GG is voltage controlled. The voltage v(7), which is proportional to the turn-on gate current, turns on the GTO. The voltage v(6), which is proportional to the turn-off current i(IGOFF), turns off the GTO. The magnitude of v(6) must be greater than the maximum anode current i(VA) so that the expression in TABLE can be negative. The value of the current i(VA) maintains the GTO on if the turn-on gate current is removed. Figure 4.30 shows the PWL relation between the TABLE input and output.
The model switch and the GG statements can be refined to account for (1) the voltage drop across the GTO if it is in the on-state and (2) the high values of the holding and latching currents, which can be 3% of the rated value of the GTO current.

A subcircuit named GT_IDEAL is found in POWER.LIB.

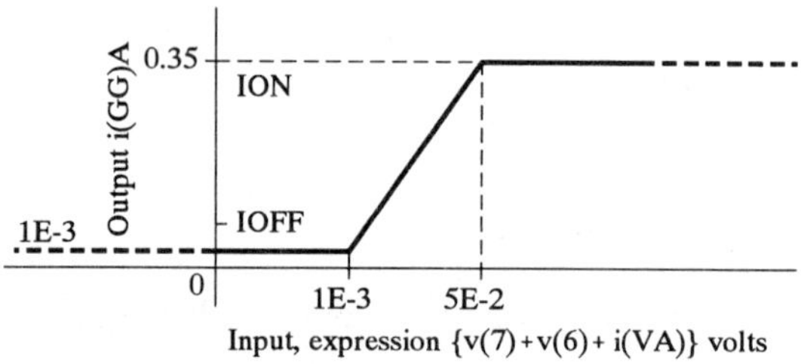

Fig. 4.30
PWL relation between input and output.

END OF EXAMPLE 4.6.1

Drill Exercise D4.6.1

Write a circuit file like EX4_6_1.CIR for the circuit diagram in Fig. 4.31. The GTO has specifications 1500V, 150A, gate turn-on current pulse 500mA and gate turn-off current pulse −50A. Use a near-ideal current switch to model the GTO. From PROBE determine the average power absorbed by the load. (Ans: 40kW.)

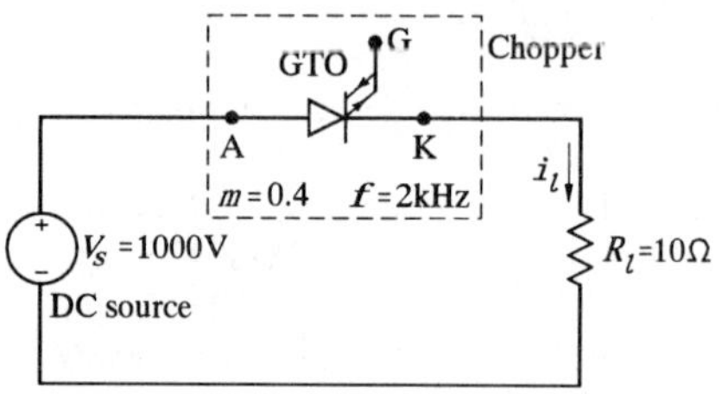

Fig. 4.31

Drill Exercise D4.6.2

Repeat Drill Exercise D4.6.1 using the subcircuit GT_IDEAL. From PROBE determine **(a)** the rms value $I_{l\,\mathrm{rms}}$ of the load current and **(b)** the average power absorbed by the load. (Ans: **(a)** 63.26A, **(b)** $I^2_{l\,\mathrm{rms}} R_l = 40.02$kW.)

There is an Extra Drill Exercise on the WEBsite.

4.7 THE IDEAL MOSFET AND IGBT SWITCHES

Physically, the MOSFET and the IGBT are different kinds of semiconductor switches. However, their external operational characteristics are similar, and their uses as switches are similar. The circuit symbols are shown in Fig. 4.32a together with the symbol of a generic switch called MOSIG that we will use to represent either the MOSFET or the IGBT. Figure 4.32b is a diagram of a simple circuit with a dc source, a resistive load, a switch and a gate-control circuit. The *I-V* characteristics of the switches are illustrated in Fig. 4.32c.

The operation of the switches is as follows.

1. The switches are voltage controlled at the gate terminal. The voltage v_{GS} turns a switch on and off.

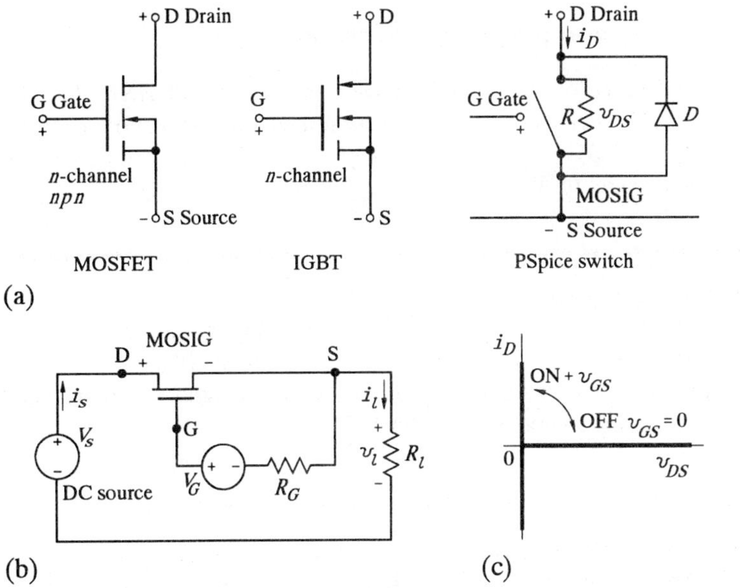

Fig. 4.32 Chopper circuit. (a) MOSFET, IGBT and PSpice switches, (b) circuit diagram, (c) switch characteristics.

2. With no voltage at the gate terminal
   (a) the switch blocks forward voltage v_{DS} and $i_D = 0$.
   (b) the switch cannot block reverse voltage $-v_{DS}$. There is a built-in reverse-connected diode that allows conduction if the switch is reverse biased. Thus, $v_{DS} \approx 0$ and i_D is negative.
3. With the application of a sufficiently positive voltage v_{GS} at the gate of an *n*-channel device, the switch will turn on and conduct current i_D if the drain is positively biased.
4. Removal of the gate voltage will cause the switch to change to the off-state. Thus, if $v_{GS} = 0$, then $i_D = 0$ and v_{DS} is positive.

We will only consider *n*-channel devices. If a *p*-channel device is to be simulated, we must reverse the polarities of the voltages and change the current directions. The actual differences between the MOSFET and the IGBT make the MOSFET useful for low power, high frequency applications, whereas the IGBT is useful for higher power, lower frequency applications.

4.7.1 CHOPPER (IDEAL MOSFET OR IGBT)

Using the actions of operation that have been described above for the MOSFET and the IGBT, we can create a near-ideal model for simulation using a PSpice switch (type VSWITCH) that is voltage controlled. We can add a near-ideal diode to account for the devices' body diodes.

EXAMPLE 4.7.1

Figure 4.33a is a circuit diagram showing the values of the circuit elements of a basic chopper. The switching device can be a MOSFET or an IGBT. Create a PSpice circuit file for simulation. From PROBE, determine the average power delivered by the source. (Ans: 194.4W.)

Solution

Three of the steps to solve this problem are described.

STEP 1 Create a PSpice configuration from the circuit diagram. A configuration is shown in Fig. 4.33b. Since we are using a PSpice voltage-controlled switch (type VSWITCH), we add a control circuit to provide the control voltage v_c. This control voltage is v(3) in the configuration. The control circuit models the gate-driver circuit of the real circuit only in so far as to provide turn-on pulses in the right sequence and right time.

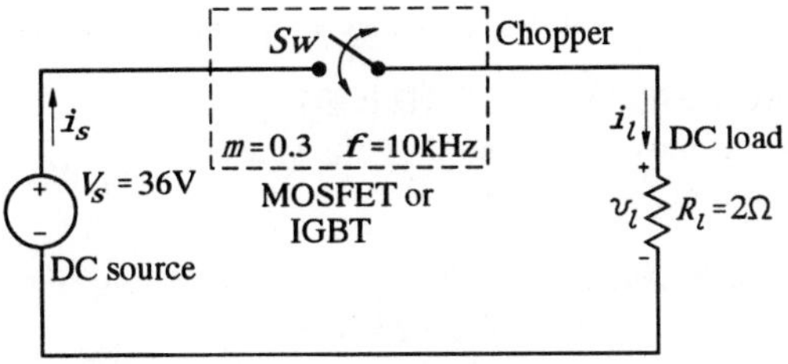

(a)

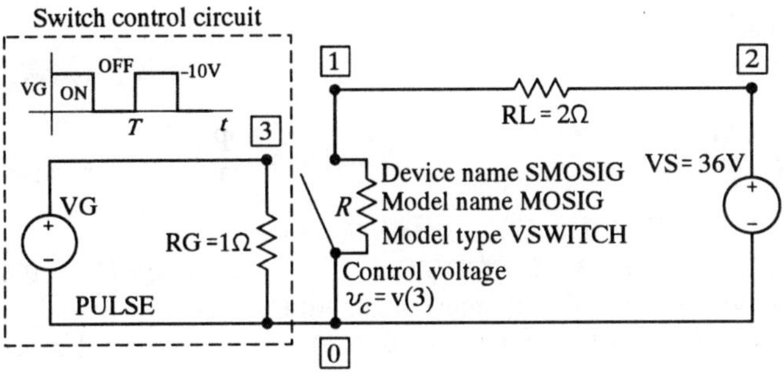

(b)

Fig. 4.33 A MOSFET or IGBT switching circuit.
(a) Circuit diagram, (b) PSpice configuration.

STEP 2

Use the PSpice configuration to formulate the circuit file, described on the next page. See POWER.LIB for the subcircuits MG_IDEAL and MG_GATE.

STEP 3

The third step is to run PSpice with the file EX4_7_1.CIR. With this particular model and without precautions such as snubbers and .OPTIONS the minimum time step for computation will be found to be about 1ps.

```
»»»»» EX4_7_1.CIR »»»»»»»»»»»»»»»»»»»»»»»»»»»»»»»»»»»»»»»»

A CHOPPER WITH AN IDEAL MOSFET OR IGBT
* To determine the average power delivered by the source.

* PARAMETERS
.PARAM FREQ=10kHz PERIOD={1/FREQ};       Switching frequency.
.PARAM DUTY=0.3      P_WIDTH={DUTY*PERIOD}
* These parameters relate to f, m and the switch on-time tON.
.PARAM TR={PERIOD/5000}  TF={TR};        Pulse rise and fall times.

* SOURCE
VS 2 0 36V;                              A dc-voltage source.

* CIRCUIT ELEMENT
RL 2 1 2ohms;                            The dc-load resistance.

* CIRCUIT DEVICE, a voltage-controlled switch.
SMOSIG  1  0  3  0  MOSIG;               A MOSFET or IGBT model.
* The device name SMOSIG must begin with S to be a switch.
* Nodes 1, 0 are for the switch. Nodes 3, 0 are for the control voltage vc.
* MOSIG is the name given to the model. This can be anything.
.MODEL MOSIG VSWITCH(RON=1E-3 ROFF=1E7 VON=10 VOFF=1E-3)
* The model statement has the name MOSIG that links the device statement.
* VSWITCH is the model type for the PSpice voltage-controlled switch.
* The switch parameters are R=RON if vc≥VON, R=ROFF if vc≤VOFF,
* and RON<R<ROFF if VON>vc>VOFF.

* CONTROL CIRCUIT, to produce vc=v(3).
VG 3 0  PULSE(0 10.1V 0 {TR} {TF} {P_WIDTH} {PERIOD})
RG 3 0  1kohm;                           An arbitrary resistance.
* v(3) is a periodic pulse at f=10kHz and pulse width tON=m/f=0.03ms.

* ANALYSIS, transient, over two periods of switching.
.TRAN   20us 0.2ms  0  50us
.PROBE;                                  To view all variables.
.END

»»»»»»»»»»»»»»»»»»»»»»»»»»»»»»»»»»»»»»»»»»»»»»»»»»»»»»»»»»»»
```

END OF EXAMPLE 4.7.1

Drill Exercise D4.7.1

The circuit diagram in Fig. 4.34 represents an ideal IGBT basic chopper. Write a circuit file like EX4_7_1.CIR for a PSpice simulation. For a duty cycle $m = 0.8$ determine the average power delivered by the supply. (Ans: 12.8kW.)

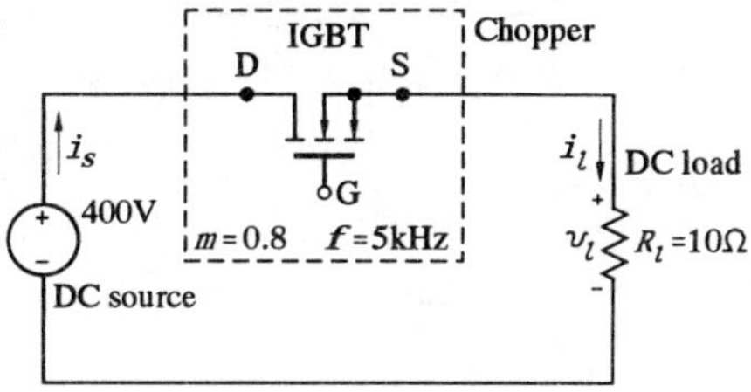

Fig. 4.34

Drill Exercise D4.7.2

Consider Fig. 4.35. Simulate the circuit using the subcircuits MG_IDEAL for the MOSFET and MG_GATE for the driver circuit. The circuit file will define the parameters FREQ for the frequency, PERIOD for the period, and P_WIDTH for the chopper pulse width. Local subcircuit parameters for the control subcircuit are DELAY for the gate pulse and PER for the period of switching. From PROBE make a trace of the load voltage and determine the average power absorbed by the load. (Ans: 1498W.)

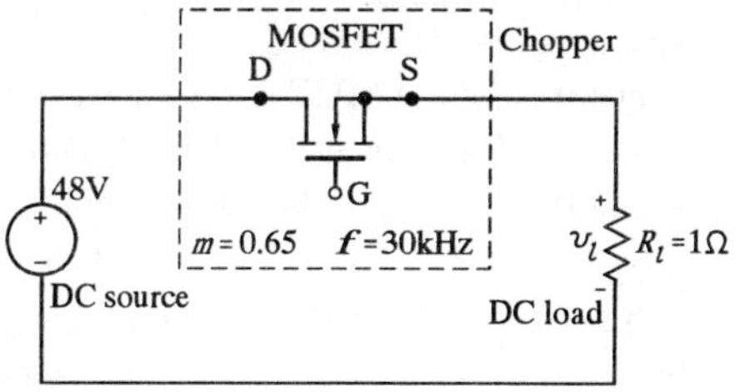

Fig. 4.35

There is an Extra Drill Exercise on the WEBsite.

4.8 THE IDEAL MCT

The MOS-controlled thyristor (MCT) is a competitor of the MOSFET and the IGBT. The MCT is voltage controlled, but, like the thyristor it latches into the on-state. Figure 4.36a, depicted on the next page, shows a circuit diagram of a basic chopper circuit with an MCT.

The operation of the MCT is as follows.

1. The MCT is normally off.
2. If the anode A is positively biased, a negative voltage pulse at the gate G with respect to the anode will cause the MCT to turn on.

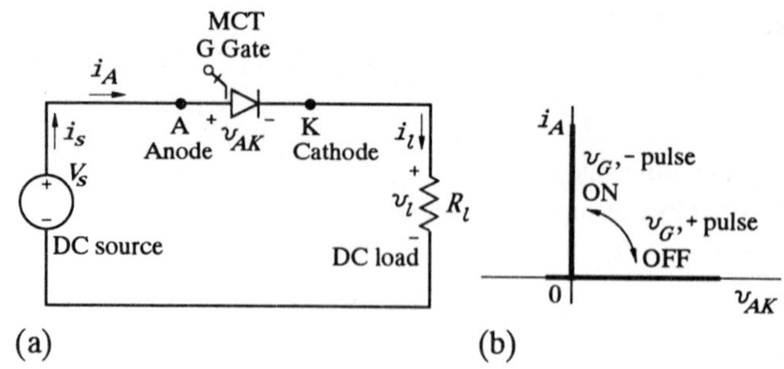

Fig. 4.36 The MCT chopper.
(a) Circuit diagram, (b) MCT *I-V* characteristic.

3. Once the MCT is on, and conducting current, internal regeneration maintains it on and the negative gate signal can be removed. This is similar to the thyristor, triac and GTO.
4. Once it is on the MCT will turn off if the anode current reduces to zero.
5. If the MCT is to be turned off before the anode current falls to zero naturally, a positive-voltage pulse can be applied to the gate G.
6. The MCT stays off until the anode is positive and the gate is negative.

In practice, designers have sacrificed reverse-voltage blocking capabilities of the MCT for fast switching and a low on-state voltage drop v_{AK}. If there is a likelihood that the MCT will be reverse biased, a diode is connected in antiparallel with the MCT's terminals A and K.

The ideal *I-V* characteristic of the MCT is shown in Fig. 4.36b.

4.8.1 CHOPPER (IDEAL MCT)

The operation of the ideal MCT is similar in principle to the operation of the GTO. The main difference is that the GTO is current controlled, whereas the MCT is voltage controlled. However, we can use the GTO as a model in order to develop the ideal MCT model for PSpice simulation.

We will use a PSpice voltage-controlled switch (model type VSWITCH) to model the MCT. For turn-on we will use a voltage pulse of negative polarity. We need a control voltage that is proportional to the MCT anode current to maintain the switch on. A voltage source will accomplish this requirement. Finally, for MCT turn-off, we will use a voltage pulse of positive polarity. The magnitude of the voltage pulse must swamp any control signal that maintains the switch in the on-state. These three control signals for the switch will satisfy those MCT operating conditions that are described above.

EXAMPLE 4.8.1

Figure 4.37a is a circuit diagram of an ideal MCT basic chopper in which the element values are shown. Create a PSpice circuit file of the ideal MCT in order to show the operation of the switch. From PROBE, record a load-voltage trace and determine the power absorbed by the load. (Ans: 33.51kW.)

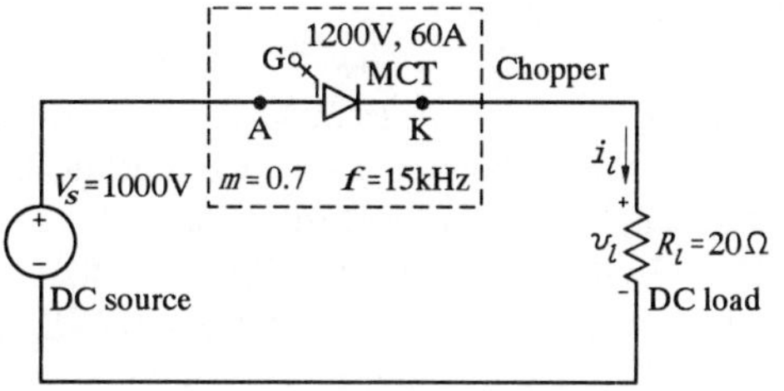

(a)

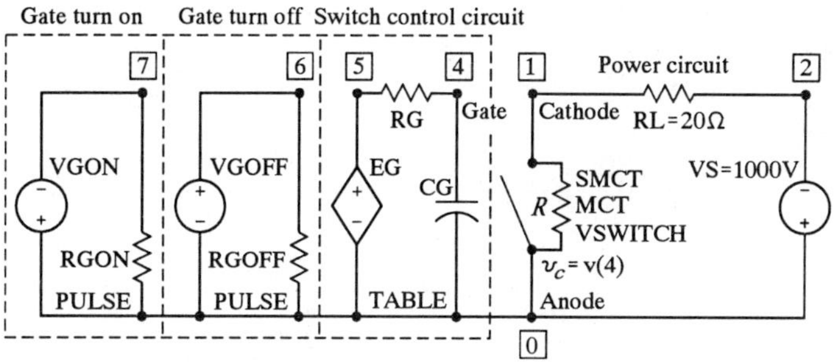

(b)

Fig. 4.37 MCT chopper.
(a) Circuit diagram, (b) PSpice configuration.

Solution

Two steps in this solution are described.

STEP 1

Create a PSpice configuration. See Fig. 4.37b. The power circuit includes the dc source, the load and the switch that models the MCT.

The control circuit has three sections. One section consists of the pulsed source VGON to provide the MCT negative turn-on gate signal. The second section comprises a pulsed source VGOFF to provide the MCT positive turn-off gate signal. The third section combines the turn-on signal v(7), the turn-off signal v(6) and the MCT anode current signal i(VS) to control the state of the PSpice switch. The turn-off signal must dominate the other two signals, in order that turn-off occurs while the MCT is conducting maximum current. The output of the dependent voltage source (VCVS EG) is used as the control voltage of the voltage-controlled switch SMCT. For the sake of numerical convergence a delay circuit (RG and CG) is added to the VCVS and the control voltage v(4) is used to determine the value of the switch resistance *R*.

STEP 2 Create a circuit file (EX4_8_1.CIR) from the PSpice configuration of Fig. 4.37b. The output voltage across the load is to have a waveform as shown in Fig. 4.38. The load voltage v(1, 2) ≈ VS is applied at the instant -VGON = VON and RON = 1E-3Ω. The load voltage becomes zero at the instant that VGOFF = VOFF and ROFF = 1E7Ω. For the interval between the two pulses VGON and VGOFF, the switch current i(VS) maintains EG > VON and *R* = RON.

Do a simulation run to find the power. A subcircuit of the MCT called MT_IDEAL is written in POWER.LIB.

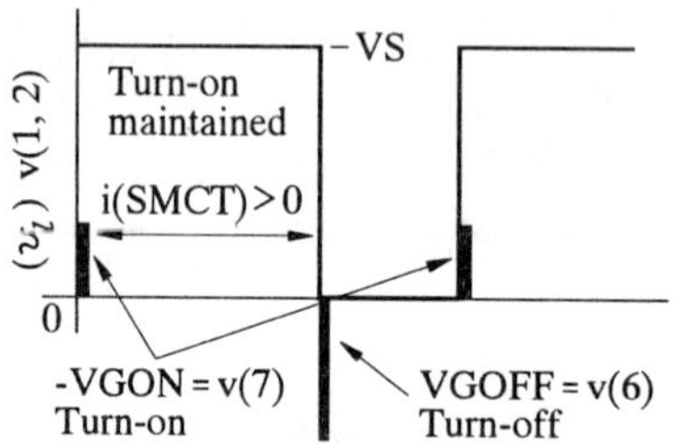

Fig. 4.38
Load voltage turn-on, maintenance, and turn-off.

» » » » » EX4_8_1.CIR »

```
AN IDEAL MCT CHOPPER     AVERAGE POWER
* SOURCE
VS  0  2   1000V;                 A 1000-V dc source at nodes(0, 2).
* CIRCUIT ELEMENT
RL  1  2   20ohms;                The dc-load resistance.
* CIRCUIT DEVICE, the voltage-controlled switch has a device name SMCT.
SMCT  0   1   4   0   MCT;        SMCT is connected to nodes (0, 1).
* SMCT's control voltage is across nodes (4, 0).
* SMCT's model name is MCT to link with the model statement.
. MODEL MCT  VSWITCH(RON=1E-3 ROFF=1E7 VON=10 VOFF=2E-3)
* If the switch control voltage v(4) > 10V, R = RON = 1E-3Ω.
* If v(4) < 2E-3V, R = ROFF = 1E7Ω.
* Otherwise, for 10V > v(4) > 2E-3V, 1E-3Ω < R < 1E7Ω.
* GATE TURN-ON CIRCUIT, a periodic pulsed gate voltage.
VGON  0 7  PULSE(0 0.1V 0 20ns 20ns 0.1us 66.667us)
RGON  7 0  10ohm;                 A nominal resistor.
* GATE TURN-OFF CIRCUIT, a periodic pulsed gate voltage.
VGOFF   6  0  PULSE(0 1.1V 44.667us 20ns 20ns 0.1us 66.667us)
RGOFF   6  0  10ohms
* CONTROL VOLTAGE, v(4) for switch SMCT, with delay
EG 5  0  TABLE {-v(7) - i(VS) - 50*v(6)}
+                    =(0,1E-3) (1E-3,1E-3) (50E-3,10.2) (0.4,10.2)
* Any current in the expression { } must be sensed by a voltage source.
RG 5  4  1ohm
CG 4  0  10nF;                    Time constant 10ns to delay v(4) rise.
* If {- v(7) - i(VS) - 50*v(6)} < 1E-3, EG = 1E-3 volts,
* That is, v(4) < VOFF, so SMCT is off.
* If {- v(7) - i(VS) - 50*v(6)} > 50E-3, EG = 10.2 volts.
* That is, v(4) > VON, so SMCT is on.
* Otherwise, 10^-3 < R < 10^7 and SMCT is neither on nor off.
* ANALYSIS, to cover the transient of 2 cycles of switching.
. TRAN   0.5us  140us  0   0.2us
. PROBE;                          To view all variables.
. END
```

» »

END OF EXAMPLE 4.8.1

Drill Exercise D4.8.1

Consider the ideal MCT chopper shown in the circuit diagram of Fig. 4.39. The chopper modulates power from the dc source of 800V to a resistive load of 10Ω. The switching frequency is f= 10kHz and the duty cycle is set to m = 0.5. Write a PSpice circuit file similar to EX4_8_1.CIR. Use PROBE to determine **(a)** the average current conducted by the MCT and **(b)** the average power absorbed by the near-ideal chopper. (Ans: **(a)** 40A, **(b)** 3.67W.)

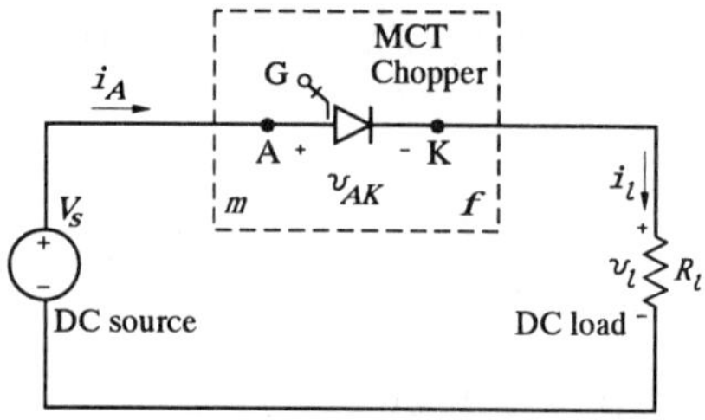

Fig. 4.39

Drill Exercise D4.8.2

Consider the ideal MCT chopper shown in the circuit diagram of Fig. 4.39. The voltage of the dc source is V_s = 600V, the load resistance is 7.1Ω, the frequency of switching is 8kHz and the duty cycle is m = 0.85. Use the subcircuit named MT_IDEAL in POWER.LIB to simulate the circuit and find the average power delivered by the supply.
(Ans: VS*AVG(i(RL)) ≈ 43.1kW.)

Drill Exercise D4.8.3

Consider an ideal MCT chopper circuit whose PSpice configuration is shown in Fig. 4.37b. The voltage of the dc source is V_s = 800V, the load resistance is 18Ω, the frequency of switching is 5kHz and the duty cycle is 0.6. Adapt the circuit file EX4_8_1.CIR so that the circuit file for this exercise has two subcircuits. One of the subcircuits is for the MCT alone and the other subcircuit is for the driver. Determine the average power delivered by the load from PROBE traces. (If the driver circuit and the power switch are to be put into separate subcircuits, the current sensor VA must be put with the driver subcircuit.)
(Ans: 21.16kW.)

Chapter 5

Simulation of Driver Circuits

5.1 INTRODUCTION

Semiconductor switches need to be turned on and off. The action of turn-on and turn-off is accomplished by low-power circuits that provide pulses at their outputs. These low-power circuits are often called gate drives, or base drives, or driver circuits, or just drivers. The output of the driver circuit is connected to the gate or base terminals of the semiconductor switch. For some switches, a signal from the driver turns on the switch and the absence of a signal from the driver maintains the switch in the off-state.

In Chapter 4 we described how PSpice can model semiconductor switches for near-ideal conditions. The models were voltage or current-controlled switches (S and W). The control voltage and control current for these switches represent the signals from the driver circuits. In this chapter we describe how PSpice can model the driver circuits.

This chapter is divided into sections that correspond to the different types of converter, chopper, controlled rectifier and inverter. In each section, appropriate PSpice driver circuits are developed and a description of each one is added to the PSpice library file called DRIVER.LIB.

5.2 DRIVERS FOR DC-DC CONVERSION

The conventional way to control voltage from a dc source is to turn a switch on and off in a fashion called chopping. A primitive chopper is illustrated in Fig. 5.1a on the next page. The switch *Swa* has the specifications that it is unidirectional, and is on if the control signal has some positive value (otherwise, it is off). By turning the switch *Swa* on and off in a particular sequence (on-time t_{ON}, off-time t_{OFF}), the load voltage v_l can be modulated as shown in the waveform of Fig. 5.1b.

The average load voltage V_{lav} is given by the expressions

$$V_{lav} = \frac{t_{ON}}{t_{ON} + t_{OFF}} V_s = \frac{t_{ON}}{T} V_s = m V_s \tag{5.2.1}$$

where T is the switching period and m is the switching duty cycle ($m = t_{ON}/T$).

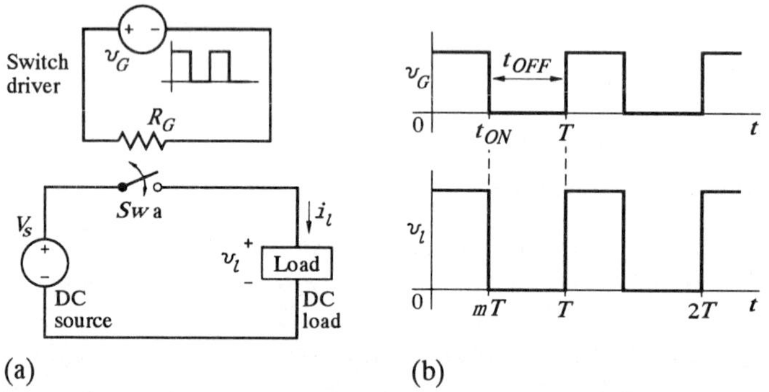

Fig. 5.1 Primitive chopper.
(a) Circuit diagram, (b) waveforms of v_G and v_l.

This type of voltage modulation is called Pulse-width Modulation (PWM). A suitable driver is required to provide a stream of controlled pulse widths to control the on/off states of the switch that enables the actual dc-dc conversion to take place. An equivalent gate driver is illustrated in Fig. 5.2a. The voltage source v_G represents a source of voltage pulses to be applied at the switch gate terminal to drive the switch *Swa*. The pulse waveform is shown in Fig. 5.2b.

EXAMPLE 5.2.1

Implement a PSpice simulation of the circuit diagram in Fig. 5.1a. The circuit values are V_s=36V, R_l=1.0Ω, switching frequency f=10kHz, duty cycle m=0.75. Choose the PSpice voltage-controlled switch (type VSWITCH). This switch is described both in EXAMPLE 1.4.1 and in the library file POWER.LIB. Plot traces of the driver control voltage v_G and the load voltage v_l over an interval of 400μs.

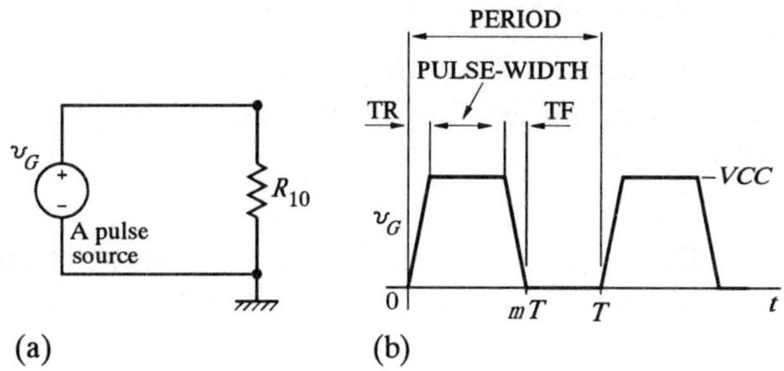

Fig. 5.2 Chopper driver.
(a) Circuit diagram, (b) periodic pulse waveform.

Solution

The waveform of the control voltage is similar to the waveform of the load voltage. From the description of the PSpice voltage-controlled switch VSWITCH, it will switch on if a control voltage $v_G \geq 1$V (VON = 1V). The switch will turn off if the control voltage $v_G \leq 0$V (VOFF = 0). The intended signal and output waveforms are in Fig. 5.1b. There are four steps in the solution.

STEP 1 The PSpice configuration in Fig. 5.3 is derived directly from the circuit diagram in Fig. 5.1a and the circuit specifications.

STEP 2 From the PSpice configuration in Fig. 5.3 and the switch operating specifications, we can write a circuit file.

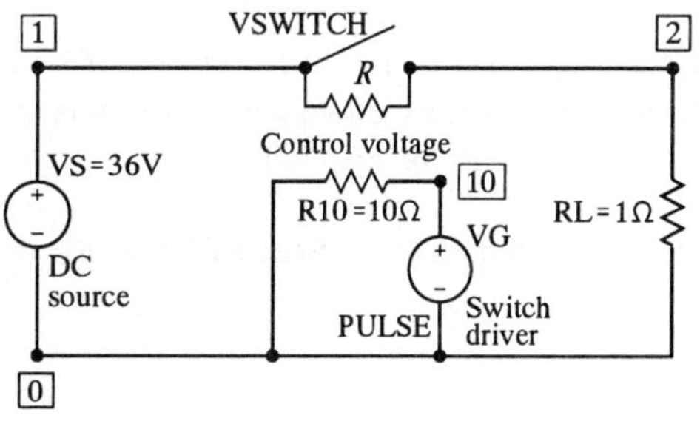

Fig. 5.3
PSpice configuration.

```
»»»»» EX5_2_1.CIR »»»»»»»»»»»»»»»»»»»»»»»»»»»»»»»»»»»»»»

A PRIMITIVE CHOPPER CIRCUIT AND DRIVER
* PSpice must know that the switch parameters are in a library file, so
.LIB     POWER.LIB

* SOURCES
VS   1   0   36V;                                    Main power supply.
VG   10  0   PULSE(0 1V 0 1ns 1ns 75us 100us);       Control voltage.

* DEVICE
S1   1   2   10   0   SWITCH;                        This is the switch Swa.
* The switch's model statement with parameter values is in POWER.LIB.
* It is described as follows.
* * A GENERIC MODEL OF A VOLTAGE-CONTROLLED SWITCH.
* * On-resistance is 0.001ohm and the off-resistance is 1Megaohm.
* . MODEL SWITCH
* + VSWITCH(RON=1E-3 ROFF=1E6 VON=1.0 VOFF=0)
* CIRCUIT ELEMENTS
R10  10  0   10ohms;       This closes the control-voltage loop.
RL   1   2   1.0ohm;       The load resistance.

* ANALYSIS, transient, up to 400μs, prints every 1μs.
.TRAN    1us    400us
.PROBE  v(2),  v(10);      We require only the load and driver waveforms.
.END

»»»»»»»»»»»»»»»»»»»»»»»»»»»»»»»»»»»»»»»»»»»»»»»»»»»»»»»»»»
```

STEP 3 Run the PSpice simulation with the file EX5_2_1.CIR.

STEP 4 Use PROBE with EX5_2_1.DAT to plot the traces of v(10), the control or driver voltage, and v(2), the load voltage. The plots are shown in Fig. 5.4. Here, we see that the load voltage v_l is a scaled replica of the driver signal v_G.

The driver circuit can be stored as a subcircuit. See VCHOPPER in DRIVER.LIB.

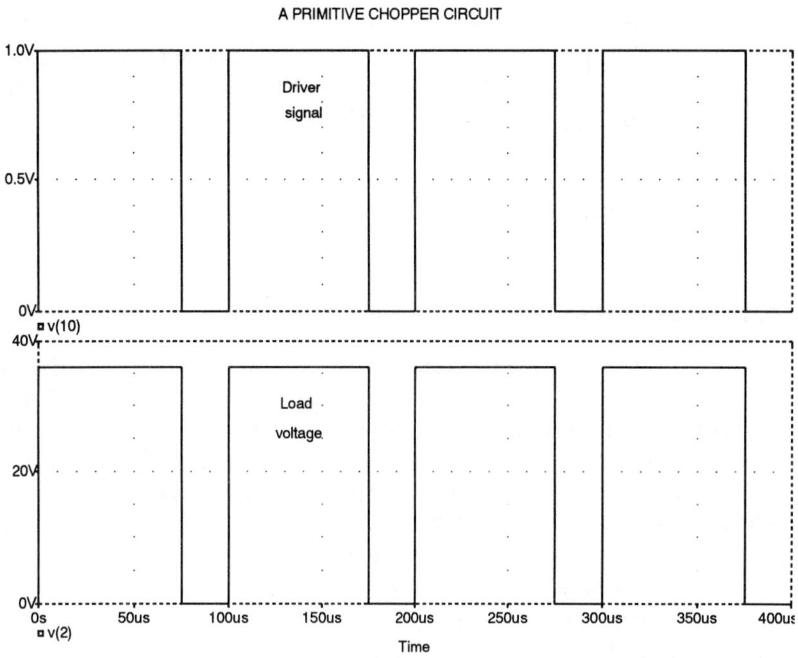

Fig. 5.4

END OF EXAMPLE 5.2.1

Drill Exercise D5.2.1

Consider a gate-drive circuit for use with a dc chopper. Use the subcircuit VCHOPPER from DRIVER.LIB. The driver specification is $VCC=1.0\text{V}$, duty cycle $m=0.5$, switching frequency $f=500\text{Hz}$. Other parameters are optional. Plot a trace of the driver output-voltage signal over an interval of 4ms.

Drill Exercise D5.2.2

Consider the circuit diagram of a primitive chopper in Fig. 5.5 depicted on the next page. Write a PSpice circuit file using the switch *Swa* from EXAMPLE 5.2.1, a diode from POWER.LIB and the driver VCHOPPER from DRIVER.LIB. Perform a transient analysis with TSTOP=4ms. Plot a trace of the load current and determine the current excursion ($\Delta I = I_{max} - I_{min}$). Note which diode models give convergence problems. (All diode models DIO_I, DIO_V, DIO_GD and

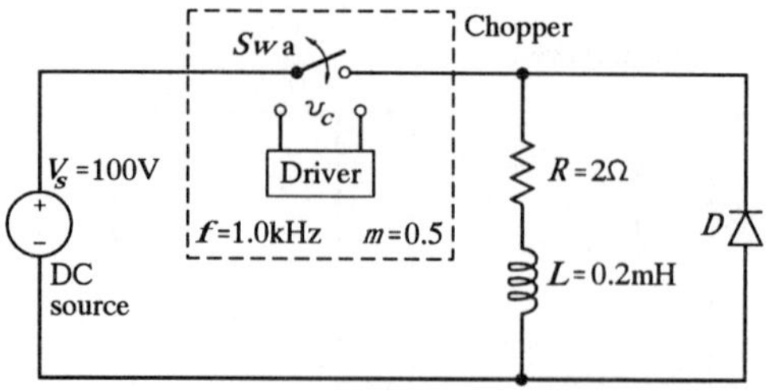

D_IDEAL work as long as the circuit time constant is less than 0.1/f. The job time is least for DIO_GD and greatest for DIO_I.)

Fig. 5.5

5.2.1 CHOPPER DUTY-CYCLE CONTROL

The driver VCHOPPER (in DRIVER.LIB) does not allow the duty cycle to change during a simulation run. In many practical cases a chopper is designed with feedback (closed-loop control) so that the driver can dynamically change its duty cycle m during operation. This adjustable duty-cycle operation can be simulated.

Let us consider the general form of a practical circuit that could act as an adjustable PWM driver. The circuit diagram is shown in Fig. 5.6. A source with a sawtooth waveform is compared with a dc reference voltage (adjustable). The result is an output whose waveform is a train of pulses.

The comparator output signal is positive, whenever the noninverting input signal is greater than the inverting input signal. Otherwise, the comparator output signal is negative or zero depending on the connections.

The comparator output v_{out} starts off positive as the magnitude v_{st} of the ramp of the sawtooth waveform begins to increase. Once the sawtooth ramp reaches the magnitude of the reference voltage v_{ref}, the comparator output switches from positive to negative (or zero). The output signal remains negative (or zero) until the sawtooth waveform suddenly becomes zero.

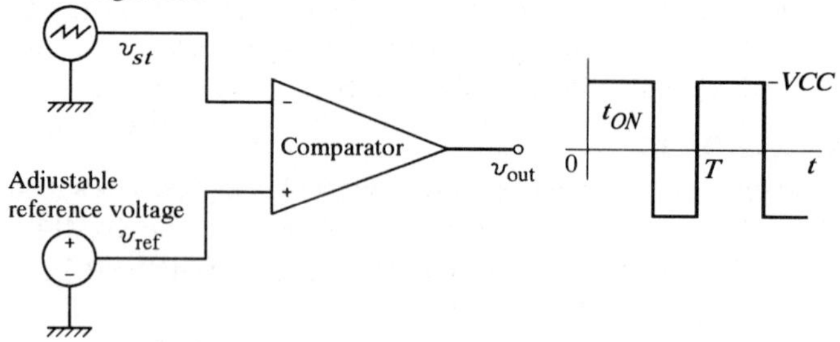

Fig. 5.6 PWM adjustment.

The cycle of the output signal repeats for each period of the sawtooth waveform. The result is a PWM signal, whose duty cycle ($m = t_{ON}/T$) is directly proportional to the reference input voltage v_{ref}, and whose frequency is equal to the frequency ($1/T$) of the sawtooth-generator waveform.

$$m = t_{ON}/T = V_{ref}/V_{st}. \qquad (5.2.2)$$

By setting the reference voltage above or below the excursion of the sawtooth waveform, a duty cycle of $m = 1$ or $m = 0$, respectively, can be achieved.

PSPICE SAWTOOTH GENERATOR

The sawtooth waveform is not a standard waveform in PSpice primitives. An approximate way to generate a sawtooth waveform is to use an independent-voltage PULSE source. The rise time would be just less than the period, for numerical reasons there must be a short pulse width, and the fall time TF would be very short. The sawtooth waveform is useful in driver circuits, so we will create a PSpice sawtooth generator in a subcircuit named VSAWTOOTH and write it in DRIVER.LIB. View PSPICE SAWTOOTH GENERATOR on the WEB.

Drill Exercise D5.2.3

Use the subcircuit VSAWTOOTH from the library file DRIVER.LIB to generate a simulated sawtooth waveform of voltage with a frequency of 50kHz and an amplitude of 5.5V. Plot a trace of the output voltage.
See WEB EXAMPLE W5.2.1.

COMPARATOR

For the duty-cycle control of a chopper, the comparator provides a gating signal that is adjusted by a reference voltage. See Fig. 5.6. The comparator is a straightforward device to use in a PSpice simulation, either by means of an analogue behavioural model (E source) or a macro model (subcircuit LM111). Reference can be made to Section 1.4 in Chapter 1. We have created subcircuits named COMPARATOR and COMPARATOR2 in DRIVER.LIB.
View COMPARATOR and the Example on the WEBsite.

Drill Exercise D5.2.4

Consider the circuit diagram in Fig. 5.7 on the next page. The input voltage is a sinusoid $v_s = 4.5\sin 12.57t$ V and the comparator has a single voltage supply, $VCC = 5$V. Use the subcircuit COMPARATOR (found in DRIVER.LIB) to simulate the circuit. Plot traces of the input and output voltage waveforms.

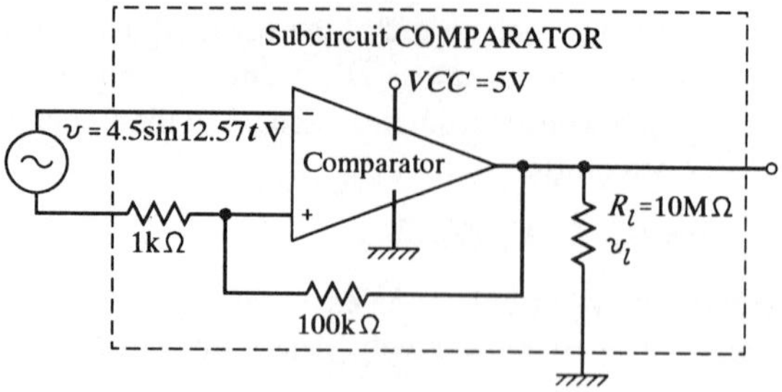

Fig. 5.7

Drill Exercise D5.2.5

The PSpice manual mentions that problems can arise in an OP analysis because of the double values in hysteresis. Repeat the problem of drill exercise D5.2.4 using the same model of the comparator but eliminating all hysteresis. Compare the simulation results. Check your circuit file and results with the subcircuit COMPARATOR2 in the library file DRIVER.LIB.
See WEB EXAMPLE W5.2.2.

GATE CONTROL

We can combine the sawtooth generator and the comparator, as shown in Fig. 5.6 to provide controlled gate signals for switch turn-on and turn-off.

EXAMPLE 5.2.2

Create a PSpice simulation that produces a train of gate pulses, suitable to control the on/off states of a switch. The pulses are to have a magnitude $V_G = 10\text{V}$, a frequency $f = 1\text{kHz}$ and a duty cycle $m = 0.5$. Plot traces of the sawtooth waveform, the reference voltage and the output gate voltage of the comparator.

Solution

The sawtooth generator was given a peak value of 1V nominally. Therefore, a reference voltage for a duty cycle of $m = 0.5$ is given by $m = V_{\text{ref}}/V_{st}$, that is, $V_{\text{ref}} = 0.5\text{V}$ (see eq. (5.2.2)). There are four steps in the solution.

STEP 1 From the schematic in Fig. 5.6, we can create a PSpice configuration of the gate driver. Figure 5.8 depicts the PSpice configuration of the required driver circuit.

STEP 2 It is from this configuration in Fig. 5.8, from the subcircuit VSAWTOOTH for the sawtooth generator, and from the subcircuit COMPARATOR2 for the comparator, that we write the circuit file EX5_2_2.CIR.

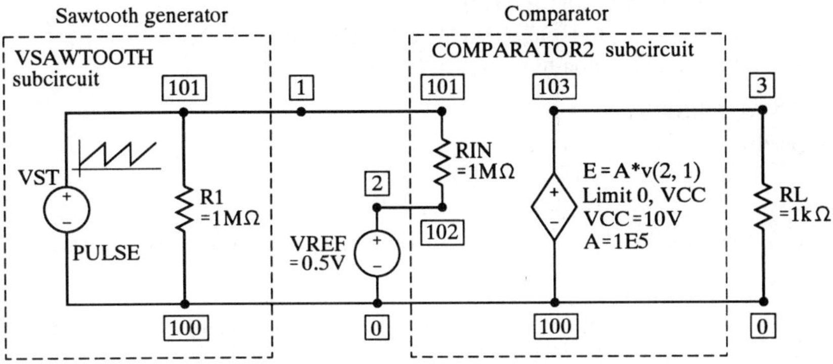

Fig. 5.8 Configuration of a controlled gate drive.

» » » » » EX5_2_2.CIR »

```
A CONTROLLED GATE-DRIVER CIRCUIT
* To plot the traces of the input and output-voltage waveforms.

* SOURCE
VREF   2   0   0.5V;              Reference voltage for a duty cycle m=0.5.

* The comparator output load (the gate impedance) is
RL  3   0   1kohm;                          An arbitrary value.
.PARAM   FREQ=1kHz   PER={1/FREQ}   VGATE=10V

* The sawtooth generator as a subcircuit named VSAWTOOTH.
.SUBCKT VSAWTOOTH 101 100 PARAMS: VM=2V PERIOD=20us
VST 101 100 PULSE(0.01 {VM-0.01} 0 {PERIOD-2ns} 1ns 1ns {PERIOD})
R1  101 100    1E6ohms
.ENDS  VSAWTOOTH

* The call for the VSAWTOOTH subcircuit is
Xgenerator 1 0 VSAWTOOTH    PARAMS: VM=1V PERIOD={PER}
* Node 1 is the inverting input of the comparator.

* The comparator as a subcircuit.
.SUBCKT  COMPARATOR2 101 102 103 100 PARAMS: VCC=10V
* 101 (inverting input). 102 (noninverting input). 103, 100 (output).
E1    103 100 VALUE={LIMIT(v(102, 101)*1E5, 0, VCC)}
RIN   102 101  1E6;                          Input resistance.
.ENDS     COMPARATOR2
```

```
* The call for the COMPARATOR2 subcircuit is
Xcomparator 1 2 3 0 COMPARATOR2 PARAMS: VCC={VGATE}
* 1 is inverting. 2 is noninverting. 3, 0 is output.

* ANALYSIS
.TRAN     40us     2.1ms
.PROBE    v(1),    v(2),    v(3)
.END
```

» »

STEP 3 The PSpice simulation is run and the results are recorded in EX5_2_2.OUT and EX5_2_2.DAT.

STEP 4 Traces of the sawtooth waveform v(1), the reference voltage v(2) and the output gate pulses v(3) are shown in Fig. 5.9. View the combined subcircuits VPWM2 and VPWM in POWER.LIB.

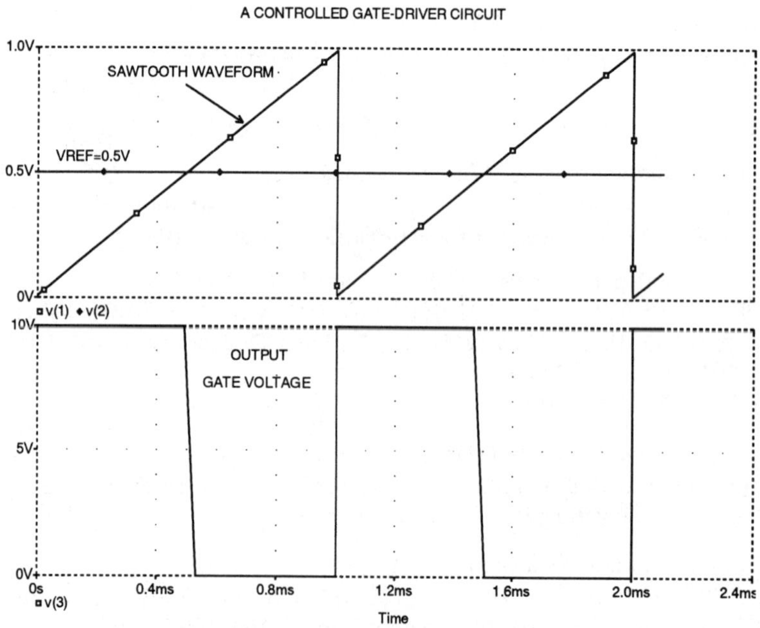

Fig. 5.9

END OF EXAMPLE 5.2.2

Drill Exercise D5.2.6

Consider the driver circuit diagram for a chopper, shown in Fig. 5.10. Create a PSpice circuit configuration to simulate a train of signals of magnitude 12V, frequency 100kHz and duty cycle 0.6. Base the configuration on the framework described in EXAMPLE 5.2.2. The comparator has no hysteresis and the load is 1kΩ. Plot traces of the sawtooth waveform v_{st}, the reference voltage v_{ref} and the output wave-form v_{gate}.

Fig. 5.10

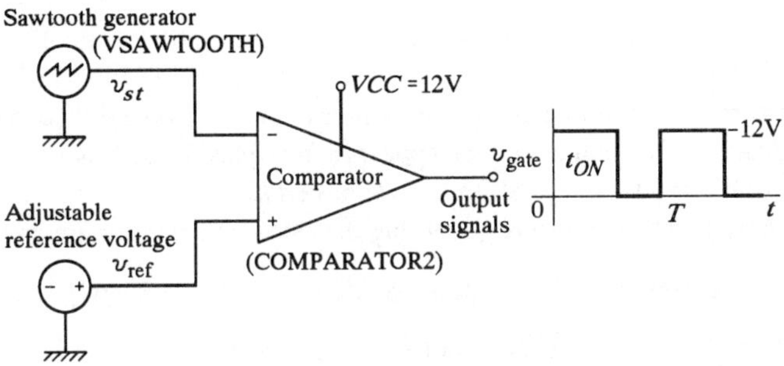

Drill Exercise D5.2.7

Consider the driver-circuit schematic shown in Fig. 5.10. Use the subcircuits named VSAWTOOTH and COMPARATOR2 in the library file DRIVER.LIB to simulate a train of signals of magnitude 13V, frequency 10kHz and duty cycle 0.7. Base the configuration on EXAMPLE 5.2.2, Fig. 5.8. Plot traces of the sawtooth waveform v_{st}, the reference voltage v_{ref} and the output waveform v_{gate}.

There are Extra Drill Exercises on the WEBsite.

DIGITAL GATE CONTROL

It is common to use digital circuits for the control of power electronics converters. PSpice has a large array of logic devices that can be incorporated in gate drivers. In this section we introduce an example of a digital circuit that can be used in a similar fashion to the driver with a comparator and a sawtooth generator. This digital circuit is derived from the 555 timer IC that is found in EVAL.LIB.

EXAMPLE 5.2.3

Design a driver for a chopper that has an astable multivibrator from a 555 timer IC. The train of pulses is to have a magnitude of 12V, a duty cycle of about 0.75 and a frequency of about 8kHz or 16kHz. From a PSpice simulation plot traces of the output voltage.

Solution

There are five steps in this solution.

STEP 1 The design of the astable multivibrator involves a fixed process. Figure 5.11 shows a block diagram of the 555 timer with its pin numbering. The source of power (voltage V_{CC}) is connected to pins 8 and 1. The output voltage appears across pins 3 and 1 as a train of pulses. The astable mode of operation follows if certain R and C elements are connected as shown. The equations governing the astable behaviour are as follows.

The on-time t_{ON} of each pulse is $t_{ON} = 0.693(R_{87} + R_{76})C_6$. (5.2.3)

The off-time t_{OFF} of each pulse is $t_{OFF} = 0.693 R_{76} C_6$. (5.2.4)

The pulse period T and frequency f are

$$T = \frac{1}{f} = t_{ON} + t_{OFF} = 0.693(R_{87} + 2R_{76})C_6. \quad (5.2.5)$$

The pulse duty cycle m is $m = \dfrac{t_{ON}}{T} = \dfrac{R_{87} + R_{76}}{R_{87} + 2R_{76}}$. (5.2.6)

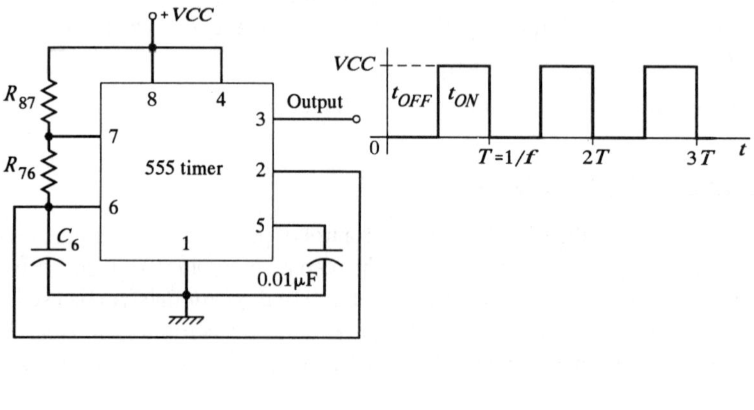

Fig. 5.11 The astable multivibrator schematic.

For this example with $m \approx 0.75$, if R_{87} is chosen to be 4.7kΩ, then $R_{76} \approx 2.2$kΩ from eq. (5.2.6). From eq. (5.2.5), for $C_6 = 20$nF, $f \approx 7.9$kHz. For $f \approx 16$kHz, then $C_6 \approx 10$nF.
Consequently, the values of the parameters external to the 555 timer are $R_{87} = 4.7$kΩ, $R_{76} = 2.2$kΩ and $C_6 = 20$nF for $f \approx 8$kHz.

STEP 2 The PSpice circuit configuration for the schematic in Fig. 5.11 is shown in Fig. 5.12. This makes use of the subcircuit named 555D that is written in the file EVAL.LIB. The subcircuit statement is

```
.SUBCKT 555D 1 2 3 4 5 6 7 8
```

where the node numbers coincide with the pin numbers of the device.

STEP 3 From the PSpice configuration in Fig. 5.12 a circuit file, EX 5_2_3.CIR, can be written. This follows on the next page.

STEP 4 The PSpice simulation is run with the circuit file EX5_2_3.CIR. The analysis runs twice to accommodate the two values of the parameter CVAL in the .STEP statement.

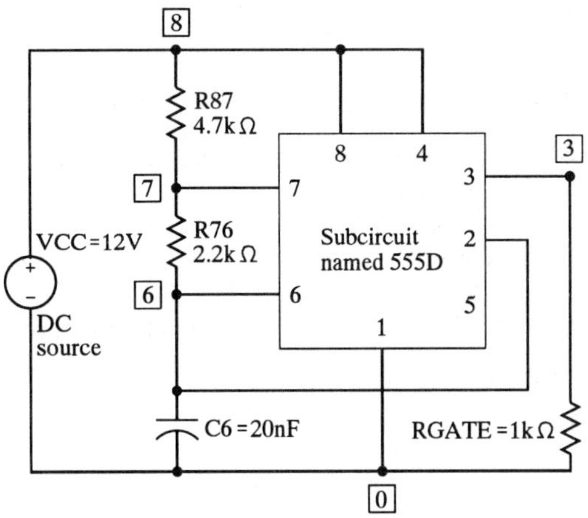

Fig. 5.12
PSpice configuration of a multivibrator.

```
»»»»» EX5_2_3.CIR »»»»»»»»»»»»»»»»»»»»»»»»»»»»»»»»»»»»»»»»»»»»»»

AN ASTABLE DRIVER FOR A CHOPPER  (555)
* To plot the driver output-voltage waveforms.
* MicroSim suggests using the options.
. OPTIONS   ITL1=500   ITL5=0
* ITL1 resets the dc iteration limit to 500 (40, default).
* ITL5=0 omits the transient iteration limit (0, default).
. PARAM   CVAL=20nF ;                    Parameter value for capacitor C6.

* SUBCIRCUIT CALL, for the 555 timer is
Xtimer   0  6  3  8  5  6  7  8  555D
*        |  |  |  |  |  |  |  |   |
*        |  |  |  |  |  |  |  |   Name of subcircuit.
*        |  |  |  |  |  |  |  Pin number 8 connection.
*        |  |  |  |  |  |  Pin number 7 connection.
*        |  |  |  |  |  Pin number 6 connection.
*        |  |  |  |  Pin number 5 connection.
*        |  |  |  Pin number 4 connection.
*        |  |  Pin number 3 connection.
*        |  Pin number 2 connection.
*        Pin number 1 connection to external circuit.

* SOURCE of power.
VCC  8  0    12V ;                     This is a dc source.

* CIRCUIT ELEMENTS in circuit external to the subcircuit.
R87       8    7     4.7kohms
R76       7    6     2.2kohms
C6        6    0     {CVAL}
C3        3    0     10pF ;            For stable operation.
RGATE     3    0     1kohms ;          The gate (load) resistance.

* The subcircuit 555D is in the library file.
. LIB   EVAL.LIB

* ANALYSIS
. TRAN    0.1us    600us
. STEP    PARAM  CVAL  LIST  20nF  10nF ;     The program is run twice.
. PROBE   v(3) ;                              Driver output voltage.
. END

»»»»»»»»»»»»»»»»»»»»»»»»»»»»»»»»»»»»»»»»»»»»»»»»»»»»»»»»»»»»»»»»»
```

STEP 5

Using PROBE with the data file EX5_2_3.DAT traces are plotted for the output gate-voltage pulses. Figure 5.13 shows the waveforms.

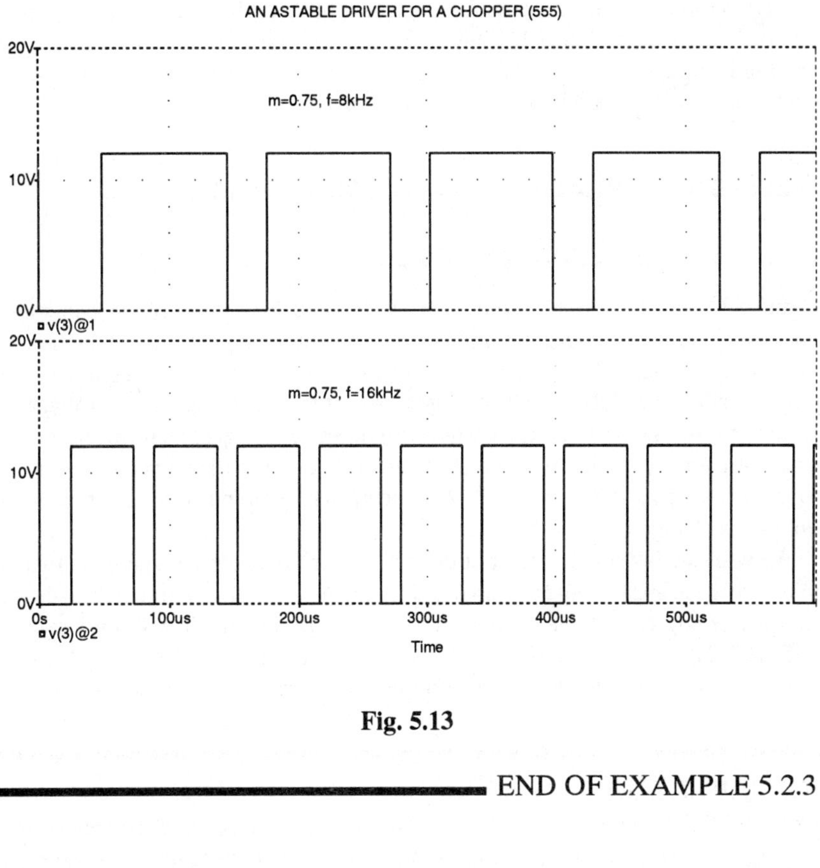

Fig. 5.13

END OF EXAMPLE 5.2.3

Drill Exercise D5.2.8

Following the pattern of EXAMPLE 5.2.3 design a gate driver from a 555 timer suitable for a chopper. The gate signals are to be a train of pulses of frequency 12kHz, of magnitude 12V and duty cycle 0.6. Do a PSpice simulation and use PROBE to plot a trace of the gate signals.

5.3 DRIVERS FOR AC-DC CONVERSION

AC-DC converters are needed if we must supply a dc load from an ac source. Controlled power switches (mainly thyristors†) are required for the converter if the load voltage is to be adjustable. We will describe drivers for two ac-dc converter techniques that involve point-on-wave switching.

5.3.1 SINGLE-PHASE HALF-WAVE CONVERSION

DELAY-ANGLE ALPHA CONTROL

Figure 5.14 depicts a circuit schematic and waveforms for a single-phase, half-wave converter††. The switch *Sw* is a unilateral device, allowing current conduction in one direction only. The driver circuit turns on the switch at an angle α in the positive half cycle of the source-voltage v_s waveform. As a result, the voltage v_s is applied across the load R_l until the current reduces to zero. At this point the switch *Sw* turns off naturally so that the voltage source is disconnected from the load. By altering the firing angle α of the switch's gate pulse the average load voltage is changed.

We will simulate a gate-driver circuit with a periodic gate voltage v_{gate}, that is initiated at the angle α and terminated at π radians of each positive half cycle of the power-source voltage waveform. This is satisfactory if a PSpice switch (VSWITCH) is used in the simulation and the load is purely resistive, or inductive. It also works if the switch is a thyristor model and the load is inductive.

EXAMPLE 5.3.1

Create a PSpice simulation of a driver circuit suitable for a single-phase, half-wave rectifier as illustrated in Fig. 5.14. Put the circuit file in the form of a subcircuit that is suitable for writing to the library file DRIVER.LIB. Use a sinusoidal voltage source of amplitude 100V and frequency $f = 100$Hz as a reference to plot a trace of the gate voltage of magnitude 6V with a trigger angle $\alpha = \pi/3$ radians.

---

† See Section 4.4 for the principle of controlled rectification and the action of the thyristor.

†† This converter is described in "Power Electronics Semiconductor Switches" (PESS) Section 1.3.2, Section 2.3.2 and Section 5.5.

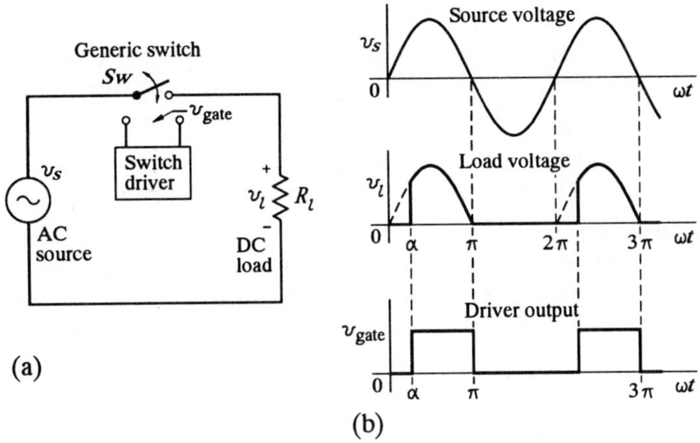

Fig. 5.14 Single-phase half-wave conversion. (a) Circuit diagram, (b) waveforms.

Solution

Three steps of the solution are described.

STEP 1 A PSpice configuration of a single-phase, half-wave rectifier is depicted in Fig. 5.15.

STEP 2 From the PSpice configuration in Fig. 5.15, the circuit file EX5_3_1.CIR can be written. The driver description is written as a subcircuit named HALFWAVE_DRVA in DRIVER.LIB.

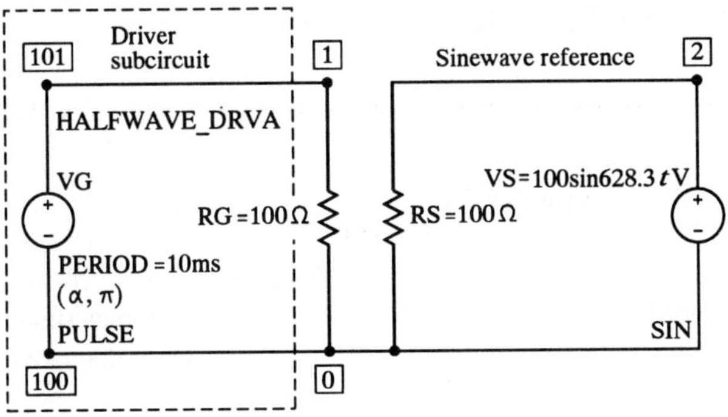

Fig. 5.15 Gate-driver configuration.

```
»»»»» EX5_3_1.CIR »»»»»»»»»»»»»»»»»»»»»»»»»»»»»»»»»»»»»»»»
DRIVER FOR POINT-ON-WAVE SWITCHING (AC-DC CONVERSION)
* To plot the trace of the gate voltage v(1) for f=100Hz and ALFA=60°.

* SINEWAVE REFERENCE SOURCE
VS 2   0    SIN(0  100V  {FREQ})
RS 2   0    100ohms;                An arbitrary resistance.

* DRIVER-PULSE characteristics.
.PARAM    VGATE=6V;                 Pulse magnitude at the gate terminals.
.PARAM    FREQ=100Hz    PERIOD={1/FREQ};      Pulse frequency.
.PARAM    ALFA=60;                  Pulse delay angle in degrees.

* A SUBCIRCUIT DRIVER NAMED HALFWAVE_DRVA.
* Nodes 101, 100 are the subcircuit terminals of the gate voltage VG.
* Nodes 1, 0 are the main circuit-file terminals for the gate voltage VG.
.SUBCKT HALFWAVE_DRVA 101    100
VG 101      100   PULSE(0 {VGATE} {PERIOD*ALFA/360} 1ns 1ns
+    {PERIOD*(180-ALFA)/360} {PERIOD});        The gate-pulse source.
* Needs global parameters VGATE, PERIOD and ALFA in the circuit file.
.ENDS    HALFWAVE_DRVA

* Call the subcircuit into main circuit, to connect nodes 101, 100 to 1, 0.
Xdriver     1     0    HALFWAVE_DRVA
RG 1        0     100ohms;      This simulates the switch's gate resistance.

.TRAN     0.05ms    31ms
.PROBE    v(1),     v(2);       Gate signal referenced to a sinusoid.
.END
»»»»»»»»»»»»»»»»»»»»»»»»»»»»»»»»»»»»»»»»»»»»»»»»»»»»»»»»»»»»»
```

STEP 3 Run PSpice simulation with the circuit file EX5_3_1.CIR. View v(1) and v(2).

END OF EXAMPLE 5.3.1

Drill Exercise D5.3.1

Following the pattern of EXAMPLE 5.3.1 design a gate circuit of a single-phase, half-wave thyristor converter. The power-source frequency is 50Hz. The gate voltage must have a magnitude of 5V. From a PSpice simulation plot traces of the source voltage (amplitude 100V) and the gate voltage over an interval of 60ms. Let the delay angle be $\alpha = 90°$ and let the driver load be 100Ω.

Drill Exercise D5.3.2

Use the subcircuit named HALFWAVE_DRVA, that is in the file DRIVER.LIB, to simulate the driver of a thyristor. The thyristor is used as a single-phase, half-wave converter that operates from a 60-Hz supply of 240V(rms). The voltage applied to the gate of the thyristor is to be 4V and the firing angle is $\alpha = 2\pi/3$ radians. It can be assumed that the driver load is 50Ω. Plot the traces of the power-source voltage waveform and the gate pulses over an interval of 50ms.

DELAY-ANGLE ALPHA AND EXTINCTION-ANGLE BETA CONTROL

If a semiconductor other than a thyristor is to perform the rectification, we can add extinction-angle (β) control to the delay-angle (α) control. That means we can turn the switch on and off at any point on the alternating voltage waveform of the ac source. This helps not only to adjust the rectified output voltage it also helps to adjust the ac-side power factor.

Figure 5.16, depicted on the next page, shows a partial circuit diagram and waveforms for a point-on-wave driver with turn-on (α) and turn-off (β) control. The switch *Sw* is a unilateral device, allowing current conduction in one direction only. The driver circuit turns on the switch after a delay angle α in the positive half cycle of the source-voltage v_s waveform. Further, the driver circuit turns off the switch after an angle β (extinction angle) in the same positive half cycle by having the gate voltage reduced to zero. The voltage v_R across the load is a dc-pulsed waveform as shown in Fig. 5.16b.

We will simulate a train of pulses, each of which is initiated at an angle α and terminated at an angle β for each positive half cycle of the sinusoidal waveform v_s. This kind of gate drive is satisfactory if the converter load is resistive. However, care has to be taken if the converter load is either inductive or capacitive.

EXAMPLE 5.3.2

Simulate a driver circuit that is suitable for a single-phase, half-wave rectifier as depicted in Fig. 5.16. Put the PSpice circuit file in the form of a subcircuit that can be written to the library file DRIVER.LIB. The sinusoidal power source has a voltage 240V(rms) at 50Hz. The parameters of the circuit are the gate resistance $R_G = 1\text{k}\Omega$, the gate pulse of magnitude 12V, and a dummy source resistance $R_s = 1\text{M}\Omega$. Let the pulsewidth of the gate voltage be such that $\alpha = 20°$ and $\beta = 160°$. Plot traces of the reference source voltage v_s and the driver pulses over an interval of 60ms.

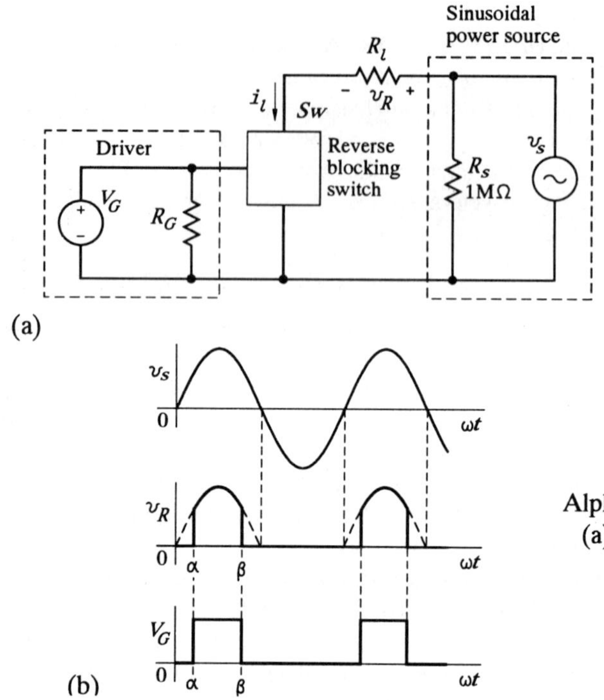

Fig. 5.16
Alpha and beta control.
(a) Circuit diagram,
(b) waveforms.

Solution
Two steps of the solution are described.

STEP 1 We need a PSpice configuration. This will contain the driver circuit and the reference power source. We do not need the converter at this stage. The driver comprises a PULSE source of magnitude 12V, a delay of α degrees (PERIOD*α/360 seconds) and a pulse width of $\beta-\alpha$ degrees (PERIOD*($\beta-\alpha$)/360 seconds) each positive half cycle of the sinusoidal source waveform.

STEP 2 From the PSpice configuration in Fig. 5.17, a circuit file, EX5_3_2.CIR, can be written as follows. Do the simulation and plot v(1) and v(2). View the subcircuit HALFWAVE_DRVAB in DRIVER.LIB.

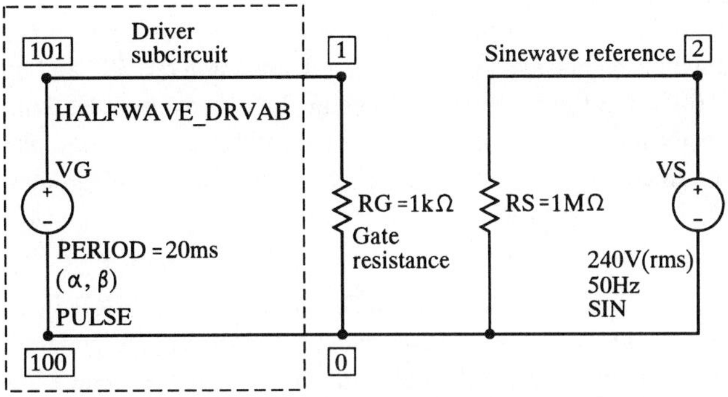

Fig. 5.17 Gate-driver configuration.

```
»»»»» EX5_3_2.CIR »»»»»»»»»»»»»»»»»»»»»»»»»»»»»»»»»»»»»»
A HALF-WAVE DRIVER WITH ALFA AND BETA CONTROL
* To plot the trace of the train of gate pulses for α=20°, β=160° and f=50Hz.
* PARAMETERS needed for the subcircuit pulse characteristics.
.PARAM    VGATE=12V    VSIN=240V    VMAX={1.414*VSIN}
.PARAM    FREQ=50Hz    PERIOD={1/FREQ}
.PARAM    ALFA=20      BETA=160;              In degrees.
* SOURCE.  Sinewave reference.
VS  2  0   SIN(0 {VMAX} {FREQ})
RS  2  0   1E6ohms
* A SUBCIRCUIT DRIVER NAMED HALFWAVE_DRVAB.
* Nodes 101, 100 are the subcircuit terminals of the gate voltage VG.
* Nodes 1, 0 are the main circuit terminals for the gate voltage VG.
.SUBCKT HALFWAVE_DRVAB    101  100
VG   101   100   PULSE(0 {VGATE} {PERIOD*ALFA/360} 1ns 1ns
+    {PERIOD*(BETA-ALFA)/360} {PERIOD});     The gate-pulse source.
.ENDS    HALFWAVE_DRVAB
* Call the subcircuit into main circuit.
Xdriver   1  0   HALFWAVE_DRVAB
RG        1  0   1E3ohms;       This simulates the switch's gate resistance.
.TRAN     0.1ms  61ms
.PROBE    v(1),  v(2);          Gate voltage referenced to a sinusoid.
.END
»»»»»»»»»»»»»»»»»»»»»»»»»»»»»»»»»»»»»»»»»»»»»»»»»»»»»»»»»»
```

END OF EXAMPLE 5.3.2

Drill Exercise D5.3.3
Following the pattern of EXAMPLE 5.3.2, simulate a single-phase, half-wave rectifier drive circuit with each pulse of 10V, having a delay angle $\alpha = 30°$ and an extinction angle $\beta = 100°$. Let the ac source voltage be 240V(rms) at 60Hz with a load of 10MΩ. Assume a gate-drive load of 2kΩ. From a PSpice simulation plot traces of the reference sinewave and the waveform of the gate-drive voltage over an interval of 50ms.

Drill Exercise D5.3.4
Use the subcircuit named HALFWAVE_DRVAB in the file DRIVER.LIB to simulate a single-phase, half-wave rectifier driver circuit. The driver is to generate a voltage with respect to a 100-V(rms), 100-Hz sinewave reference. The 15-V gate voltage is to be turned on at $\alpha = 40$ degrees and turned off at $\beta = 130$ degrees every positive half cycle of the reference voltage. The driver load is 3kΩ. Plot traces of the reference sinewave voltage and the gate voltage over an interval of 30ms.

Section 5.3.2 about full-wave drivers is to be found on the WEBsite. Also there are EXAMPLES and Drill Exercises.

5.4 DRIVERS FOR DC-AC CONVERSION

5.4.1 INTRODUCTION

A power electronics converter, that modulates power from a dc source and supplies ac power to a load, is an inverter. The parameters that are controlled by the inverter† are the magnitude of the output voltage and the frequency of alternation.

5.4.2 A SINGLE-PULSE DRIVER

A common technique for inverter control is to *chop* the dc-supply voltage waveform so that the load experiences rectangular signal waveforms that change polarity periodically.

A single-phase bridge configuration of an inverter is illustrated in Fig. 5.18. There are four switches. Bridge configurations are normally operated such that

† For inverter operation see "PESS" sections 1.3.2 and 2.4.

pairs of switches ($Sw1$, $Sw2$ and $Sw3$, $Sw4$) are driven in unison. However, because some of the switches in the bridge are not referenced to ground, isolated driver outputs are normally required in practice. That is, four gate signals are required. With a PSpice simulation this problem does not arise. Isolation between the driver and switch always exists. We need only use two gate voltages for four switches.

From an inspection of the circuit schematic in Fig. 5.18, it can be surmised that a finite value of control voltage (gate voltage) will keep a PSpice switch in the on-state. No gate voltage means that the switch is off. In practice, such a representation would hold for BJTs, MOSFETs and IGBTs.

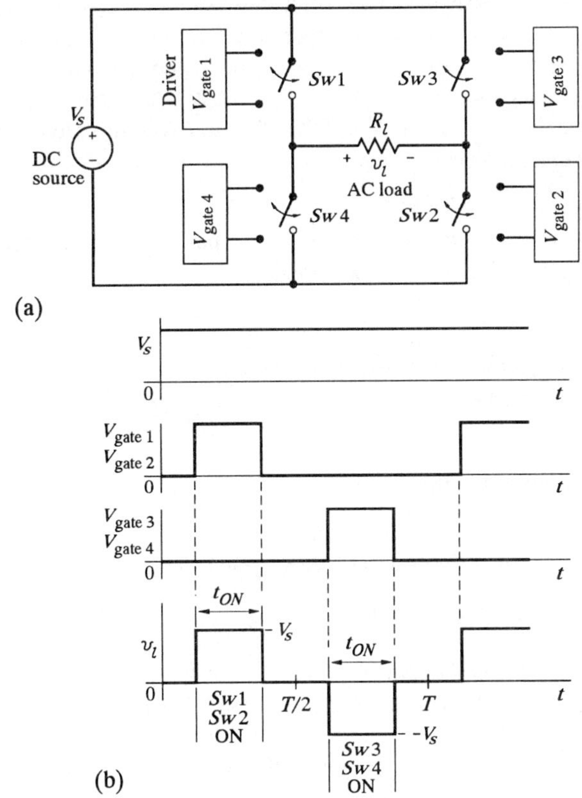

Fig. 5.18 A single-phase bridge inverter. (a) Circuit diagram, (b) waveforms.

Each pair of switches is turned on by gate signals for an interval t_{ON} ($t_{ON}<T/2$) over each period T. First, the $Sw1$ and $Sw2$ pair of switches is turned on (at $t=(T/2-t_{ON})/2$). This will result in a positive voltage ($v_l = V_s$) across the load resistance. Second, after these switches are turned off (at $t=(T/2+t_{ON})/2$), the pair of switches $Sw3$ and $Sw4$ is turned on (at $t=(3T/2-t_{ON})/2$). This results in a negative voltage ($v_l=-V_s$) across the load resistance. The load voltage waveform is shown in Fig. 5.18b.

EXAMPLE 5.4.1

Consider the single-phase bridge inverter, illustrated in Fig. 5.18. The dc source has a voltage $V_s=400\text{V}$ and the inverter output frequency is $f=1\text{kHz}$. Make a PSpice simulation of a 15-V driver circuit, general enough to use as a subcircuit, but particular enough for an average power of 160W to be dissipated in the inverter load resistance, whose value is $R_l=800\Omega$. Plot traces of the waveforms of V_{gate1} and $V_{\text{gate 3}}$ over an interval of 3ms.

Solution

The output waveform period is 1ms. This is the period of the gate signals. From Fig. 5.18 the rms value of the load voltage is

$$V_{l\,\text{rms}} = \left[\frac{1}{T/2}\int_0^{t_{ON}} V_s\,dt\right]^{1/2} = V_s\left(\frac{2t_{ON}}{T}\right)^{1/2}. \tag{5.4.1}$$

The average power P in the load is $P = \dfrac{V_{l\,\text{rms}}^2}{R_l}$. (5.4.2)

Therefore, $t_{ON}=0.4\text{ms}$. This gives a duty cycle $m=t_{ON}/T=0.4$.

There are four steps in the solution.

STEP 1 A PSpice circuit configuration of the gate driver can be drawn from an inspection of Fig. 5.18. The configuration is shown in Fig. 5.19.

STEP 2 From the configuration in Fig. 5.19 the circuit file can be written.

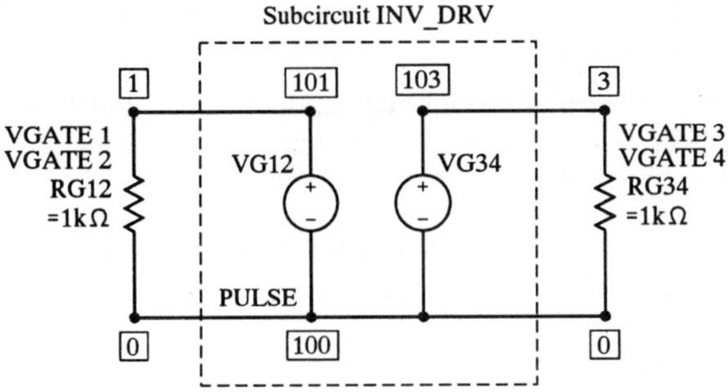

Fig. 5.19
PSpice configuration for an inverter drive.

```
»»»»» EX5_4_1.CIR »»»»»»»»»»»»»»»»»»»»»»»»»»»»»»»»»»»»»»»»
DRIVER FOR A SINGLE-PHASE BRIDGE INVERTER

* PARAMETERS used in the subcircuit.
.PARAM   FREQ=1kHz   PERIOD={1/FREQ};      Inverter frequency.
.PARAM   VGATE=15V;                        The gate voltages.
.PARAM   TON=0.4ms;               The pulse width of the gate signal.

* SUBCIRCUIT NAMED INV_DRV for the driver circuit in DRIVER.LIB.
.SUBCKT  INV_DRV 101 103 100
VG12 101 100   PULSE(0 {VGATE} {PERIOD/4-TON/2}
+   1ns 1ns {TON} {PERIOD});  Gate pulses VGATE1 and VGATE2.
VG34 103 100   PULSE(0 {VGATE} {3*PERIOD/4-TON/2}
+   1ns 1ns {TON} {PERIOD});  Gate pulses VGATE3 and VGATE4.
.ENDS  INV_DRV
* Nodes 101, 100 are for the gate-drive outputs for Sw1 and Sw2.
* Nodes 103, 100 are for the gate-drive outputs for Sw3 and Sw4.

* Calling the subcircuit configuration into the main circuit requires.
Xdriver 1 3  0  INV_DRV
* Nodes 1, 3, 0 in the main circuit relate to 101, 103 and 100 in the subcircuit.

* CIRCUIT ELEMENTS
RG12    1  0   1E3ohms;                         Gate driver load.
RG34    3  0   1E3ohms;                         Gate driver load.

.TRAN   50us   2.5ms
.PROBE  v(1),  v(3);      Driver pulses to the gates of the switches.
.END
»»»»»»»»»»»»»»»»»»»»»»»»»»»»»»»»»»»»»»»»»»»»»»»»»»»»»»»»»»»»
```

STEP 3 A PSpice simulation can be run with EX5_4_1.CIR.

STEP 4 From the simulation results, PROBE can be used to plot traces of the gate-driver voltages. These traces are shown in Fig. 5.20.

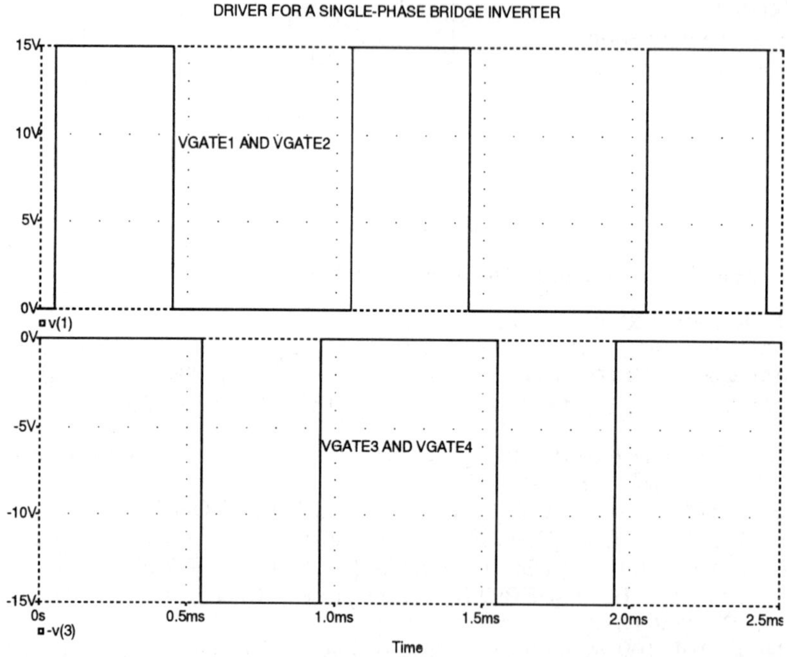

Fig. 5.20

END OF EXAMPLE 5.4.1

Drill Exercise D5.4.1

Following the pattern of EXAMPLE 5.4.1 simulate a driver for a single-phase bridge inverter as illustrated in Fig. 5.18. The gate voltage is to have pulses of magnitude $V_{gate} = 12V$. The inverter output is to operate at 50kHz with rectangular pulse widths of $5\mu s$ duration. Plot traces of the gate voltages over an interval of $60\mu s$.

Drill Exercise D5.4.2

Use the subcircuit INV_DRV in the file DRIVER.LIB to simulate the driver for a single-phase bridge inverter. The specifications are that the gate signals are to be pulses of magnitude 10V, of duration 15μs and frequency 20kHz. Plot the traces of the gate voltages over an interval of 0.15ms.

5.4.3 A UNIFORM MULTIPLE-PULSE DRIVER

A variation of the single-pulse driver, described in Section 5.4.2 is the multiple-pulse driver. We note that the driver signals to the gates of the inverter switches are reflected in amplified form in the output-voltage waveform. A single gate pulse of duration t_{ON} each half cycle of the output waveform enables a single pulse ($v_l = V_s$) of duration t_{ON} each half cycle to appear across the inverter load. Equally well, if the gate signal was a train of pulses over the interval t_{ON}, then the inverter output would have a simple train of pulses, with the same pulse width, the same duration t_{ON}, but with a magnitude equal to the power dc-source voltage V_s.

The main-circuit schematic for an n-pulse inverter remains the same as is shown in Fig. 5.18. The differences are associated with the details of the driver circuit. A driver schematic is shown in Fig. 5.21 on the next page. The variables are the number of pulses n, the pulse width t_{ON}, the pulse magnitude and the inverter-output frequency f (or period T). Once the pulse width and the number of pulses have been chosen, the output of the inverter is symmetrical. The output rms value of the load voltage is

$$V_{l\,\mathrm{rms}} = \left[\frac{1}{T''}\int_0^{t_{ON}} V_s^2\,dt\right]^{1/2} = V_s\left[\frac{t_{ON}}{T''}\right]^{1/2} = V_s\sqrt{m} = \left[2nt_{ON}\,f\right]^{1/2} \qquad (5.4.3)$$

where m is the duty cycle t_{ON}/T''. Consequently, the driver can be designed in terms of the number of pulses and pulse width. A subcircuit to produce multiple pulses as shown in Fig. 5.21 is named INV_PWM_DRV in DRIVER.LIB.

Drill Exercise D5.4.3

Consider the single-phase bridge inverter depicted in Fig. 5.18. The dc source has a voltage $V_s = 400$V and the inverter output frequency is $f = 1$kHz. Make a PSpice simulation of a 12-V driver circuit, using the INV_PWM_DRV subcircuit. An average power of 120W is to be delivered to the inverter load resistance, whose value is $R_l = 750\Omega$. Let the number of gate pulses per half cycle be two. Plot traces of the waveforms of $V_{\text{gate 1}}$ and $V_{\text{gate 3}}$. See the WEBsite for Extra Drills.

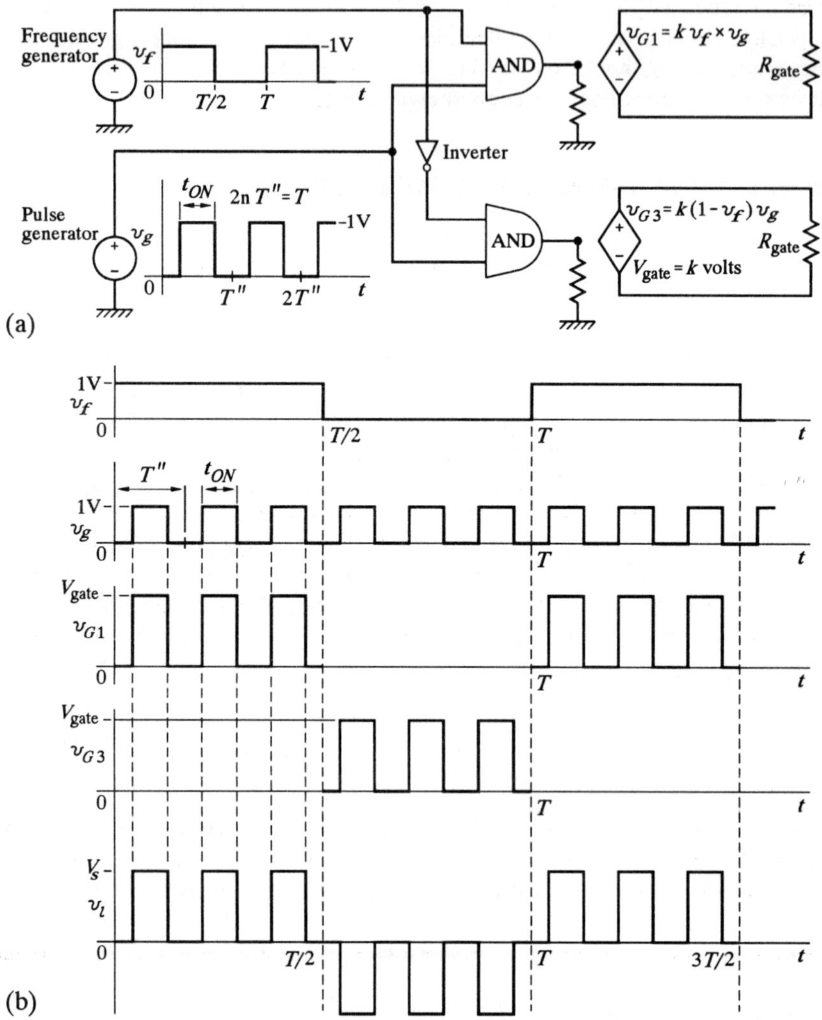

Fig. 5.21 A uniform *n*-pulse driver.
(a) Driver schematic, (b) waveforms.

5.4.4 A MODULATING PWM DRIVER (SPWM)

Many driver switching schemes have been devised in the quest to obtain an inverter ac waveform that closely resembles a sinewave. One technique is to extend the method of equal multiple pulses described in Section 5.4.3. The extension is to modulate the width of the pulses over the half cycle of the gate signal. This can be called a modulating PWM driver. We will consider the single-phase bridge inverter whose schematic diagram is shown in Fig. 5.18. The switches are driven in pairs $Sw1$, $Sw2$ and Sw3, $Sw4$. Our aim in this section is not to design switching strategies for a PWM driver, it is to exemplify a technique that can be simulated using PSpice.

Figure 5.22 shows a sinewave whose frequency is to be the same as that required for the inverter. Superimposed on the same plot is a sawtooth waveform, whose frequency is greater than that of the sinewave. The strategy is that, during the intervals that the sinewave-voltage magnitude is greater than the sawtooth-voltage magnitude, there is a gate voltage of magnitude V_{gate} as shown. Otherwise the gate voltage is zero. This method of modulation is called *Sinewave Pulse-width Modulation* or SPWM.

Referring to Fig. 5.22, as the sinewave cycle begins, the gate pulses (for $v_s > v_{st}$) are relatively narrow. These gate pulses increase gradually in width until the peak of the ac waveform is reached. Then the pulse widths decrease until the zero-crossing point of the sinewave is reached. The rms value of the inverter output voltage is determined by the ratio of the amplitudes of the sinewave and the sawtooth waveforms. This latter ratio $M = \hat{V}_s / \hat{V}_{st}$ is known as the modulation index.

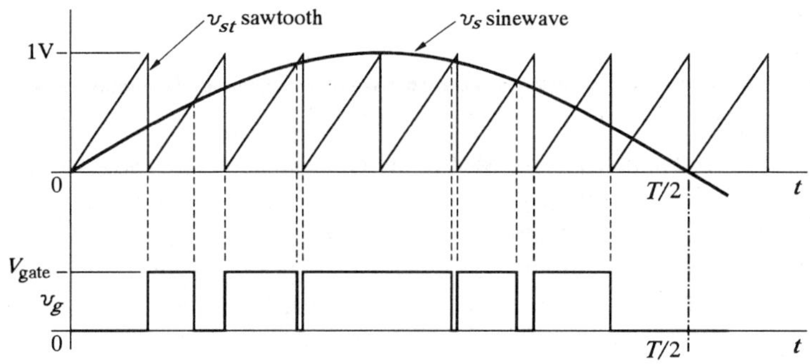

Fig. 5.22 Gate pulse if $v_s > v_{st}$.

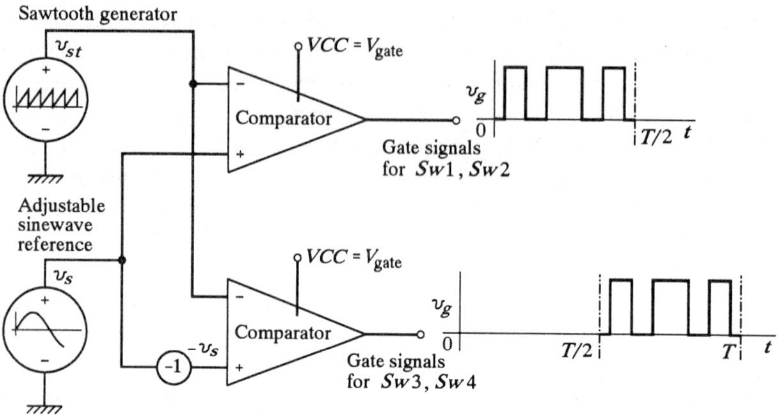

Fig. 5.23 Implementation of a modulating PWM driver (SPWM).

The implementation of this control strategy for modulating the pulse widths of the gate signals is shown in the schematic of Fig. 5.23. For a PSpice simulation we can use existing subcircuits as building blocks. The triangular-wave generator (VSAWTOOTH) and the comparator (COMPARATOR2) can be used to generate a PWM drive similar to the chopper generator in Section 5.2.1.

EXAMPLE 5.4.2

Create a PSpice simulation of the SPWM driver, illustrated in Fig. 5.23. Formulate a general subcircuit for the driver, but make the circuit specific to the following requirements by using parameter values. The single-phase bridge inverter frequency is 50Hz. Let the number of gate pulses per half cycle of the inverter-output waveform be five, and let the gate-pulse magnitude be 15V.

Solution

This particular solution is accomplished in four steps.

STEP 1 From the schematic of the driver in Fig. 5.23 a PSpice configuration can be drawn. See Fig. 5.24. Each comparator is chosen to be the subcircuit named COMPARATOR2 and the sawtooth generator is chosen to be the subcircuit named VSAWTOOTH, both from DRIVER.LIB. The sinewave reference is a voltage SIN source and its inverse is obtained by using a voltage-controlled voltage source EVS.

STEP 2 The information given in the PSpice configuration of the driver in Fig. 5.24 can be written to a circuit file named EX5_4_2.CIR.

STEP 3 The PSpice simulation can be run with the file EX5_4_2.CIR.

STEP 4 Using PROBE, we can plot v(12) and v(34) which are the signal outputs of the driver to the gates of the inverter switches *Sw*1, *Sw*2, *Sw*3 and *Sw*4. Figure 5.25, shown on page 175, shows these traces.

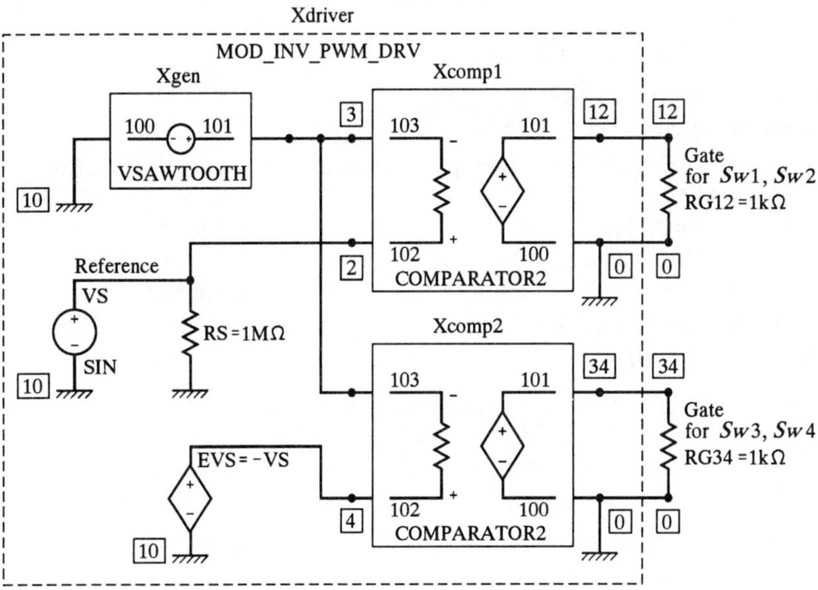

Fig. 5.24 PSpice configuration of an SPWM modulator.

```
»»»»» EX5_4_2.CIR »»»»»»»»»»»»»»»»»»»»»»»»»»»»»»»»»»»»»»»»»
INVERTER DRIVER WITH SPWM MODULATION
* To plot the traces of the gate-voltage waveforms for Sw1, Sw3.
* We will use subcircuits already used for drivers so we call
.LIB   DRIVER.LIB

* We will create a driver subcircuit called MOD_INV_PWM_DRV
.SUBCKT  MOD_INV_PWM_DRV 12 34 10;     Found in DRIVER.LIB.
** We will call two comparator subcircuits entitled
**.SUBCKT COMPARATOR2 103 102 101 100 PARAMS: VCC=10V
Xcomp1  3  2  12  10   COMPARATOR2   PARAMS:   VCC={VGATE}
Xcomp2  3  4  34  10   COMPARATOR2   PARAMS:   VCC={VGATE}
** We call the sawtooth-generator subcircuit entitled
**.SUBCKT VSAWTOOTH 101 100 PARAMS: VM=2V    PERIOD=20us
Xgen      3  10  VSAWTOOTH   PARAMS:  VM={VMST}  PERIOD={PER}
** The reference sinewave is described by
VS         2  10  SIN(0  {VMAX} {FREQ})
RS         2  10  1E6ohms;                    Artificial load for VS source.
** The source VS needs to be inverted for the negative half cycle pulses.
EVS        4  10 2  10   -1;                  EVS=-1*v(2,10)=-VS.
.ENDS   MOD_INV_PWM_DRV

* PARAMETERS needed in the main circuit file.
.PARAM VGATE=15V;        This is the magnitude of the gate pulses.
.PARAM VMST=1V;          This is the amplitude of the sawtooth waveform.
.PARAM MOD=1;            This is the modulation index. Ratio VMAX/VMST.
.PARAM VMAX={MOD*VMST};          This is the amplitude of the sinewave .
.PARAM FREQ=50Hz;        This is the required inverter frequency.
.PARAM NUM=6;            (6-1)≈number of pulses per half cycle.
.PARAM FREQST={2*NUM*FREQ};  Frequency of sawtooth waveform.
.PARAM PER={1/FREQST};      This is the period of the sawtooth waveform.

* We need to call the driver subcircuit with the X statement.
Xdriver 12    34    0   MOD_INV_PWM_DRV
RG12   12    0     1E3ohms;              Gate resistance for Sw1, Sw2.
RG34   34    0     1E3ohms;              Gate resistance for Sw3, Sw4.

.TRAN        0.1ms   40ms;               Two inverter output cycles.
.PROBE       v(12),   v(34);             These are the driver outputs vg.
.END

»»»»»»»»»»»»»»»»»»»»»»»»»»»»»»»»»»»»»»»»»»»»»»»»»»»»»»»»»»»»»»»
```

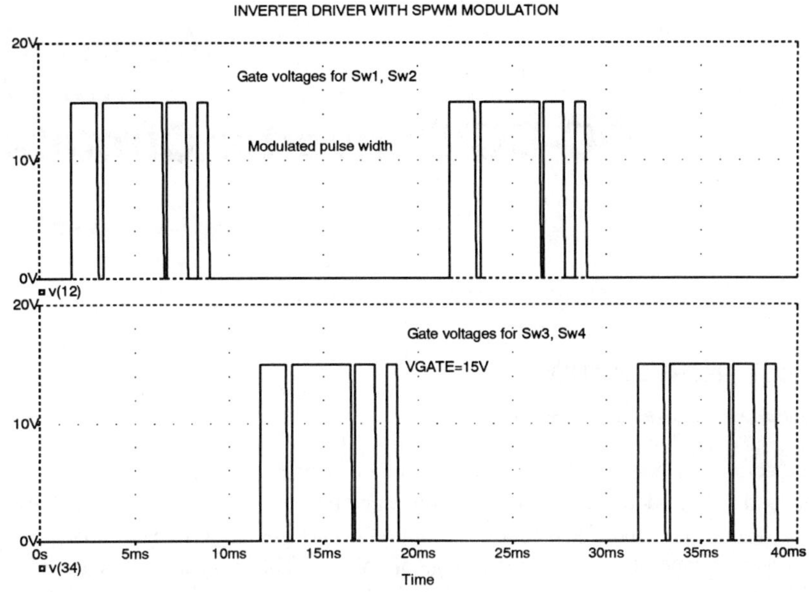

Fig. 5.25

END OF EXAMPLE 5.4.2

Drill Exercise D5.4.4

Follow the pattern of EXAMPLE 5.4.2 to simulate a modulating-multipulse driver for a single-phase bridge inverter, depicted in Fig. 5.18. The driver schematic is shown in Fig. 5.23 with its PSpice configuration in Fig. 5.24. Figure 5.22 illustrates the form of a modulated train of gate pulses to the inverter. For a 60-Hz inverter that requires 12-V gate pulses, plot traces of the gate voltages if the number of pulses of the carrier sawtooth waveform is 10 per half cycle of the inverter output waveform.

Drill Exercise D5.4.5

Use the subcircuit named MOD_INV_PWM_DRV to simulate a driver for a single-phase bridge inverter. The number of pulses of the carrier signal (sawtooth generator) is 8 per half cycle of the inverter output waveform which has a frequency of 100Hz. The gate pulses are to have a magnitude of 10V. Plot traces of the driver-output voltages.

A digital SPWM driver is to be found in Section 5.4.4 on the WEBsite.

Chapter 6

DC-DC Converter Circuits

6.1 INTRODUCTION

The dc-dc converter produces an adjustable dc voltage at the load. The value of the load voltage depends on the duty ratio *m* and the value of the dc source voltage.

Elements *L* and *C* and a diode are found in chopper circuits. The capacitor *C* acts as a voltage filter to limit the change of voltage. The inductor acts as a current filter to limit the change of current, and the diode is included to limit the voltage across the inductor due to the switching action. These elements play a part to modify the chopper modulation. By simulation we will determine how the power modulation is affected.

Some of the examples and problems that follow are similar to those in the text "Power Electronics Semiconductor Switches" by R. S. Ramshaw, (PESS) Chapter 2, Section 2. This PSpice guide to the simulation of power electronics circuits is independent of this text. However, the normal analysis is found in the above text and the Solutions Manual provides numerical answers.

6.2 THE BUCK CONVERTER

A dc-dc converter that utilizes a chopper to produce an adjustable output voltage that is less than the dc supply voltage is called a buck converter. One is depicted in Fig. 6.1a. Figure 6.1b shows the ideal waveforms for the condition that the capacitor value is large.

The average voltage V_{lav} across the load is ideally expressed by $V_{lav}=mV_s$ where m is the duty cycle ($m=t_{ON}/T$). From a mathematical analysis the inductance L that must be in the circuit so that its current falls to zero only once per cycle ($i = I_{min} = 0$) is $L = R_l(1-m)T/2$, where $T = 1/f$ and f is the frequency of switching. We would like to test this for a practical case.

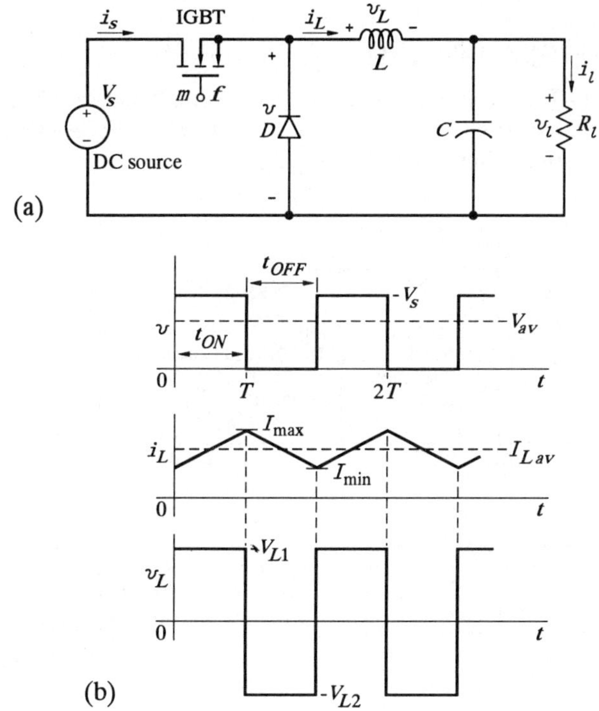

Fig. 6.1 Buck converter.
(a) Circuit diagram, (b) ideal waveforms.

EXAMPLE 6.2.1

An IGBT buck converter† has been designed so that for a duty cycle $m = 0.707$ the average power to the load is ideally 5kW and the inductor current just falls to zero once per cycle. See Fig. 6.1. The converter parameters are $V_s = 100$V, $f = 1$kHz, $L = 147\mu$H, $R_l = 1\Omega$, and $C = 350\mu$F. Use PSpice to simulate the chopper circuit

† This is EXAMPLE 2.1 in PESS ("Power Electronics Semiconductor Switches").

for these conditions. Use PROBE to create traces of **(a)** the diode voltage and current, **(b)** the inductor current i_L and **(c)** the inductor voltage v_L. Initialize the capacitor voltage to 70V so that steady state is reached early in the transient analysis. Treat the elements V_s, IGBT and D as near-ideal devices.

Solution

There are four steps in this solution.

STEP 1 — The first step involves the creation of a PSpice configuration from the circuit diagram in Fig. 6.1 and the given element values. A configuration is shown in Fig. 6.2.

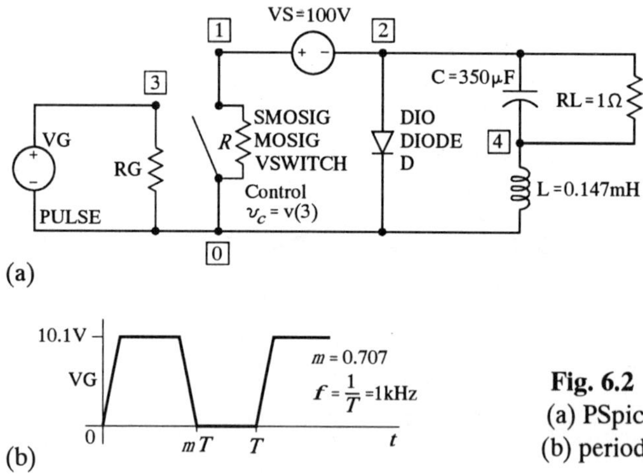

Fig. 6.2 Buck converter. (a) PSpice configuration, (b) periodic gate pulse VG.

The IGBT model, described in Section 4.7, EXAMPLE 4.7.1, or the subcircuits MG_IDEAL and MG_GATE, is the one used here.

The near-ideal diode is described in Section 4.2, EXAMPLE 4.2.3. The default value of the series parasitic resistance is RS = 0. The default values of the junction capacitance is CJO = 0 and the transit time TT = 0, both associated with charge storage effects. With RS = 0 there may be no limit to the current and with CJO = 0 (TT = 0) there is zero switching time. To prevent numerical convergence problems it is better to have small values RS = 1mΩ, CJO = 0.1pF and TT = 0.1ns.

STEP 2 — The second step involves writing the circuit file EX6_2_1.CIR from Fig. 6.2. This file description is as follows.

```
»»»»» EX6_2_1.CIR »»»»»»»»»»»»»»»»»»»»»»»»»»»»»»»»»»»»»»»»

BUCK CONVERTER WITH IDEAL IGBT
.OPTIONS    RELTOL=0.01;        This option is described below.

VS    1  2   100V;              DC source voltage.
RL    4  2   1;                 The load resistance.
L     0  4   0.147mH;           Limits the current excursion.
C     4  2   350uF;             Tends to keep the load voltage constant.
.IC   V(4, 2)=70V;              Average load voltage = mVS = 0.707×100.
* The .IC statement is for initial conditions across C, and steady state.
* There is a choice of adding UIC to the .TRAN statement.

DIO   2   0   DIODE;            Device, diode.  See EXAMPLE 4.2.3.
.MODEL  DIODE  D(RS=0.001  CJO=0.1pF  TT=0.1ns)
* Model parameters chosen to aid convergence, see Section 7.2.

SMOSIG  1   0   3   0   MOSIG;  Device, IGBT.  See EXAMPLE 4.7.1.
.MODEL  MOSIG    VSWITCH(RON=1E-3 ROFF=1E6 VON=10 VOFF=2)

* GATE CONTROL OF SWITCH
VG    3   0   PULSE(0  10.1V  0  6ns  6ns  0.707ms  1ms)
* Pulsewidth = tON = mT = m/f = 0.707ms.        Period = T = 1/f = 1ms.
* TR = TF = 6ns were chosen to aid numerical analysis.
RG    3   0   1kohm;        An arbitrary resistance to close the voltage loop.

* ANALYSIS, transient, to cover two cycles of switching.
.TRAN   10us  2.1ms  0   80us  UIC;        UIC:- Use Initial Conditions.
* TMAX = 80us is used to eliminate diode voltage oscillation.
* Try using TMAX=20μs and plot the trace of v(2).
.PROBE  v(2),  i(L),  v(0, 4)
.END

»»»»»»»»»»»»»»»»»»»»»»»»»»»»»»»»»»»»»»»»»»»»»»»»»»»»»»»»»»»»
```

EXPLANATION OF CIRCUIT FILE EX6_2_1.CIR

A transient analysis of a switching circuit with diodes is prone to numerical convergence problems and oscillations. For stability it may be required to change the simulator parameters. In this example we added a .OPTIONS statement. Further, we tuned the gate signal pulse VG and used a large value TMAX = 80μs for the maximum step size in the numerical analysis.

.OPTIONS

In order to aid convergence, accuracy can be sacrificed to varying degrees. PSpice provides options that can be included in a .OPTIONS statement. There are 32 options that are described in the manual. For the analysis of power electronics circuits there are at least three important options. These three options are listed in Table 6.1. An option statement in the circuit file has the form

```
.OPTIONS  option=value ...
```

If an analysis has convergence problems, one of the first steps to find a solution is to increase the value of the relative tolerance RELTOL. The default tolerance of RELTOL is 1 part in 1000 (0.001 or 0.1%). At low values of the variables (approaching zero) this high tolerance can produce a problem. The relative tolerance can be increased systematically until there is no convergence problem. In some cases the absolute values of the voltage and current tolerances, VNTOL and ABSTOL, are too tight. In a power electronics circuit, where the variables may be at the level of kilovolts and kiloamperes, a minimum voltage of 1mV and a minimum current of 1μA are more practical than the default values.

Table 6.1 Some options.

Options	Effect
RELTOL	Sets the relative error tolerance of the voltages and currents. Default 0.001 (0.1%)
ABSTOL	Sets the absolute error tolerance of the current. Default 1pA.
VNTOL	Sets the absolute error tolerance of the voltage. Default 1μV.

INITIALIZATION

In this example there is an initialization statement

```
.IC  V(0, 4)=70V
```

This indicates that the voltage v(0, 4) across the capacitor C is given the initial condition $v_C(0)=70$V. The value of the initial condition is important. Time can be saved to reach steady state. If the initial conditions are not known, relax the options RELTOL, VNTOL and ABSTOL. Find the operating point, reset the initial conditions, and tighten the options again. Then, the transient analysis can be performed. For this example an operating point of 68V was found for the capacitor voltage. This is a deviation from the theoretical analysis.

The .IC statement is one of three methods of initializing a circuit before the required analysis is performed. The other two are (i) IC=value in an element statement with UIC in the .TRAN statement, and (ii) a .NODESET statement†.

PSpice can act in two ways if the .IC statement is used. UIC placed in the .TRAN statement makes PSpice Use Initial Conditions stipulated in the .IC statement

† See Section 7.2 of this book, or the MicroSim manual.

directly in the transient analysis, by omitting a bias-point calculation. No UIC in the .TRAN statement makes PSpice do an initial bias-point calculation with the .IC conditions held by voltage sources. After the bias-point calculation, all values are transferred to the transient analysis as initial conditions (see the output file for values) and then the values are relaxed.
The general form for the .IC statement is

.IC node_pair_voltage = value ...

This ends the explanation of circuit file EX6_2_1.CIR.

STEP 3 The third step involves using PSpice to run the circuit file EX6_2_1.CIR. Errors and convergence problems can be tackled at this stage. In this example different values of RELTOL, snubber circuits, diode parameters, gate-voltage pulse parameters, and .TRAN parameters were all tried and adjusted.

STEP 4 The fourth step involves invoking PROBE to obtain traces of the diode voltage v(2), the diode current i(DIO), the inductor current i(L) and the inductor voltage v(0,4) = – v(4). These traces are shown in Fig. 6.3. The waveforms are not quite linear like those used in a theoretical analysis.

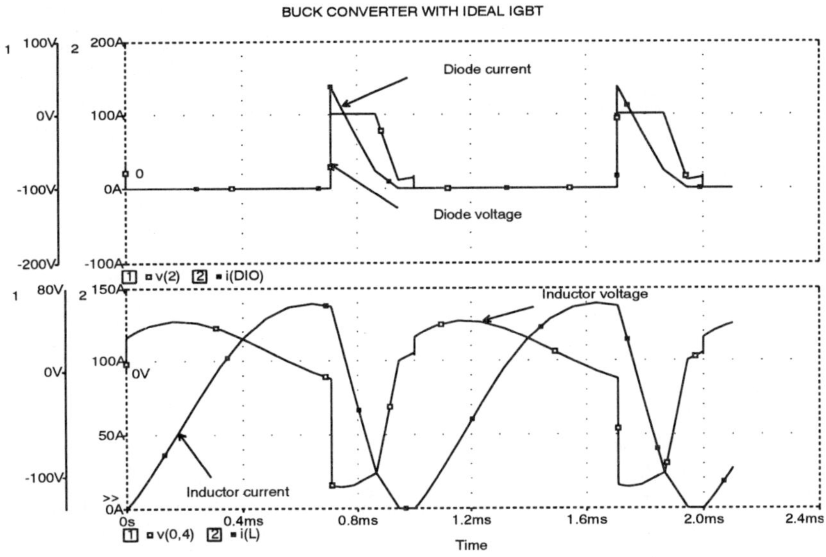

Fig. 6.3

END OF EXAMPLE 6.2.1

Drill Exercise D6.2.1

A MOSFET buck converter is used with a duty cycle $m = 0.707$ and a switching frequency $f = 1$kHz. The converter parameters are $V_s = 100$V, $L = 147\mu$H, $R_l = 1\Omega$ and $C = 350\mu$F. Determine the average power **(a)** delivered by the supply or **(b)** absorbed by the load. The circuit diagram is shown in Fig. 6.1a. Set an initial voltage across the capacitor at 68V. The diode parameters can be chosen to be the same as those in the circuit file EX6_2_1.CIR.
(Ans: 6.42kW. If (a) < (b), steady conditions have not been reached.)

Drill Exercise D6.2.2

A MOSFET chopper has its circuit diagram in Fig. 6.4. The operating frequency of the chopper is $f = 10$kHz and the duty cycle is $m = 0.6$. From a PSpice simulation for steady-state conditions **(a)** plot a trace of the load current, **(b)** determine the load-current excursion, **(c)** determine the average load current and **(d)** determine the average power delivered by the supply. (Hint. (i) Use the switch and diode models from EX6_2_1.CIR. (ii) Determine the initial steady-state conditions from a transient analysis over many cycles. As a check, the average inductor voltage is zero in the steady state.) (Ans: **(b)** 2.97A, **(c)** 5.93A, **(d)** 181.3W.)

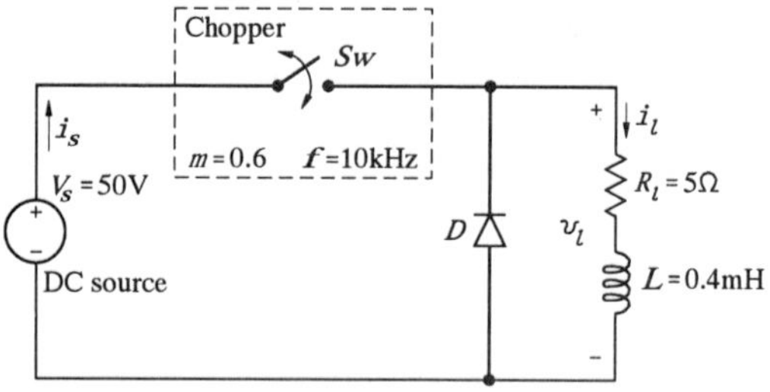

Fig. 6.4

There are Extra Drill Exercises on the WEBsite.

6.3 THE BOOST CONVERTER

A dc-dc converter that produces an adjustable output voltage that is greater than the dc supply voltage is called a boost converter. Figure 6.5 shows the circuit diagram of a boost converter. The ideal waveforms are shown for the case that the value of the capacitor C is very large (a constant load voltage).

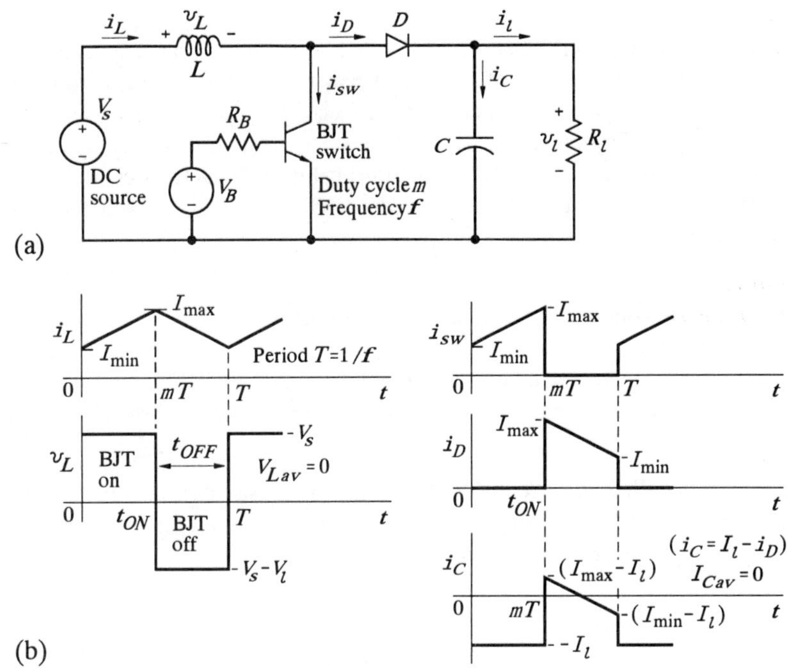

Fig. 6.5 Boost converter.
(a) Circuit diagram, (b) ideal waveforms.

A theoretical analysis of the circuit with ideal elements and a very large value of C provides the average value of the output voltage V_l in terms of the source voltage V_s and the chopper duty cycle m. The results are

$$V_l = V_s/(1-m) \tag{6.3.1}$$

and

$$V_s I_{L\,av} = V_l I_l . \tag{6.3.2}$$

We can check these with the results of a simulation study.

EXAMPLE 6.3.1

An ideal boost converter[†] utilizes a BJT switch to modulate power from a 100-V dc source to a resistive load of 3Ω. The frequency of modulation is 10kHz and the duty cycle is $m = 0.667$. Circuit elements (see Fig. 6.5a) have values $L = 0.2\text{mH}$ and $C = 600\mu\text{F}$. From a simulation plot the steady-state traces of i_L, i_D, i_C and i_l and compare them with the ideal waveforms in Fig. 6.5b. Determine the average power absorbed by the load under steady conditions. The theoretical value is ideally 30kW.

Solution

Two of the steps in this solution are described below.

STEP 1 — Create a PSpice configuration from the circuit diagram in Fig. 6.5 and the given element values. A configuration is shown in Fig. 6.6.

STEP 2 — Write the circuit file from Fig. 6.6. A base current pulse of about 10A is chosen to satisfy a BJT collector current of about 200A. Now run the simulation.

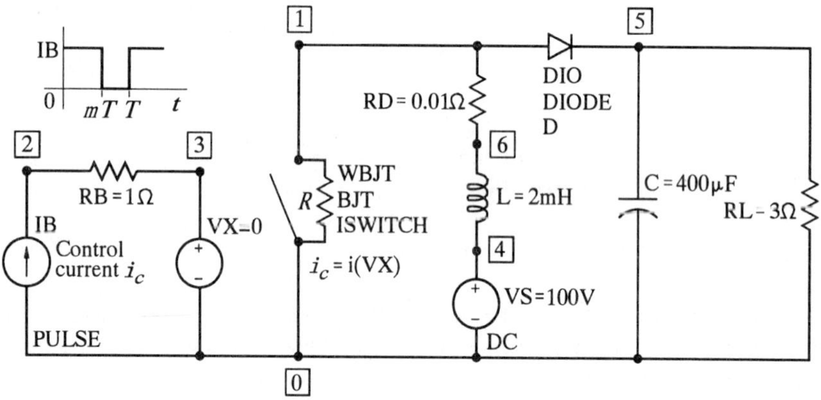

Fig. 6.6 Boost converter configuration.

† Taken from PESS, EXAMPLE 2.2.

» » » » » EX6_3_1.CIR »

```
BOOST CONVERTER WITH IDEAL BJT
* To determine the average load power.
. OPTIONS      RELTOL=0.01

VS  4  0  100V;                          The dc-source voltage.
RL  5  0  3ohms;                         The load resistance.
L   4  6  2mH    IC=276.6A;              Initial conditions.
RD  6  1  0.01;                          Damper; inductor resistance.
C   5  0  400uF  IC=282.7V;              Initial conditions.

DIO 1  5  DIODE;                         Device, diode.  See EXAMPLE 6.2.1.
. MODEL   DIODE D(RS=0.001  CJO=0.1pF  TT=0.1ns); Diode parameters.

WBJT  1  0  VX  BJT;                     Device, BJT.  See EXAMPLE EX4.3.1.
. MODEL   BJT   ISWITCH(RON=1E-2  ROFF=1E6  ION=9.8A  IOFF=1A)

* CONTROL CIRCUIT FOR BJT SWITCH ON/OFF STATE
IB   0  2  PULSE(0  10A  0  6ns  6ns  66.7us  100us)
* m=0.667, tON=pulse width =mT=m/f=66.7µs.  T=100µs.
RB  2  3  1;                     An arbitrary loop resistance.
VX  3  0  0;                     Current sensor for controlling WBJT switch.

* ANALYSIS, transient.
. TRAN     2us     24ms 20ms  UIC;    240 cycles for SS operation initially.
. PROBE;                         Data for all variables stored in EX6_3_1.DAT.
. END
```

» »

END OF EXAMPLE 6.3.1

Drill Exercise D6.3.1

Consider a boost converter whose circuit diagram is given in Fig. 6.5. The BJT can operate at a switching frequency of 8kHz and requires a base current of 10A to be in the saturation region for turn-on. The values of the circuit elements are $V_s = 100\text{V}$, $R_l = 4\Omega$, $L = 2.3\text{mH}$ (with resistance 0.01Ω) and $C = 300\mu\text{F}$. The average load voltage is to be 200V nominally ($m = 0.5$). **(a)** Run a PSpice simulation to find the time to reach steady conditions. Modify the values of L and C if necessary. (Use .PROBE i(L) etc.) **(b)** Use the results from part (a) to initialize the inductor current and capacitor voltage for a steady-state simulation. Determine the average power loss in the power circuit for steady conditions. (Find the difference between the source and load powers.) **(c)** What is the average load voltage for $m = 0.5$? (Ans: **(a)** About 20ms, **(b)** 196.2W, **(c)** 195.95V.)

There is another Drill Exercise on the WEBsite.

6.4 THE BUCK-BOOST CONVERTER

The buck converter and the boost converter can be combined to produce a single buck-boost converter. In this case the load voltage can be either less than or greater than the supply voltage by adjusting the duty cycle m of the chopper (a unidirectional near-ideal switch). The circuit diagram of a buck-boost converter is shown in Fig. 6.7. Also shown are the ideal waveforms associated with a very high value of capacitance C for virtually constant load voltage. For such a situation in the steady state the average voltage across the inductor is zero. Therefore, for continuous current in the inductor the input-output relations are

$$V_l = \frac{m}{1-m} V_s \tag{6.4.1}$$

and

$$V_l I_l = V_s I_{s\,av}. \tag{6.4.2}$$

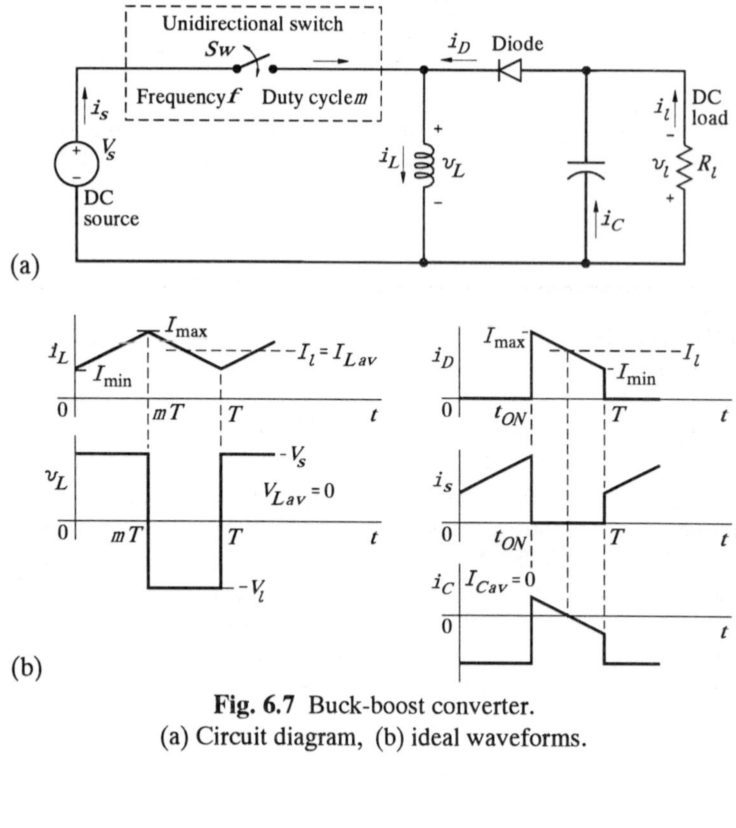

Fig. 6.7 Buck-boost converter.
(a) Circuit diagram, (b) ideal waveforms.

For $m < 0.5$ the chopper is a buck converter, and for $m > 0.5$ the chopper is a boost converter. The PSpice simulation of a buck-boost converter will take into account the finite value of the capacitor C. Accordingly, the simulation results will be more like those that would be found in practice.

EXAMPLE 6.4.1

Consider the buck-boost converter illustrated in Fig. 6.7a. The parameter values are $V_s = 24$V, the switch is a MOSFET operating at 30kHz, $L = 500\mu$H, $C = 250\mu$F and $R_l = 10\Omega$. For a duty cycle $m = 0.38$ execute a PSpice simulation. **(a)** Plot a trace of the inductor current i_L over the interval $t=0$ to $t=5$ms. **(b)** From the trace of i_L determine the inductor current excursion, $\Delta I = I_{max} - I_{min}$. **(c)** Plot a trace of the load voltage and determine approximately the average value.
(Ans: **(b)** 0.6A, **(c)** 12V.)

Solution

Two steps in the solution, PSpice circuit configuration and circuit-file creation, are described below.

STEP 1

A PSpice configuration of the circuit is shown in Fig. 6.8. A PSpice switch models the MOSFET and a dependent voltage source models the diode. See Chapter 4.

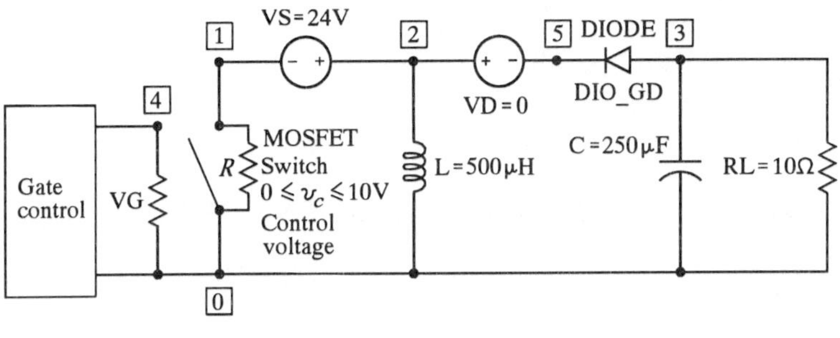

Fig. 6.8 PSpice configuration of a buck-boost converter.

STEP 2

From the PSpice configuration in Fig. 6.8 we can write a circuit file. In this example no comments are written in the circuit file. It is left as an exercise for you to interpret the statements. We have chosen subcircuits MG_IDEAL and DIO_GD from POWER.LIB (and Section 4.7 and

4.2.3) to represent the MOSFET switch and the diode respectively. The driver circuit for the model MOSFET could be the subcircuit MG_GATE (in POWER.LIB and Chapter 5). We have arbitrarily chosen VCHOPPER. The parameters needed are PERIOD ($T=1/f$), PULSE_WIDTH (mT) and VCC (the gate voltage VG in Fig. 6.8). The circuit file EX6_4_1.CIR follows. Run the simulation. Plot the inductor current and the load voltage.

»»»»» EX6_4_1.CIR »»»»»»»»»»»»»»»»»»»»»»»»»»»»»»»»»»»»»»»

```
BUCK-BOOST CONVERTER
.LIB   POWER.LIB
.LIB   DRIVER.LIB
.OPTIONS  RELTOL=0.01
VS     2    1    24V
VD     5    2    0
L      2    0    500uH      IC=0A
C      3    0    250uF      IC=0V
RL     3    0    10ohms
.PARAM    FREQ=30kHz   PERIOD={1/FREQ}
.PARAM    DUTY=0.38    PULSE_WIDTH={DUTY*PERIOD}
.PARAM    VCC=10.2V
XMOS      1    0    4      MG_IDEAL
XDRIVER   4    0    VCHOPPER
XDIODE    3    5    DIO_GD
.TRAN     1us  5ms   UIC
.PROBE    i(VD),     v(3)
.END
```

»»»

END OF EXAMPLE 6.4.1

Drill Exercise D6.4.1

Consider the circuit diagram of the buck-boost converter in Fig. 6.7. Let the values of the elements and parameters be as follows. $V_s=12\text{V}$, $L=12\mu\text{H}$, $C=120\mu\text{F}$ and $R_l=6\Omega$. The chopper frequency is 100kHz with a duty cycle $m=0.7$. Use any of the appropriate devices and drivers described in Chapters 4 and 5. For example, let *Sw* be the near-ideal MOSFET (subcircuit MG_IDEAL in POWER.LIB), let *D* be the near-ideal diode (subcircuit DIO_GD in POWER.LIB), and let the MOSFET driver circuit be a pulse circuit (subcircuit VCHOPPER in DRIVER.LIB). Determine from a PSpice simulation **(a)** the peak diode current during the initial transient interval, **(b)** the time to reach steady state, **(c)** the steady-state inductor current excursion, $\Delta I = I_{max} - I_{min}$ and **(d)** the steady-state average diode current. (Hint. Use .TRAN 0.1us 14.01ms 13ms 0.1us UIC. Store only the one variable i_D for economy.) (Ans: **(a)** 95A, **(b)** <13ms, **(c)** 38 - 24.2 = 14.2A, **(d)** 9.3A.)

Drill Exercise D6.4.2

The buck-boost converter illustrated in Fig. 6.9 is known as a Ćuk regulator†. The switching frequency is 20kHz with a duty cycle $m = 0.8$. Do a PSpice simulation and, for steady conditions, plot traces of the voltage v_{Sw} across the switch, the load current i_l, the capacitor current i_{C2} and the input current i_s. Show, from the results, that the output voltage $V_{lav} \approx -V_s m/(1-m)$. Use any appropriate device model from Chapter 4. (Ans: After a transient interval of 90ms, there is still an oscillation (period 50μs) on an oscillation (period 0.9ms) on an oscillation (period 14ms). $V_{lav} = 89.7$V (ideally 96V).)

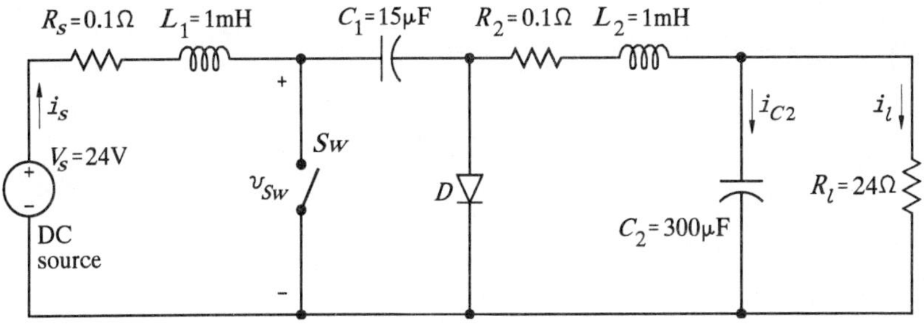

Fig. 6.9

There is another Drill Exercise on the WEBsite.

6.5 THE FLYBACK CONVERTER (TRANSFORMER)

The flyback converter is a buck-boost converter (see Section 6.4) with an isolating transformer replacing the inductor. The principle of operation is the same for both converters. We use the simulation of the flyback converter in order to demonstrate a PSpice model of a transformer. Figure 6.10, depicted on the next page, shows the circuit diagram and the waveforms of an ideal flyback converter whose capacitor value has a very high value. The output voltage V_l is given by

† S. Ćuk, R. D. Middlebrook, "Advances in Switched-mode Power Conversion", (Teslaco, Pasadena, 1981).

$$V_l = \frac{V_s}{n} \frac{m}{1-m} \tag{6.5.1}$$

where n is the transformer turns ratio N_1/N_2.

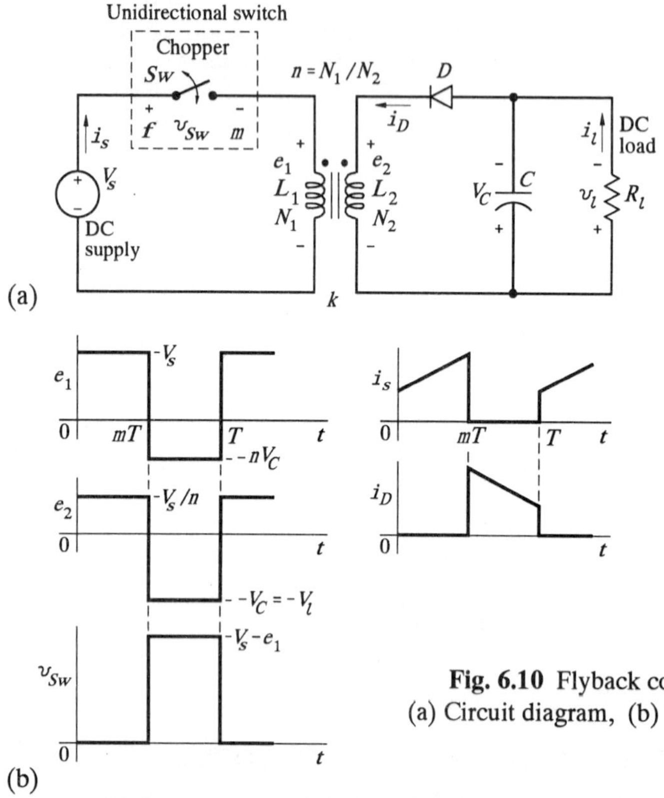

Fig. 6.10 Flyback converter. (a) Circuit diagram, (b) waveforms.

EXAMPLE 6.5.1

A flyback converter is to modulate voltage from a 200-V dc supply with a source resistance of 0.1Ω to a 1.0-Ω resistive load. The ideal transformer has a turns ratio $n = N_1/N_2 = 4$ and the inductance of the primary winding is 200μH. The filter capacitor has a value of 200μF. The switch (IGBT) has a switching frequency of 50kHz with a duty cycle of 0.4. Do a PSpice simulation to determine **(a)** the transient interval, **(b)** the peak transient load voltage, **(c)** the steady-state load voltage excursion and **(d)** the steady-state average load voltage.

Solution

We will treat the transformer as a lossless ideal linear device. That is, the transformer comprises two inductors with perfect coupling. PSpice handles mutual coupling with a K-statement

K L1 L2 *k*

where k is the coefficient of coupling. For perfect coupling ($k = 1$), the relation between the inductors is

$$L_1/L_2 = (N_1/N_2)^2 = n^2.$$

There are four steps in the solution.

STEP 1 The circuit diagram in Fig. 6.10 and the given data allow a PSpice configuration to be made. This is shown in Fig. 6.11.

STEP 2 A circuit file, EX6_5_1 .CIR, can be written from an inspection of the configuration in Fig. 6.11, and a reference to Chapter 4 for the diode and the IGBT.

STEP 3 Run PSpice with the circuit file EX6_5_1.CIR.

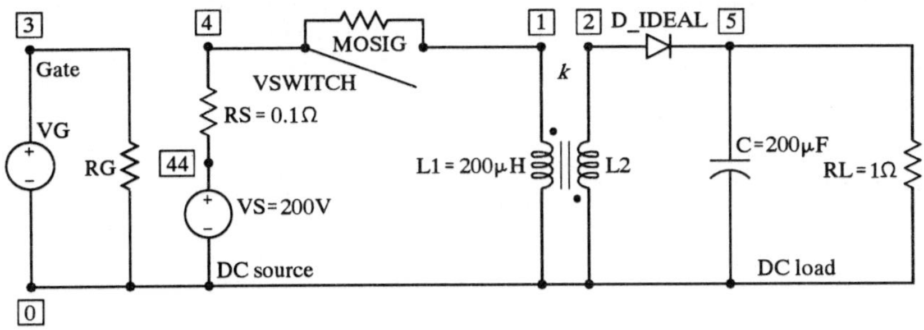

Fig. 6.11 PSpice configuration of a flyback converter.

```
»»»»» EX6_5_1.CIR »»»»»»»»»»»»»»»»»»»»»»»»»»»»»»»»»»»»»»»»»»

A FLYBACK CONVERTER     (MOSIG AND D_IDEAL)
* To plot the transient waveform of the load voltage.

* PARAMETERS
.PARAM  NUM=4           NUM2={NUM*NUM};          Turns ratio n.
.PARAM  PRIM=200uH      SEC={PRIM/NUM2};         Inductance values.
* SOURCE
VS   44   0    DC   200V;                        Main dc source.
RS   44   4    0.1ohm;                           Source resistance.

* DEVICE, IGBT switch.
SMOSIG  4   1   3   0   MOSIG;        From Chapter 4, Section 4.7.
.MODEL  MOSIG   VSWITCH(RON=1E-3 ROFF=1E6 VON=10 VOFF=2)
* GATE DRIVER, f=50kHz, m=0.4.
VG  3    0    PULSE(0 10.1V 0 10ns 10ns 8us 20us)
RG  3    0    1k;                                Closes the voltage loop.
* DEVICE, diode, Section 4.2.4, modified for convergence (Section 7.5.1.)
DIO 2    5    D_IDEAL
.MODEL  D_IDEAL  D(RS=1m CJO=0.1pF N=0.001)

* DEVICE, the transformer.
L1   1    0    {PRIM};                The primary-winding inductance.
L2   0    2    {SEC};                 The secondary-winding inductance.
K    L1   L2   1;                     Coefficient of coupling, k=1.

* CIRCUIT ELEMENTS
C    5    0    200uF;                 Filter capacitor.
RL   5    0    1ohm;                  The load resistance.

* ANALYSIS, 100 cycles.
.TRAN    3us   2ms
.PROBE   v(5);                        Viewing only the load voltage.
.END
»»»»»»»»»»»»»»»»»»»»»»»»»»»»»»»»»»»»»»»»»»»»»»»»»»»»»»»»»»»»»
```

STEP 4 Invoke PROBE with the data file EX6_5_1.DAT. A trace of the load voltage v(5) is shown in Fig. 6.12.

Part (a) of Solution. From the trace of the load voltage the transient interval is about 2ms.

Part (b) of Solution. Using the **Cursor**, the **Max** value of the voltage transient is 51.2V.

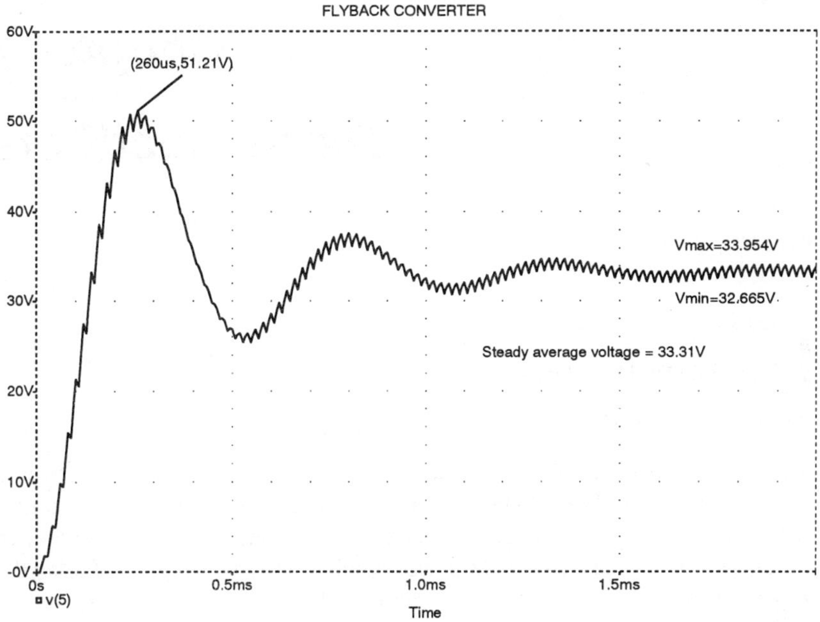

Fig. 6.12

Part (c) of Solution. Using the **Cursor**, the **Peak** and **Trough** values of the load voltage in steady operation are 33.954V and 32.665V respectively. Thus, the voltage excursion is $\Delta V = 1.289$V.

Part (d) of Solution. The average load voltage is estimated from part (c) to be 33.31V.

END OF EXAMPLE 6.5.1

Drill Exercise D6.5.1
Do a PSpice simulation of the flyback converter, described in EXAMPLE EX6.5.1. To be close to steady operation set the capacitor initial condition to IC=33.954V. Plot traces of the primary and secondary winding currents and the primary winding voltage for comparison with the waveforms in Fig. 6.12. Determine **(a)** the peak withstand voltage of the IGBT and **(b)** the average current in the diode. (Ans: **(a)** 332.98V, (332V calculated), **(b)** 32.72A.)

There is another Drill Exercise on WEBsite.

Chapter 7

Diode Rectifiers

7.1 INTRODUCTION

The uncontrolled rectification from alternating current to direct current is accomplished using diodes in the converter. We will look at a number of rectifier problems† to see how well PSpice simulation can solve them. Rectifying converters take many forms depending on the power requirement and the type of filtering that is needed. We will choose different forms of rectifier circuit to exemplify the use of PSpice simulation. The diode models used are from Chapter 4 and are near-ideal models.

We use this chapter to discuss transformer modelling and consolidate our experience with nonconvergence problems.

7.2 SINGLE-PHASE HALF-WAVE RECTIFICATION

A simple rectifier comprises a single diode between the ac source and the dc load. The load is usually a combination of resistance and inductance. An effect on the circuit response is made if a filter is connected to smooth the current waveform or the voltage waveform at the output.

† Some of the problems are taken from "Power Electronics Semiconductor Switches" by R. S. Ramshaw, Sections 2.3 and 3.8.2, referred to as PESS.

EXAMPLE 7.2.1 (PESS EXAMPLE 3.5)
Consider the circuit diagram in Fig. 7.1a. This single-phase, half-wave rectifier circuit has the ideal waveforms as shown in Fig. 7.1b. The source voltage is 240V at 50Hz and the load comprises resistance $R_l = 1\Omega$ and inductance $L_l = 3.18$mH. From a PSpice simulation plot the source voltage v_s, the voltage across the diode and the load current. Determine the current rating of the diode D. Use the diode model DIO_I (Section 4.2). (Ans: >27A.)

Solution
The mathematical analysis of this circuit response is

$$i_l = \frac{V_m}{Z}\sin(\omega t - \theta) + \frac{V_m}{Z}\sin\theta\, e^{-t/\tau} \tag{7.2.1}$$

where $Z = (R_l^2 + \omega^2 L_l^2)$, $\theta = \tan^{-1}\omega L_l/R_l$ and $\tau = L_l/R_l$.

The diode rating is the average value of the expression in eq. (7.2.1). A PSpice simulation can find the average value for us. Two steps of this simulation are described below.

STEP 1	From the circuit diagram in Fig. 7.1a draw a PSpice con-figuration as depicted in Fig. 7.2 on the next page.
STEP 2	From the PSpice configuration in Fig. 7.2 write a PSpice circuit file. Now do a simulation run.

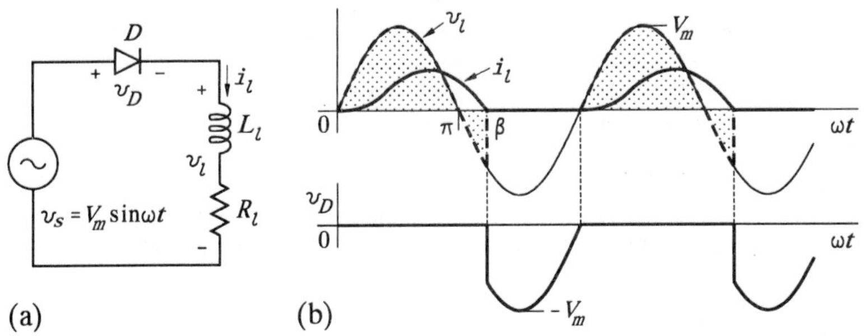

Fig. 7.1 Single-phase, half-wave rectifier.
(a) Circuit diagram, (b) waveforms.

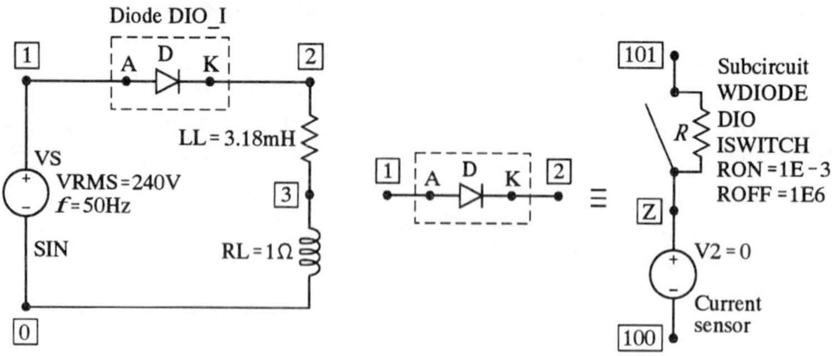

Fig. 7.2 PSpice configuration of a rectifier.

» » » » » EX7_2_1.CIR »

```
SINGLE-PHASE HALF-WAVE RECTIFIER     (DIO_I)
* To plot traces of the diode voltage and current.
. PARAM VRMS=240V  VMAX={1.414*VRMS}
. PARAM FREQ=50Hz

* SOURCE
VS  1  0  SIN(0  {VMAX}  {FREQ});                  The ac source.

* CIRCUIT ELEMENTS
LL  2  3  31.8mH;                                  Load inductance.
RL  3  0  1ohm;                                    Load resistance.
* DIODE, see Section 4.2.  Embedded in a subcircuit.
. SUBCKT  DIO_I  101  100
WDIODE 101  Z  V2  DIO; Current-controlled switch modelling the diode.
. MODEL DIO  ISWITCH(RON=1E-3  ROFF=1E6  ION=0  IOFF=-1E4)
V2  Z   100     0;                 Sensor for the switch's control current.
* Nodes 101, Z are for the diode anode and cathode.
* Nodes Z, 100 are for the zero-voltage source VZ.
* Nodes 101, 100 are for the subcircuit diode model.
. ENDS  DIO_I
Xdiode  1  2  DIO_I
* X statement calls the diode subcircuit named DIO_I into the main circuit.
* Nodes 1, 2 are the main-circuit connections to the diode model.

* ANALYSIS, over two cycles of the supply waveform.
. TRAN   0.2ms  35ms    UIC
. PROBE v(1),   v(1, 2),   i(RL);   Supply voltage.  Diode variables.
. END
```

» »

END OF EXAMPLE 7.2.1

Drill Exercise D7.2.1

Follow the pattern set out in EXAMPLE 7.2.1 to simulate a single-phase, half-wave rectifier. Use the same diode model DIO_I. The specifications of the circuit are that the ac source voltage is 120V(rms) at 60Hz, and the dc load is R_l=4.5Ω and L_l=20.7mH. Plot traces of the diode current and voltage, and determine the current rating of the diode. (Ans: Average current ~8.6A.)

Drill Exercise D7.2.2

A single-phase, half-wave rectifier has a 240-V(rms), 60-Hz supply and a dc load of resistance R_l=6.1Ω and inductance L_l=33.5mH. Carry out a PSpice simulation using the subcircuit named DIO_I from POWER.LIB for the diode. Plot traces of the diode voltage and current, and determine the current rating of the diode. (Ans: Average current ~11.8A.)

There are Extra Drill Exercises on the WEBsite.

7.3 SINGLE-PHASE FULL-WAVE RECTIFICATION

An ideal rectifier using a single-phase, full-wave bridge of diodes is illustrated in Fig. 7.3a. The average value of the load voltage V_{lav} for the ideal circuit is given by

$$V_{lav} = 2\hat{V}_s/\pi . \qquad (7.3.1)$$

Figure 7.3b shows the waveforms of the input and output variables.

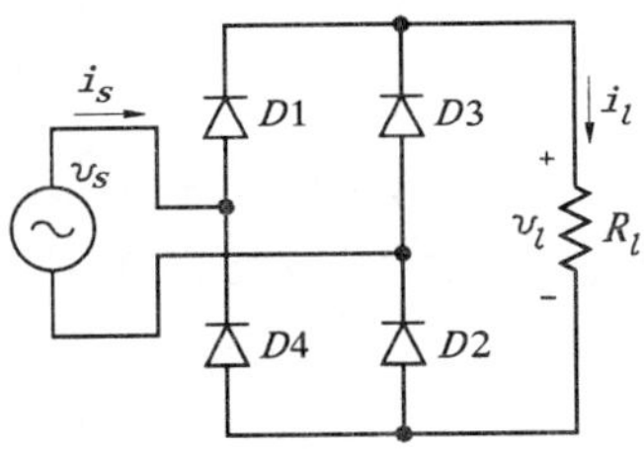

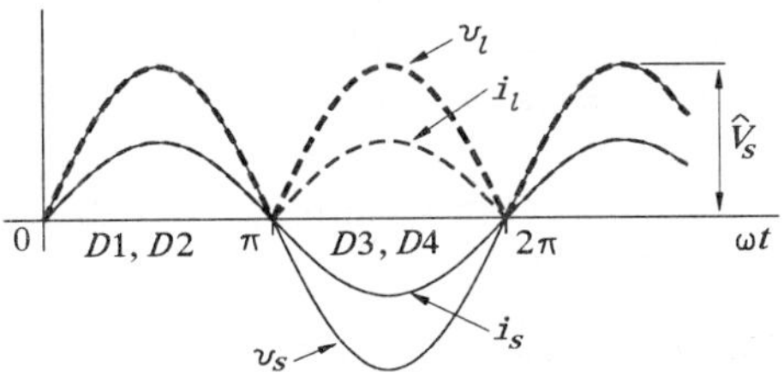

Fig. 7.3 A single-phase, full-wave bridge rectifier.
(a) Circuit diagram, (b) waveforms of input and output.

Drill Exercise D7.3.1

Consider the circuit diagram in Fig. 7.3a. The diode bridge rectifies an ac source of 240V(rms) at 60Hz to provide direct current to a load whose resistance is $R_l = 1.2\Omega$. Plot traces of the load current and the supply voltage using a PSpice simulation with the model diode DIO_GD (see Section 4.2). Determine from a current trace the average power absorbed by the load. To prevent nodes from floating, connect a high-valued resistance across each diode. (From a hand analysis the average power is ideally V_s^2/R_l, that is 48kW.)

Drill Exercise D7.3.2

A single-phase, full-wave, diode bridge converts power from an ac source of 240V(rms) at 50Hz to a dc load whose resistance is $R_l = 11.3\Omega$ and whose inductance is $L_l = 9.1$mH. See Fig. 7.4. Plot traces of the voltage and current of the load and determine the average load power. Use the diode model DIO_I. (See Section 4.2.) (Ans: 4.85kW.)

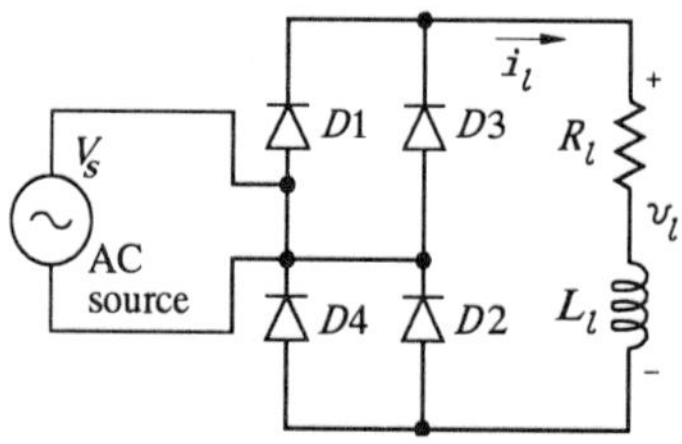

Fig. 7.4

There are Extra Drill Exercises on the WEBsite.

7.4 BIPHASE RECTIFICATION

A biphase rectifier is associated with a centre-tapped transformer, as shown in the circuit diagram in Fig. 7.5a. The output is the same as for a full-wave bridge rectifier. Waveforms are shown in Fig. 7.5b.

In order to do a PSpice simulation we must be able to model transformers. A description of two transformer models appears in the following subsection.

7.4.1 TRANSFORMER MODEL

Transformers give isolation or provide a more appropriate alternating voltage. A single-phase transformer can be considered to be two inductors which are mutually coupled. The mutual coupling can be accommodated in PSpice by a K-statement.

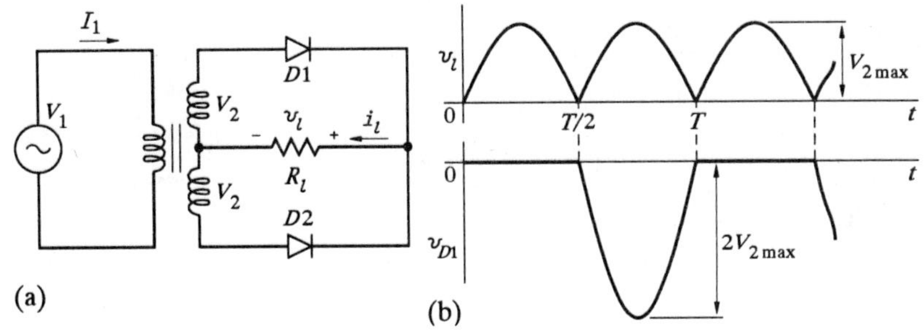

Fig. 7.5 A biphase, full-wave rectifier.
(a) Circuit diagram, (b) waveforms.

The self-inductance of an ideal element is proportional to the square of the effective number of turns N on the winding. Therefore, the effective turns ratio n of a two-winding transformer is related to the inductances L_1 and L_2 by

$$L_1/L_2 = (N_1/N_2)^2 = n^2. \tag{7.4.1}$$

The linear transformer, depicted in Fig. 7.6, on the next page, has the relations

$$V_1/V_2 = I_2/I_1 = N_1/N_2 = n \tag{7.4.2}$$

for sinusoidal excitation. The voltages and currents are rms values†. The mutual inductance M between the two coils of a transformer is given by

$$M = k\sqrt{L_1L_2} \tag{7.4.3}$$

where k is the coefficient of coupling between the two coils. Standard transformers with iron cores can have a coefficient k as high as 0.99. If the transformer has an equivalent air core, then $k \le 0.5$.

The information above is sufficient to create a PSpice model of a linear transformer. There are two points to note.

1. A loop with only an independent voltage source and an inductor cannot be tolerated in PSpice. A resistor of small value $10^{-6}\Omega$ or less can be added in series without upsetting the performance of the device.
2. An isolated winding cannot be tolerated in PSpice. Each element must have a dc path to ground. This is achieved by creating a path of high resistance ($10^8\Omega$) between the windings. Since simulation does not complain if isolation is ignored, both sides of the transformer could be grounded to give the dc path to ground.

† PSpice accommodates rms values or amplitudes.

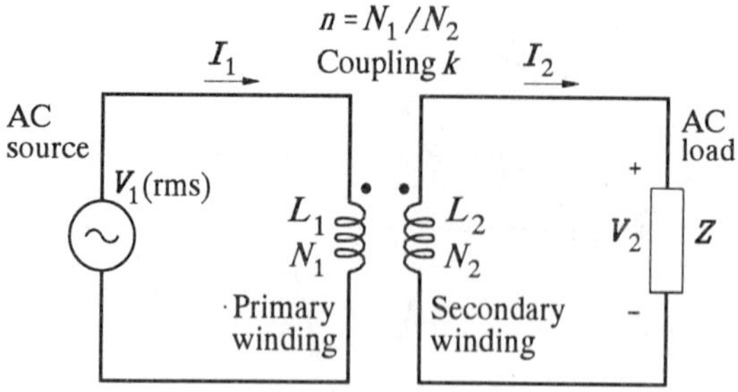

Fig. 7.6
A linear transformer.

There are two features to address in the steady-state ac analysis of a transformer by PSpice. One is the ac analysis and the other is the mutual coupling between windings of the transformer.

We will treat ac analysis in PSpice by using the command statement

```
.AC LIN number freq_start freq_stop
```

.AC is the dot command that invokes ac analysis. *Number* is the total number of frequencies in the linear (LIN) sweep between the frequency values *freq_start* and *freq_stop*, inclusive. There are two similar statements with logarithmic sweeps. See Appendix C.

The mutual inductance between the windings is represented by a data statement

```
Kxxx Lyyy Lzzz k
```

The statement must begin with the letter K, but *xxx* is an arbitrary name. The two inductors are L*yyy* and L*zzz*, that must also be described separately in the circuit file as inductances. Finally, *k* is the coefficient of coupling between the two inductors. In a three-winding transformer, the K-statement has the form

```
Kxxx Lyyy Lzzz Lwww k
```

In any PSpice steady-state ac analysis, there is a number of ways that the results can be printed in the output file. The command statement has the form that begins with the . PRINT command

```
.PRINT AC variables ...
```

AC stipulates that the results are from the ac analysis. *Variables* are any of the circuit voltages or currents listed in the statement. The variables V and I can have suffixes, as listed in Table 7.1 to demarcate real and imaginary parts.

The phase angle (in degrees) of the phasor is taken with respect to the ac source voltage. The source voltage statement has the form

```
Vxxx N+ N- AC ac_volts phase_angle
```

Table 7.1 Variable suffixes for ac analysis.

Suffix	Meaning	Example
M	Magnitude (default)	VM(2, 3)
P	Phase (degrees)	IP(RL)
R	Real part	VR(4)
I	Imaginary part	II(VS)
DB	20 log (value) (decibels)	VDB(1)

The magnitude of the voltage source V*xxx* can be written as volts in either rms or amplitude form. Whichever is chosen, the results are printed in the same form. In the third field of the voltage statement AC signifies an alternating voltage source.

In a steady-state ac analysis (phasor analysis), PSpice does not allow independent sources to be described as a sinusoid in the SIN form.

The print statement can request a maximum number of eight variables. However, many print statements can be written to cover all requirements. Less than eight variables per print statement is better with respect to the output format.

EXAMPLE 7.4.1

Consider the circuit diagram in Fig. 7.6. The ac source has a voltage of 240V(rms) at 50Hz. The load has a resistance R_l=4.1Ω and a series inductance L_l=13.06mH. Specifications of the single-phase transformer are as follows: 2kVA, 240/120V, linear, power losses are negligible, coupling coefficient k=0.999; no-load test on the low-voltage side, 120V(rms), 0.89A(rms), 60Hz. From a PSpice ac, steady-state simulation determine (a) the magnitude, the phase, the real part and imaginary part of each of the secondary winding phasor variables, current I_2 and voltage V_2 (b) Check the phase difference between I_2 and V_2 with the load-impedance angle. (c) Calculate the average power absorbed by the load using PSpice results.

Solution

There are five steps in this solution.

STEP 1 It is necessary to calculate the self-inductances L_1 and L_2 of the primary and secondary windings, respectively. From the open-circuit test data, the inductive reactance of the secondary winding is given by

$$X_{2OC} = \omega L_2 = V_{2OC}/I_{2OC}, \quad \text{where} \quad \omega = 2\pi f.$$

Hence, the self-inductance L_2 of the secondary winding is

$L_2 = V_{2OC}/(\omega I_{2OC}) = 120/(2\pi \times 60 \times 0.89) = 357.65$ mH.

From eq. (7.4.1) the self-inductance L_1 of the primary winding is

$L_1 = n^2 L_2 = 4 \times 0.3577 = 1.43$ H.

The secondary-winding voltage is $V_2 = V_1/n = 240/2 = 120$V (ideally).

The load impedance $\boldsymbol{Z}$ is given by

$$\boldsymbol{Z} = R_l + jX_l = R_l + j\omega L_l = R_l + j2\pi f L_l = 4.1 + j314 \times 13.06 \times 10^{-3}\Omega .$$

That is, $\boldsymbol{Z}_l = 4.1 + j4.1 = 5.799\angle 45°\,\Omega$.

If we take $\boldsymbol{V}_2 = 120\angle 0$V as the reference phasor, the load current $\boldsymbol{I}_2$ is

$$\boldsymbol{I}_2 = \frac{\boldsymbol{V}_2}{\boldsymbol{Z}} = \frac{120}{5.799\angle 45°} = 20.69\angle -45°\text{ A}.$$

This can be checked by using a PSpice model.

STEP 2 From Fig. 7.6 the PSpice configuration can be drawn. See Fig. 7.7. Very small resistances R1, R2 are added to the windings to satisfy the PSpice requirements.

STEP 3 A circuit file (EX7_4_1.CIR) can be written from an inspection of the configuration in Fig. 7.7.

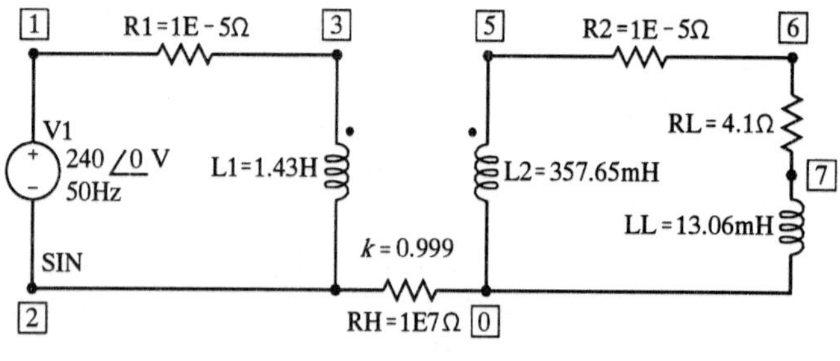

Fig. 7.7 A PSpice configuration of a transformer.

```
»»»»» EX7_4_1.CIR »»»»»»»»»»»»»»»»»»»»»»»»»»»»»»»»»»»»»»»»
PSPICE MODEL OF A TRANSFORMER
* To print out the steady-state output response.

* SOURCE
V1  1   2    AC    240    0
* AC source, rms value for rms responses. 0 degree phase shift.

* CIRCUIT ELEMENTS
R1  1   3    1E-5ohms;        Must have resistance in an inductive loop.
R2  5   6    1E-5
RL  6   7    4.1ohms;         Load resistance.
LL  7   0    13.06mH;         Load inductance.
RH  2   0    1E7;             A dc path to ground.
* DEVICE, the transformer.
L1  3   2    1.43H;           The primary winding.
L2  5   0    0.35765H;        The secondary winding.
K12 L1  L2   0.999;   K-statement that provides the value of the coefficient
* of coupling (0.999) between the two windings (inductors L1 and L2).
* L1, L2 can be in any order.
* According to the dot convention for mutual coupling, the first node of the
* inductor statement has the dot.

* ANALYSIS
.AC    LIN     1     50Hz    50Hz
* In this ac analysis there is to be a linear (LIN) range of 1 frequency between
* 50Hz and 50Hz. That is, 50Hz is the frequency of the ac analysis.
* Print the magnitude and phase of the voltages and currents.
.PRINT  AC   VM(1, 2),   VP(1, 2),    IM(V1),    IP(V1)
.PRINT  AC   V(3, 2),    VP(3, 2),    I(R1),     IP(R1)
.PRINT  AC   V(5),       VP(5),       I(RL),     IP(RL)
* Print the real and imaginary parts of the secondary voltage and current.
.PRINT  AC   VR(6),      VI(6),       IR(RL),    II(RL)
.END
»»»»»»»»»»»»»»»»»»»»»»»»»»»»»»»»»»»»»»»»»»»»»»»»»»»»»»»»»»
```

STEP 4 Run a PSpice simulation for the single frequency of 50Hz with the circuit file EX7_4_1.CIR. The results are printed in EX7_4_1.OUT, shown on the next page. Since the source voltage is an rms value, all results are rms values. The job time is short.

STEP 5 Part (a) of Solution. The output voltage and current phasors from the PSpice results are

$$V_2 = 116.7\angle -1.5°\ \text{V} \quad \text{and} \quad I_2 = 20.1\angle -46.5°\ \text{A}.$$

Part (b) of Solution. The phase angle difference between V_2 and I_2 is 45°.

Part (c) of Solution. The average power P absorbed by the load is

$P \approx V_2 I_2 \cos\theta \approx \mathrm{VM(5)*IR(RL)} = 116.7 \times 13.83 = 1.614$ kW.

* * * * * * * * * * * Evaluation PSpice *

```
PSPICE MODEL OF A TRANSFORMER
* OUTPUT FILE EX7_4_1.OUT
* * * * CIRCUIT DESCRIPTION
* CIRCUIT FILE EX7_4_1.CIR
* AC source with 0 deg phase shift.  RMS value gives rms response.
V1  1   2   AC    240     0
R1  1   3   1E-5
R2  5   6   1E-5
RL  6   7   4.1
LL  7   0   13.06mH
RH  2   0   1E7
L1  3   2   1.43H
L2  5   0   0.35765H
K12     L1  L2    0.999
.AC     LIN 1     50  50
.PRINT      AC    V(1,2),  VP(1,2),    I(V1),     IP(V1)
.PRINT      AC    V(3,2),  VP(3,2),    I(R1),     IP(R1)
.PRINT      AC    V(5),    VP(5),      I(RL),     IP(RL)
.PRINT      AC    VR(6),   VI(6),      IR(RL),    II(RL)
.END
* * SMALL SIGNAL BIAS SOLUTION  TEMPERATURE=27.000 DEG C
* * AC ANALYSIS
FREQ        V(1,2)        VP(1,2)        I(V1)        IP(V1)
5.000E+01   2.400E+02     0.000E+00      1.044E+01    1.314E+02
----------------------------------------------------------------
FREQ        V(3,2)        VP(3,2)        I(R1)        IP(R1)
5.000E+01   2.400E+02     1.869E-05      1.044E+01    -4.856E+01
----------------------------------------------------------------
FREQ        V(5)          VP(5)          I(RL)        IP(RL)
5.000E+01   1.167E+02     -1.526E+00     2.011E+01    -4.655E+01
----------------------------------------------------------------
FREQ        VR(6)         VI(6)          IR(RL)       II(RL)
5.000E+01   1.166E+02     -3.107E+00     1.383E+01    -1.460E+01
TOTAL JOB TIME .82s
```

* *

END OF EXAMPLE 7.4.1

Drill Exercise D7.4.1
Consider the circuit diagram in Fig. 7.8. The transformer is rated at 24/10V and 10kHz. Using the method set out in EXAMPLE 7.4.1 determine, from a PSpice simulation, the real, reactive and apparent power supplied by the source for steady-state conditions. (Hint. Assume that the primary winding has a nominal inductance of 0.1H.)
(Ans: 2.1405mW, 48.63mVAR, 48.69mVA.)

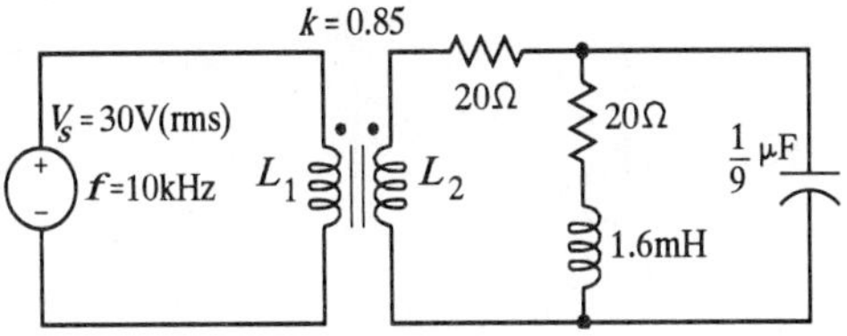

Fig. 7.8

It must be noted that an ideal transformer can have its primary modelled by an F source (CCCS) and the secondary modelled by an E source (VCVS) because of the voltage and current ratios. See the WEBsite Section 7.4.2.

We can now return to the simulation of a biphase, full-wave diode rectifier.

EXAMPLE 7.4.2

A step-down, centre-tapped transformer, as illustrated in Fig. 7.5a, has a turns ratio 2:1. The ac source has a voltage of 240V(rms) at 50Hz. Consider the biphase, full-wave rectifier to be near-ideal. For a load of resistance 11.1Ω and series inductance 18.3mH, plot traces of the load current and voltage. Determine **(a)** the average load current I_{lav} and **(b)** the ac power factor.

Solution
There are four steps in this solution.

STEP 1

The circuit diagram in Fig. 7.5a is used to create a PSpice configuration of the biphase rectifier. See Fig. 7.9 on the next page. The value of the primary winding inductance L1 is taken to be 1H nominally. The transformer turns ratio is $n = 2$ so that the secondary windings' inductances are given by $L2 = L3 = L1/n^2 = 1/4 = 0.25\,H$.
The secondary side of the circuit is connected to the primary side by a short circuit. Isolation of the windings is not a concern of simulation. The short circuit satisfies the PSpice requirement that each node must have a dc path to ground.

STEP 2

From the PSpice configuration in Fig. 7.9, a circuit file EX7_4_2.CIR can be written.

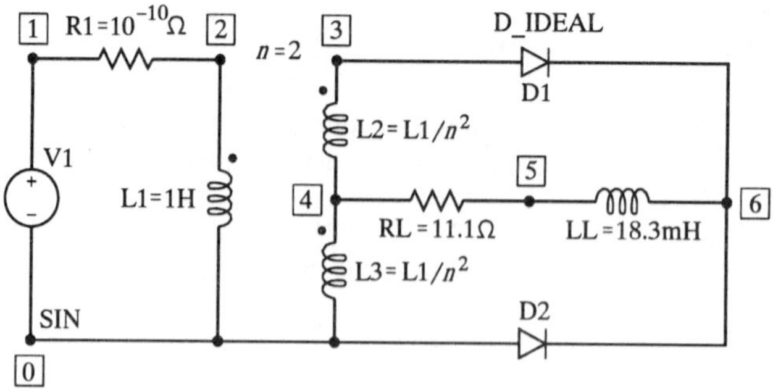

Fig. 7.9 PSpice configuration of a biphase rectifier.

» » » » » EX7_4_2.CIR »

```
BIPHASE RECTIFIER WITH DIODES D_IDEAL
* To determine the average load current and input PF.
. OPTIONS  RELTOL=0.01 ;   To ensure analysis convergence.
* PARAMETERS
. PARAM   LPRIM=1H;                          Nominal inductance of the primary winding.
. PARAM   RATIO=2      NSQ={RATIO*RATIO};    Turns ratio n=N1/N2.
. PARAM   VRMS=240V    VMAX={1.414*VRMS};    Source voltage.
. PARAM   FREQ=50Hz
* SOURCE
V1   1    0    SIN(0 {VMAX} {FREQ});     AC source.
* CIRCUIT ELEMENTS
R1   1    2    1E-10;                    Resistance for the primary loop.
RL   5    4    11.1ohms;                 The load resistance.
LL   6    5    18.3mH;                   The load inductance.
* TRANSFORMER
L1   2    0    {LPRIM};                  Primary winding #1.
L2   3    4    {LPRIM/NSQ};              Secondary winding #2.
L3   4    0    {LPRIM/NSQ};              Secondary winding #3.
K    L1   L2   L3 1;                     Coupling factor k=1.
* DEVICES, the diodes and subcircuit named D_IDEAL.
. SUBCKT  D_IDEAL  101 100;      101, 100 are anode and cathode nodes.
D_RECT    101  100  D_IDEAL
. MODEL   D_IDEAL  D(N=0.001);   PSpice diode, minimum parameters.
. ENDS    D_IDEAL;               This subcircuit is written in POWER.LIB.
* Call for the diode subcircuit
```

```
XD1  3   6   D_IDEAL;                Nodes 3, 6 are the anode, cathode of D1.
XD2  0   6   D_IDEAL;                Nodes 0, 6 are the anode, cathode of D2.

.TRAN    0.1ms  80ms    UIC;         Adjust these to suit steady conditions.
.PROBE v(1),    I(R1),   I(RL);      For input and output.
.END
```

» »

STEP 3 The simulation can be run with the file EX7_4_2.CIR and the results stored in EX7_4_2.DAT. With inductors and diodes care has to be taken to achieve convergence. In this example, the option RELTOL was increased from 0.001 to 0.01.

STEP 4 Plots are shown in Fig. 7.10. The load-current trace i(RL) is close to steady state at time $t = 80\text{ms}$.

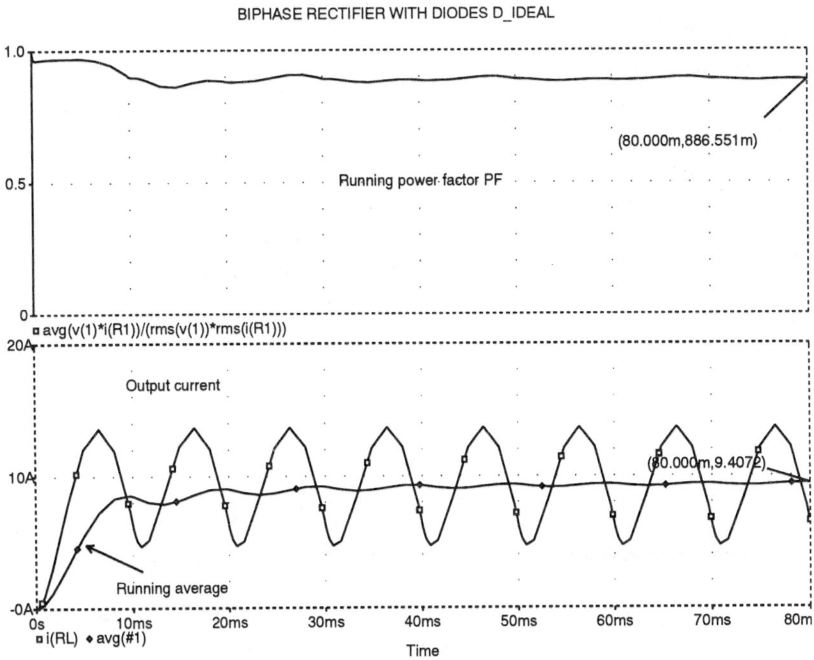

Fig. 7.10

Part (a) of Solution. From the running average value of the load current, the periodic average value at t=80ms is I_{lav}=9.4A with a peak-to-peak ripple (excursion) of about 9A.

Part (b) of Solution. The power factor is given by the expression

$$PF = \cos\theta = P_{av}/(V_{s\,rms} I_{s\,rms}).$$

The running value of this expression was plotted by PROBE. Using the cursor, the periodic value at t=80ms gives the value PF ≈ 0.887.

END OF EXAMPLE 7.4.2

Drill Exercise D7.4.2

Follow the pattern of EXAMPLE 7.4.2 to simulate a biphase rectifier. The source is 240V at 60Hz and the load comprises a resistance of 2.2Ω in series with an inductance of 3.7mH. Each turns ratio of the centre-tapped transformer is 3:1. Using the same diode model D_IDEAL and the option RELTOL=0.01, do a simulation and plot traces to determine **(a)** the average load current and **(b)** the input PF. Plots over an interval of 100ms would be appropriate.
(Ans: **(a)** 30.96A, **(b)** 0.912.)

Drill Exercise D7.4.3

A biphase, full-wave, diode rectifier converts power from an ac source of 240V(rms) at 50Hz with an equivalent reactance of 0.628Ω. The dc load has a pure resistance of 0.9Ω. See Fig. 7.5 and Fig. 7.9. The centre-tapped transformer has a turns ratio $n = N_1/N_2 = 3$. Carry out a PSpice simulation incorporating the diode model D_IDEAL whose subcircuit is described in POWER.LIB. Using PROBE, plot traces of the load voltage for the situations with and without source impedance. Determine for both cases the average load voltage.
(Ans: 70.69V, 71.32V.)

There are Extra Drill Exercises on the WEBsite.

7.5 THREE-PHASE HALF-WAVE RECTIFICATION

For economy at medium power levels it is possible to use only three diodes in a three-phase half-wave rectifier. Figure 7.11 illustrates a rectifier circuit.

The circuit diagram is more complex than previous rectifier circuits. That is because this diagram contains many, but not all, of the effects found in practice. It is assumed that the system is balanced and symmetrical. The three-phase sources V_{sa}, V_{sb} and V_{sc} are considered to be ideal. The source and line impedance is mainly inductive and is modelled by an inductance L_s.

The three-phase transformer provides a different set of values of ac voltages applied at the rectifier terminals. Here, the transformer is treated as being linear and lossless. Modelling of the transformer will be similar to that used in Section 7.4.1. The connections can change to be Y-Y, Y-Δ, Δ-Y, or Δ-Δ according to design needs. The turns ratio $n = N_1/N_2$ is the phase relation V_{1a}/V_{2a}.

Three diodes constitute the three-phase, half-wave rectifier that forces the load current to be unidirectional.

An *LC* filter tends to minimize the current and voltage ripple at the load, which is modelled by a series resistance R_l and inductance L_l. The conversion performance of ac power to dc power depends on the circuit elements as well as the load. Performance is measured in terms of the quality of both the input and output variables. Performance can be determined by PSpice.

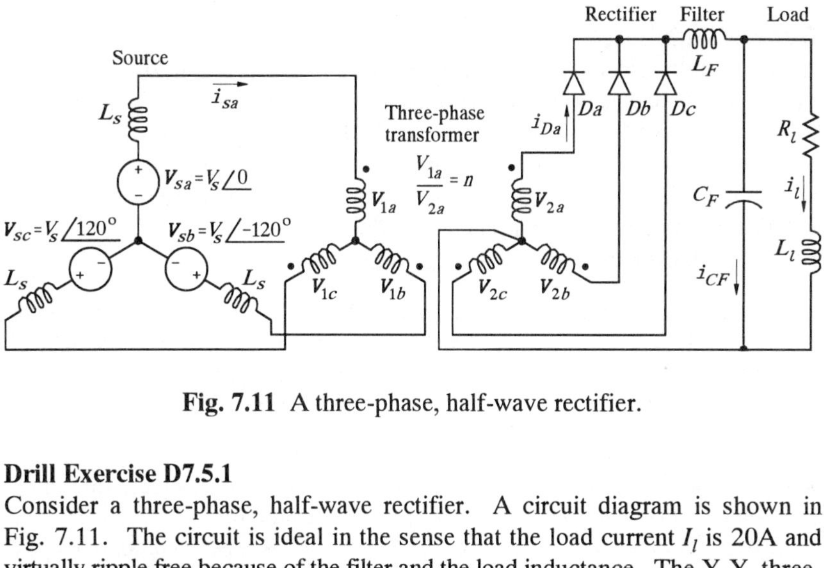

Fig. 7.11 A three-phase, half-wave rectifier.

Drill Exercise D7.5.1

Consider a three-phase, half-wave rectifier. A circuit diagram is shown in Fig. 7.11. The circuit is ideal in the sense that the load current I_l is 20A and virtually ripple free because of the filter and the load inductance. The Y-Y, three-phase transformer is ideally linear and has a rating 6.3kV/208V(line), 30kVA at 60Hz. The supply has a line voltage of 6.3kV(rms) with negligible impedance. Carry out a PSpice simulation and use PROBE to plot traces of the load voltage v_l, the diode current i_{Da}, the diode voltage v_{Da} and the input currents. Determine **(a)** the average load voltage, **(b)** the diode ratings, **(c)** the average output power and **(d)** the average input power.

(Ans: **(a)** 140V, **(b)** 6.7A, 294V, **(c)** 2.8kW, **(d)** 2.8kW.)

7.5.1 SIMULATION CONVERGENCE

Up to now the simulation of power-electronics circuits has not given much trouble. If nonconvergence of a simulation analysis has occurred, the problem has been solved by adjusting some of the error tolerances in the program. Transient analysis of circuits with inductors and multiple semiconductor devices is prone to nonconvergence problems. Drill Exercise D7.5.1 showed problems. These were fixed with resistors and an increased value of relative tolerance error (RELTOL) in the numerical analysis algorithm. Further, the load current limit for the circuit file is 20A for convergence.

We need a systematic method† to attempt to overcome the majority of nonconvergence problems. If the systematic method works for a particular circuit simulation, then some of the convergence options can be relaxed to obtain more accurate results. The nonconvergence solutions are described here in four categories, general aids, dc bias analysis, dc sweep analysis and transient analysis.

General Aids

General aids to help avoid nonconvergence in a simulation analysis are provided as .OPTIONS statements in the PSpice circuit file. In particular, they are the error tolerances RELTOL, VNTOL and ABSTOL. During the Newton-Raphson iterative process to determine node voltages and branch currents, the value found is compared with the previous iteration value. If the difference in the values is within a specified error tolerance, the solution has been reached. Otherwise, the program is halted and an error message is given.

RELTOL is a parameter that defines a relative tolerance error. The default setting is 0.001 (0.1%). For a 1A branch circuit, the iterative process of analysis will end if the branch current is $0.001 \times 1 = 1\text{mA}$, or less, within the value of the previous iteration. To help ensure convergence for power-electronics circuits set RELTOL=0.01.

VNTOL, ABSTOL are absolute values of error tolerances. They are for voltage and current respectively. VNTOL is set by default to $1\mu\text{V}$. ABSTOL is set by default to 1pA. Convergence is achieved if a per-iteration change Δ satisfies the expression
$\Delta \leq$ (relative error tolerance + absolute error tolerance).
This criterion is important for the case where the variable (voltage or current) has a value near zero. The relative error tolerance is almost zero, so convergence would not take place without the absolute error tolerance determining when iteration will stop.

† See "Inside SPICE" by Ron Kielkowski, McGraw-Hill Inc., 1994, Chapter 3.

The default values of VNTOL and ABSTOL are low for power circuits.
If the minimum circuit voltage $V_{\min} \approx 0.1$ volt, and if
VNTOL $\approx$ RELTOL $\times V_{\min}$,
we can set VNTOL = 1mV to aid convergence.
If the minimum circuit current $I_m \approx 0.1\mu A$, and if
ABSTOL $\approx$ RELTOL $\times I_{\min}$,
we can set ABSTOL = 1nA to aid convergence.

GMIN is a conductance parameter. It is set to a default value of 1pS. Care must be taken that the range of resistance values in a circuit is not greater than 1/GMIN. In the analysis algorithm the change of voltage ΔV per iteration has the form f(v)/G, where G is conductance. If conductance G is zero the program crashes. To prevent this happening, PSpice places GMIN across every *pn* junction because semiconductors have zero conductance in *I-V* regions of constant current. Further, GMIN is connected across every dependent source and every independent current source.
It aids convergence to have GMIN as large as possible, but the number of iterations increases. Set GMIN = 0.1nS in a .OPTIONS statement. Take care that the current in GMIN is less than ABSTOL, otherwise, the analysis will not converge.

RS is the series resistance parameter in PSpice diodes and BJTs. The default value is set to zero. To counteract against a large value of conductance and many iterations, make the value of RS finite without disturbing the circuit responses by a noticeable amount.
Set RS = 1mΩ.

DC Bias-point Analysis

This is the first analysis carried out by PSpice. Inductors and capacitors are treated as short circuits and open circuits respectively. The dc bias voltage at each node and the source currents are determined. These values are used as initial values for subsequent analyses. There are aids for convergence associated with this bias-point analysis.

ITL1 There is a default iteration limit of ITL1 = 40. If convergence has not been reached after 40 iterations, then a source-stepping algorithm is automatically used. This involves the source values being stepped from zero to full value. At each new step the previous node voltages are used as initial values. Increasing ITL1 gives a greater chance of finding the bias points. For the majority of cases success is expected if we set ITL1 = 500.

.NODESET The initial guess by PSpice for the node voltage values before analysis begins may be off sufficiently that too many iterations are required to find a solution. We can aid the process of the initial guess by setting some of the initial conditions with the .NODESET statement.

Example:

.NODESET V(12) = 200V, V(7, 10) = 20V

This aid is usually used after a failure to converge on the operating points. After a failure, the output file can be inspected. Those node-voltage values which are unrealistic can be made the subject of the .NODESET statement. More realistic values can be inserted in the statement.

SAVEBIAS, LOADBIAS commands (see MicroSim's manuals) saves bias point voltages from one simulation and loads them in another to save iteration time. We do not use these commands in this text.

DC Sweep Analysis

In analyses that involve DC SWEEP, a bias-point analysis is done first, then a dc analysis follows for the first value of the source parameter that is being *swept*. In the following analyses of the sweep, the previous results are used as initial conditions.

.STEP Device model discontinuities can produce nonconvergence problems if the iteration limit is exceeded due to oscillations at a discontinuity. The discontinuity arises from two equations describing adjacent regions of an *I-V* characteristic (eg. linear and saturation regions). An imperfect *join* of the two equations causes a discontinuity at the intersection. If iteration steps are small for accuracy, there is a greater chance of *falling into* a discontinuity. The first attempt to overcome a failure is to increase the increment in the .STEP statement.

The second attempt is to create an offset in the .STEP statement.

Example with offset and increased step.

.STEP VS 6V 10V 0.2V might become .STEP VS 6.1 10.3 0.3

A third attempt to avoid a nonconvergence problem is to create a new set of parameters for the device model.

ITL2 Rapid voltage transitions at a node may be a natural characteristic of a circuit. Such changes of voltage between steps may be too large for the Newton-Raphson algorithm for the limited number of iterations. The default value is 20 iterations. This can be changed by ITL2 in the .OPTIONS statement. The problem can often be solved by setting ITL2 = 200

Transient Analysis

Like dc sweep analysis, transient analysis can be plagued by a combination of rapid voltage transitions and discontinuities in the model of a device. These can produce nonconvergence problems. In transient analysis, if a nonconvergence occurs at the iteration limit, PSpice discards that time point, reduces the time step and retries. Large voltage transitions, often accompanied by a discontinuity, still offer a likely trap for nonconvergence.

ITL4 For transient analysis ITL4 sets the limit of the number of iterations before abandoning a time point and reducing the time step. The larger the time step, the less likely the discontinuity trap will be triggered. It is worth having a large number of allowed iterations to increase the possibility of convergence. The default value of ITL4 is set to 10.
Set ITL4 = 40.

ITL5 To limit the number of iterations for a transient analysis solution at any particular time point. The default value is ITL5 = 0, which suppresses any limit. We have left it at ITL5 = 0 in this text.

Capacitance A capacitive current can help to bridge the discontinuity of a model's characteristics and aid convergence in analysis. The default setting of the capacitance parameter in a semiconductor-device model is zero. All devices have capacitance, so it is reasonable to give the model's capacitances finite values.
For a diode set the parameter CJO = 0.1pF.

Summary of Convergence Aids for Power Electronics Circuits

General Aids

.OPTIONS RELTOL = 0.01 VNTOL = 1mV ABSTOL = 1nA
.OPTIONS GMIN = 1E-10S
Set diode parameter RS = 1mV.

DC Bias

.OPTIONS ITL1 = 500
.NODESET V(10) = 150V; Initial condition guess, two bias-point calculations.

DC Sweep

.OPTIONS ITL2 = 200; For large voltage transitions.
Increase step size or use offset, for model discontinuity.
Change model parameter values, for a discontinuity.

Transient Analysis

.OPTIONS ITL4 = 40; ITL5 = 0 by default.
Set CJO = 0.1pF as a diode-model parameter.

General Failure

Find another device model if the output file pinpoints the device failing to converge.

General Success

Relax the aids for greater accuracy or a faster solution time.

If we add the above aids to convergence to the circuit file of D7.5.1, a simulation will converge even with load currents of 500A. It must be noted that the program is run with UIC in the .TRAN statement. This sets all initial conditions to zero since none is specified. Had we not included UIC, the program would have done a dc-bias analysis first, then would have crashed early (300μs) in the transient analysis. This is because this special circuit comprises sources and inductors mainly. The dc-bias currents from the sources are greater than 10^9A. Such initial values provide nonconvergence problems in the transient analysis, since the actual source currents have amplitudes of no more than tens of amperes. The current transition is too large. The use of UIC in the .TRAN statement is common for ac circuits with transformers. Transformers have high impedance, low dc-resistance primary windings.

If problems were experienced with nonconvergence in the exercises of Chapter 6, we can redo them now with the new options set.

Drill Exercise D7.5.2

Figure 7.12 shows the circuit diagram of a three-phase, half-wave rectifier. The three-phase source can be represented by an ideal voltage source of 60V(rms) per phase at 50Hz together with a series inductance of 1mH. The dc load can be modelled by a constant-current source of 100A. Do a PSpice simulation using the diode model name D_IDEAL (described in POWER.LIB). Using PROBE, plot traces of the diode currents and the running average load power over an interval of 61ms. Determine **(a)** the commutation overlap angle of diode conduction and **(b)** the average power absorbed by the load.

(Ans: **(a)** 56.9°, **(b)** 5.45kW.)

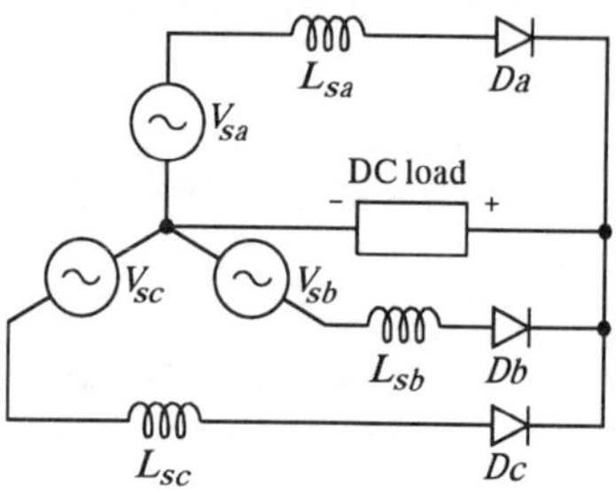

Fig. 7.12

There are Extra Drill Exercises on the Website.

7.6 THREE-PHASE FULL-WAVE RECTIFICATION

Figure 7.13 shows the circuit diagram of a full-wave bridge rectifier. This is common for high power conversion. The three-phase source could represent the ac supply on the secondary side of a transformer. The impedance (L_s, R_s) models the source, line and stray impedance.

The bridge rectifier comprises six diodes. These uncontrolled switches turn on when their anodes are positively biased and turn off when their currents reduce to zero. The sequence of switching follows a pattern of six pulses per cycle. Conduction is in pairs D6, D1 and D1, D2 and D2, D3 and D3, D4 and D4, D5 and D5, D6, cyclically repeating.

The dc load comprises resistance R_l and inductance L_l. Since the input is alternating, the output will have an ac component, that is the ripple on top of the constant dc component. To keep the output current and voltage ripple to a low value, filter elements L_f and C can be added to the circuit.

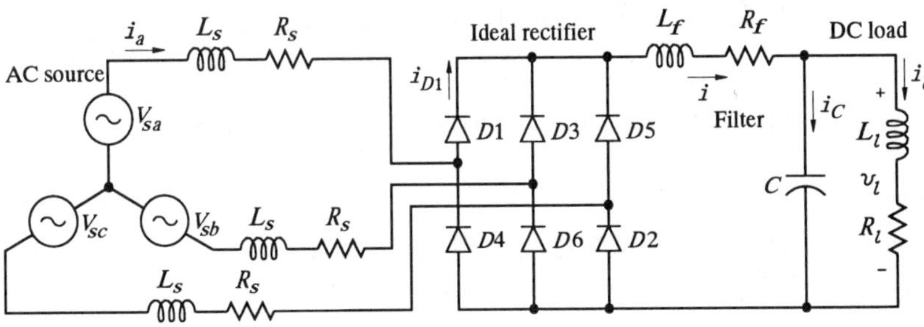

Fig. 7.13 A full-wave diode bridge rectifier.

EXAMPLE 7.6.1

Consider the circuit diagram of a three-phase, full-wave rectifier in Fig. 7.13. The circuit has the specifications

$$V_{sa} = 100\text{V(rms) at 50Hz},\ L_s=0,\ R_s=0,\ L_f=0,\ R_f=0,$$
$$C=0,\ L_l=0 \text{ and } R_l=0.9\Omega.$$

Carry out a PSpice simulation using the diode model named D_IDEAL in the file POWER.LIB. Plot traces of the three-phase line voltages, the output voltage and the output current. Determine **(a)** the average output power, **(b)** the current rating and peak current of a diode and **(c)** the peak reverse voltage of each diode.

Solution

There are four steps in the solution of this basic rectifier circuit.

STEP 1 From the circuit diagram in Fig. 7.13 and from the data supplied, a PSpice configuration can be drawn. See Fig. 7.14.

STEP 2 From the PSpice configuration in Fig. 7.14, the circuit file named EX7_6_1.CIR can be written. Note that a general set of .OPTIONS statements from Section 7.5.1 has been added for convergence stability during the analysis.

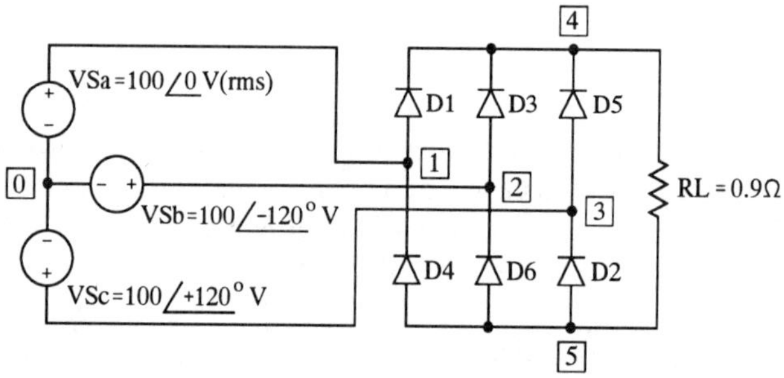

Fig. 7.14 PSpice configuration of a full-wave rectifier.

»»»»» EX7_6_1.CIR »»»

```
THREE-PHASE FULL-WAVE DIODE BRIDGE D_IDEAL
* To plot waveforms and find diode ratings and output power.

* OPTIONS;                          See Section 7.5.1.
. OPTIONS   RELTOL=0.01 ;           General convergence aid.
. OPTIONS   ABSTOL=1nA;             General convergence aid.
. OPTIONS   VNTOL=1mV;              General convergence aid.
. OPTIONS   ITL1=500;               Aid dc bias convergence.
. OPTIONS   ITL2=200;               DC sweep convergence.
. OPTIONS   ITL4=40;                Transient convergence.
. OPTIONS   GMIN=1E-10;             For general convergence.

* PARAMETERS
. PARAM     VRMS=100V   VMAX={1.414*VRMS}   FREQ=50Hz

* SOURCES, three-phase ac source.
VSa 1 0  SIN(0 {VMAX} {FREQ} 0 );             Phase a.
VSb 2 0  SIN(0 {VMAX} {FREQ} 0 -120);         Phase b.
VSc 3 0  SIN(0 {VMAX} {FREQ} 0 -240);         Phase c.
```

```
* DIODE, also found in file POWER.LIB.
.SUBCKT  D_IDEAL  101  100;              Node 101 is the diode anode.
D_RECT   101 100  D_IDEAL
.MODEL   D_IDEAL  D(RS=1m  CJO=0.1pF  N=0.001)
.ENDS    D_IDEAL

* DIODE BRIDGE
XD1  1  4  D_IDEAL
XD2  5  3  D_IDEAL
XD3  2  4  D_IDEAL
XD4  5  1  D_IDEAL
XD5  3  4  D_IDEAL
XD6  5  2  D_IDEAL

* LOAD, dc output.
RL    4  5  0.9ohm

* ANALYSIS
.TRAN  0.2ms  21ms  0   0.1ms;           If this has problems, add UIC.
.PROBE
.END
```

» »

STEP 3 A PSpice simulation can be run with the circuit file EX7_6_1.CIR. There were no problems with convergence.

STEP 4 The use of PROBE with the data file EX7_6_1.DAT enables traces of all the variables to be plotted. For this example the traces are limited to those shown in Fig. 7.15 on the next page. Traces of the output voltage and current give the following results.

Peak load voltage $\hat{V}_l \approx 244.4\,\text{V}$, excursion $\Delta V_l \approx 31.3\,\text{V}$, average $V_{lav} \approx 233.4\,\text{V}$.

The theoretical average value is $V_{lav} = 3\sqrt{3}\,\hat{V}_s/\pi = 233.87\,\text{V}$. (No diode voltage drop.)

Peak load current $\hat{I}_l = 271.52\,\text{A}$, excursion $\Delta I_l \approx 34.8\,\text{A}$, average $I_{lav} \approx 260.7\,\text{A}$.

The theoretical average load current is $I_{lav} = V_{lav}/R_l = 233.87/0.9 = 259.9\,\text{A}$.

Part (a) of Solution. The trace of the running average load power in Fig. 7.15 gives a value of the periodic average power at $t = 20\text{ms}$ as $P \approx 60.6\text{kW}$ from avg(v(4,5)*i(RL)). This could have been obtained from RMS(i(RL))*RMS(i(RL))*0.9 where $R_l = 0.9\Omega$, or PWR(rms(i(RL)),2)*0.9.

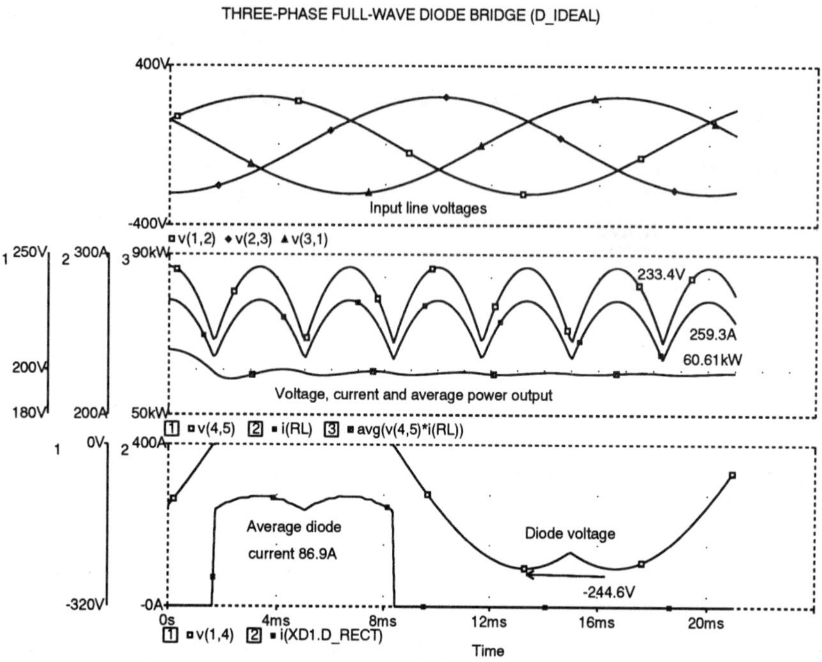

Fig. 7.15

Part (b) of Solution. There is a trace of diode D1 current in Fig. 7.15. The peak diode current is 271.52A. From a plot of avg(i(XD1.D_RECT)) the periodic average value of the diode current at $t = 20$ms is $I_{D\,av} = 86.89$A. From the trace of the diode voltage v(1, 4), the peak diode voltage is $V_{D\,\mathrm{peak}} = -244.6$V.

END OF EXAMPLE 7.6.1

Drill Exercise D7.6.1

Repeat the three-phase rectifier problem in EXAMPLE 7.6.1. In turn, use the diodes named DIO_V, DIO_I and DIO_GD from the file POWER.LIB. Check the average output voltage for each circuit configuration. (Note: DIO_GD needs a high resistance in parallel so that the load does not *float*.)
(Ans: D_IDEAL job 9.12s, $V_{lav} = 233.4$V. DIO_V job 18.95s, $V_{lav} = 233.34$V. DIO_I job 18.95s, $V_{lav} = 233.34$V. DIO_GD job 11.15s, $V_{lav} = 233.82$V.)

Drill Exercise D7.6.2

Consider the circuit diagram of the bridge rectifier in Fig. 7.13. The circuit has the following specifications.

$$V_{sa}=60\text{V(rms) at 60Hz},\ L_s=0,\ R_s=0,\ L_f=0,\ R_f=0,$$
$$C=0,\ L_l=3.0\text{mH and } R_l=0.4\Omega.$$

Use the diode model D_IDEAL. **(a)** Determine the approximate time it takes for steady conditions to be reached. For steady state, plot traces of the input power factor, a diode's current and voltage, and the load current. Determine **(b)** the input power factor, **(c)** the average current rating of a diode, **(d)** the average load current and **(e)** the load current excursion ($\Delta I = I_{l\max} - I_{l\min}$).
(Ans: **(a)** 50ms, **(b)** 0.955, **(c)** 116.2A, **(d)** 349.1A, **(e)** 2.33A.)

Drill Exercise D7.6.3

A three-phase bridge rectifier is depicted in Fig. 7.13. The circuit has the following specifications.

$$V_{sa}=60\text{V(rms) at 60Hz},\ L_s=0,\ R_s=0,\ L_f=0,\ R_f=0,$$
$$C=200\mu\text{F},\ L_l=0\text{ and } R_l=0.4\Omega.$$

Do a PSpice simulation using the diode model D_IDEAL. **(a)** Determine the approximate time it takes for steady conditions to be reached. For steady state, plot traces of the input power factor, a diode's current and voltage and the load current. Determine **(b)** the input power factor, **(c)** the average current rating of a diode, **(d)** the average load current and **(e)** the load current excursion ($\Delta I = I_{l\max} - I_{l\min}$). (Ans: **(a)** Negligible transient, **(b)** 0.9556, **(c)** 116.311A, **(d)** 349.06A, **(e)** 48.56A.)

Drill Exercise D7.6.4

Consider the circuit diagram of the full-wave bridge rectifier in Fig. 7.13. The circuit has the following specifications.

$$V_{sa}=60\text{V(rms) at 60Hz},\ L_s=0,\ R_s=0,\ L_f=20\text{mH},\ R_f=0.004\Omega,$$
$$C=350\mu\text{F},\ L_l=0\text{ and } R_l=0.4\Omega.$$

Do a PSpice simulation using the diode model D_IDEAL from POWER.LIB (and Section 4.2.4). **(a)** Determine the approximate time it takes for steady conditions to be reached. For steady state, plot traces of the input power factor, a diode's current and voltage, and the load current. Determine **(b)** the input power factor, **(c)** the average current rating of a diode, **(d)** the average load current and **(e)** the load current excursion.
(Ans: **(a)** 350ms, **(b)** 0.955, **(c)** 114.9A, **(d)** 345.63A, **(e)** 0.33A.)

There is an Extra Drill Exercise on the WEBsite.

Chapter 8

Controlled Rectification

8.1 INTRODUCTION

AC-DC conversion can be made to be controllable if the converter switches are thyristors. The amount of dc power delivered to a load from an ac supply can be adjusted by altering the trigger angle (phase angle) α of the thyristor. The configurations of thyristor converters are similar to those uncontrolled rectifiers described in Chapter 7. In this chapter we will use the added degree of control afforded by gate drivers.

8.2 SINGLE-PHASE HALF-WAVE CONTROLLED RECTIFICATION

The circuit diagram of a single-phase, half-wave controlled rectifier is shown in Fig. 8.1a, together with the waveforms of the load voltage v_l and load current i_l. From PSpice simulations we can determine the degree of influence of the circuit elements added for improved performance.

EXAMPLE 8.2.1 (PESS EXAMPLE 2.3)
A supply delivers a rectangular-wave ac voltage of amplitude 200V at 100Hz. A thyristor modulates the power delivered to a load resistor of value 2 ohms. For a trigger angle $\alpha = 90°$, determine **(a)** the rms value of the load current, **(b)** the average load power and **(c)** the average current in the thyristor. Neglect all circuit inductance.

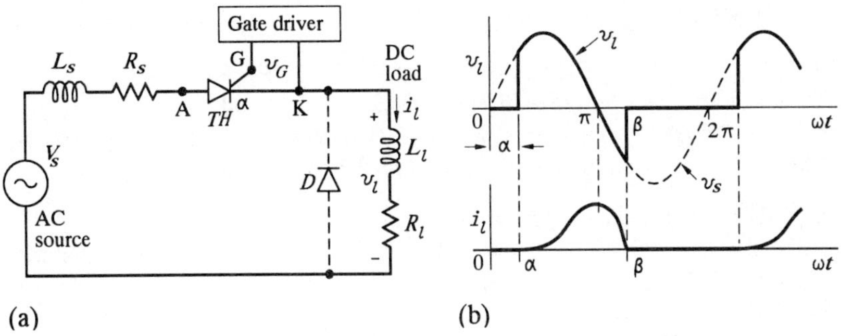

Fig. 8.1 Half-wave controlled rectification.

Solution

There are four steps in this solution.

STEP 1 From the circuit diagram in Fig. 8.1a and from the given data a circuit configuration can be drawn. See Fig. 8.2. The gate-driver circuit is from Section 5.3.1 and the thyristor model is essentially from Section 4.4.

STEP 2 The circuit file EX8_2_1.CIR can be written from an inspection of the PSpice configuration of the half-wave rectifier in Fig. 8.2.

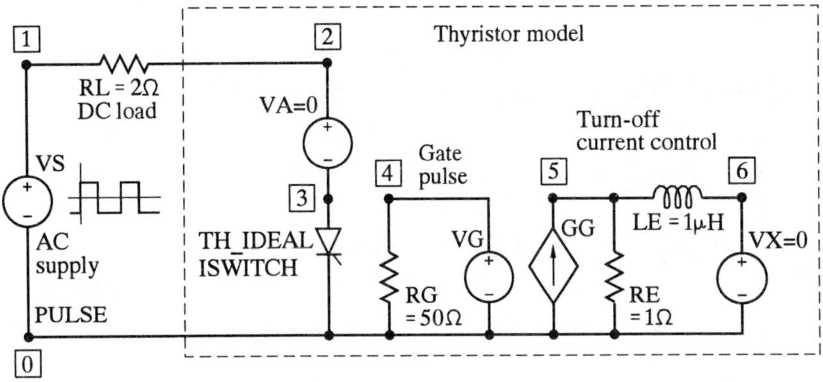

Fig. 8.2 PSpice configuration of a half-wave, controlled rectifier.

```
»»»»» EX8_2_1.CIR »»»»»»»»»»»»»»»»»»»»»»»»»»»»»»»»»»»»»»

THYRISTOR SINGLE-PHASE HALF-WAVE RECTIFIER
* To determine the average load power for a resistive load.
.INC    OPTIONS;   See Section 7.5.1, EXAMPLE 7.6.1, and Appendix A.
* The OPTIONS file must be in the subdirectory of the circuit file.

* SOURCE, squarewave alternating voltage.
.PARAM  V1=-200V       V2=200V;                Initial and peak values.
.PARAM  TR=10ns        TF={TR};                Pulse rise and fall times.
.PARAM  FREQ=100Hz  PERIOD={1/FREQ};           Source frequency.
.PARAM  P_WIDTH={PERIOD/2};                    Source pulsewidth.
.PARAM  TD=0;                    Delay time before pulse rise.
VS 1 0  PULSE({V1} {V2} {TD} {TR} {TF} {P_WIDTH} { PERIOD})

* LOAD
.PARAM  RLOAD=2ohms
RL 1 2  {RLOAD};                 Load resistance, "parameterized".

* DEVICE, the near-ideal thyristor.
WTH   3 0 VX  TH_IDEAL;    Thyristor as a current-controlled switch.
* The switch is connected between nodes 3, 0.
* VX is the zero-voltage current sensor that determines switch on/off.
.MODEL  TH_IDEAL ISWITCH(RON=1E-3 ROFF=1E6 ION=0.1 IOFF=0)
* GATE CIRCUIT, to provide pulses for thyristor turn on.
.PARAM  V_P=2V;                  The gate pulse voltage for turn-on.
.PARAM  ALFA=90deg   DELAY={ALFA*PERIOD/360};   Trigger angle α.
.PARAM  TON=1ms;                 The on-time of the gate pulse.
VG 4 0  PULSE(0 {V_P} {DELAY} {TR} {TF} {TON} {PERIOD})
RG 4 0  1ohm;                    A nominal value for simulation.

* CURRENT CONTROL, to have the thyristor on if a gate signal exists or
* if the thyristor current is finite (above a holding value).
VA 2 3  0;                       Current sensor for the thyristor.
GG 0 5  TABLE {v(4)+i(VA)}=(1E-3, 0) (1.5E-3, 0.11)
RE 5 0  1ohm
LE 5 6  1uH;              RE, LE time delay for current rise and fall in VX.
VX 6 0  0;                Current sensor for TH_IDEAL on/off condition.

* ANALYSIS, transient for 3 cycles of switching.
.TRAN    0.05ms 30ms
.PROBE   i(RL),   v(1);          Load current and source voltage.
.END

»»»»»»»»»»»»»»»»»»»»»»»»»»»»»»»»»»»»»»»»»»»»»»»»»»»»»»»»»»»
```

STEP 3 A PSpice simulation can be run with the circuit file EX8_2_1.CIR.

STEP 4 Using PROBE, traces of the source voltage v(1), the load current i(RL), and the running rms and average values of the load current can be plotted. The results are shown in Fig. 8.3. An important observation is that for 4μs while the source has reversed polarity the thyristor is still on. A reverse thyristor current of 100A is conducted. This result is due to the delay created by RE and LE in the switch control circuit. The elements RE and LE were added to give the analysis stability. Without the delay in the rise or fall of the switch's control current a nonconvergence problem exists (the transition of i(GG) itself is too great). The circuit file can be altered to give a lower value of the thyristor reverse current and for a shorter interval of time.

1. Decrease the slope of the PWL function of GG.
2. Reduce the time constant of the control-current circuit.
3. Make the source voltage changes less rapid.
4. Add a diode in series with the switch that models the thyristor. This prevents reverse current of any significance.

The first three points are incorporated in a driver subcircuit named TH_GAT in DRIVER.LIB.

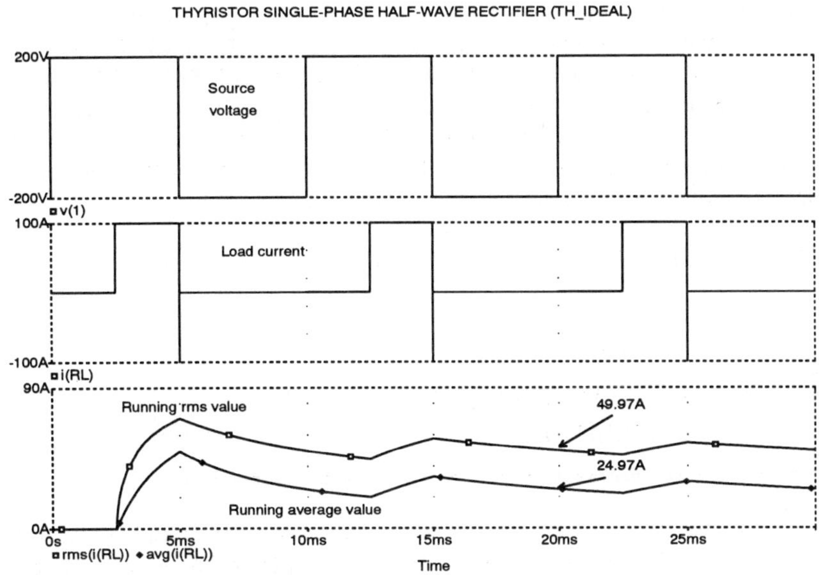

Fig. 8.3

Part (a) of Solution. The periodic rms value of the load current can be read from the trace of rms(i(RL)) at $t=30$ms. That is, $I_{lrms}=49.97$A (theoretically 50A).

Part (b) of Solution. The average power absorbed by the load is
$P = I_{lrms}^2 R_l = (49.97)^2 \times 2 = 4994$ W.

Part (c) of Solution. The periodic average value of the load current can be read from the trace of avg(i(RL)) at $t=30$ms. That is, $I_{lav}=24.97$A (theoretically 25A).

END OF EXAMPLE 8.2.1

Drill Exercise D8.2.1
Follow the pattern of EXAMPLE 8.2.1 to simulate a single-phase, half-wave, controlled rectifier. The source supplies a rectangular-wave ac voltage of amplitude 100V at 50Hz. A thyristor modulates the power delivered to a load of resistance $R_l=1\Omega$ and inductance $L_l=1.6$mH. For a trigger angle $\alpha=70°$, determine the change of average load current due to the effect of the inductance. (Use the .STEP statement.) (Ans: 15.27A, 13.89A.)

Drill Exercise D8.2.2
Consider the circuit diagram in Fig. 8.1a. The ac source provides a rectangular waveform of amplitude 30V at 50Hz. The thyristor is triggered at a firing angle $\alpha=30°$ to deliver power to a dc load whose resistance is $R_l=0.3\Omega$. Do a PSpice simulation adapting the model TH_IDEAL (from EXAMPLE 4.4.1) for the thyristor, and the driver subcircuit named HALFWAVE_DRVA (from POWERLIB and EXAMPLE 5.3.1). A gate voltage of 5V will turn the thyristor on. Plot a trace of the running average power absorbed by the load and determine the value of the periodic average power delivered to the load. (Ans: 1.24kW.)

There are Extra Drill Exercises on the WEBsite.

8.3 SINGLE-PHASE FULL-WAVE CONTROLLED RECTIFICATION

For adjustable low to medium power modulation from an ac supply to a dc load, the bridge configuration of thyristors is common. Figure 8.4 shows the circuit diagram of a single-phase converter.

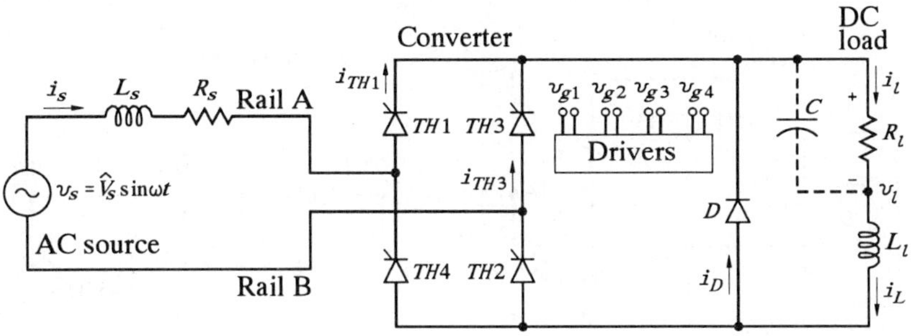

Fig. 8.4 Full-wave bridge rectifier.

Operation of the converter involves switching the thyristors in pairs. Thyristors *TH*1, *TH*2 are switched on together at some point (α) in the source waveform while rail A is positive. Thyristors *TH*3, *TH*4 are switched on together at the same point of the waveform while Rail B is positive. All four thyristors turn off if they are reverse biased and if their currents are zero. In this way, the current in the load is unidirectional. The diode, which is optional, provides a freewheeling path for the load current if all four thyristors turn off. The ac supply usually has a source impedance and this affects the converter performance by what is known as commutation overlap u.

EXAMPLE 8.3.1

Consider a fully-controlled, single-phase, full-wave, bridge converter as depicted in Fig. 8.4. The circuit specifications are as follows.

$$V_s = 120\text{V(rms) at 60Hz},\ R_s = 0,\ L_s = 0,\ \text{no diode},\ R_l = 8\Omega,$$
$$C = 0,\ L_l = 5\text{mH},\ \alpha = 30°.$$

Do a PSpice simulation and determine **(a)** the output voltage form factor FF, **(b)** the output voltage ripple factor RF, **(c)** the rectification ratio RR, **(d)** the first 13 harmonics of the input current, **(e)** the input current harmonic distortion THD, **(f)** the input power factor PF, **(g)** the input displacement power factor DPF and **(h)** the input distortion factor DF.

(Ans: **(a)** 1.2, **(b)** 0.66, **(c)** 0.72, **(e)** 8.34%, **(f)** 0.95, **(g)** 0.957, **(h)** 0.99.)

Solution

By definition: Form Factor $= V_{l\,\text{rms}}/V_{l\,av}$, Ripple Factor $= (V^2_{l\text{rms}} - V^2_{l\text{av}})^{1/2}/V_{lav}$, Rectification Ration $= V_{l\,av}I_{l\,av}/(V_{l\text{rms}}I_{l\text{rms}})$, Displacement Power Factor $= \cos\theta_1$, Input Distortion Factor $= I_{s\,1\text{rms}}/I_{s\text{rms}}$.

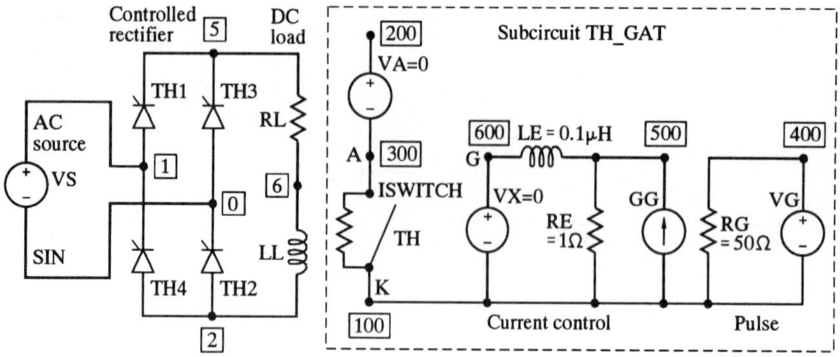

Fig. 8.5 PSpice configuration of a bridge rectifier.

STEP 1 We can use the circuit diagram in Fig. 8.4 and the data specified to create a PSpice configuration as shown in Fig. 8.5. The driver-circuit and thyristor configuration is the same as that which was used in EXAMPLE 8.2.1. We will use it as a subcircuit named TH_GAT from DRIVER.LIB.

STEP 2 The circuit file EX8_3_1.CIR can be composed from the configuration in Fig. 8.5. Now run the simulation.

»»»»» EX8_3_1.CIR »»

```
SINGLE-PHASE FULL-WAVE THYRISTOR BRIDGE RECTIFIER
* To determine the converter performance on the dc and ac sides.
.INC    OPTIONS;                See Section 7.5.1. For stable analysis.
* PARAMETERS
.PARAM  V_P=2V;                         Gate voltage for each thyristor.
.PARAM  TR=1us     TF={TR};             Gate pulse rise and fall times.
.PARAM  TON=1ms;                                    Gate pulse width.
.PARAM  LLOAD=5mH;                                  Load inductance.
.PARAM  RLOAD=8ohms;                                Load resistance.
.PARAM  FREQ=60Hz    PERIOD={1/FREQ};               Supply frequency.
.PARAM  VRMS=120V    VMAX={1.414*VRMS};             Supply voltage.
.PARAM  ALFA=30deg   DEL={ALFA*PERIOD/360}
* ALFA is the thyristor trigger angle α.
```

```
* SOURCE and LOAD
VS   1  0   SIN(0 {VMAX} {FREQ});          Sinusoidal voltage source.
RL   5  6   {RLOAD}
LL   6  2   {LLOAD}

*CONTROLLED RECTIFIER, thyristor and driver subcircuits.
. LIB DRIVER .LIB
XTH1    1   5  TH_GAT   PARAMS:  DELAY={DEL}
XTH2    2   0  TH_GAT   PARAMS:  DELAY={DEL}
XTH3    0   5  TH_GAT   PARAMS:  DELAY={DEL+PERIOD/2}
XTH4    2   1  TH_GAT   PARAMS:  DELAY={DEL+PERIOD/2}

* ANALYSIS
. FOUR     60Hz     13     i(VS);           Fourier analysis of source current.
. TRAN     0.05ms   50ms
. PROBE
. END
```

» »

END OF EXAMPLE 8.3.1

Drill Exercise D8.3.1 (PESS EXAMPLE 2.6)
Consider a fully-controlled, single-phase, full-wave, bridge converter as depicted in Fig. 8.4. The circuit specifications are as follows.
$V_s = 120$V at 60Hz, $R_s = 0$, $L_s = 0$, no diode, $R_l = 8\Omega$, $C = 0$, $L_l = \infty$, $\alpha = 30°$.
Do a PSpice simulation to determine **(a)** the output voltage form factor FF, **(b)** the rectification ratio RR, **(c)** the first 13 harmonics of the input current, **(d)** the input current harmonic distortion THD and **(e)** the input power factor.
(Notes: (1) Ideally the average load voltage is $V_{lav} = (2\hat{V}_s/\pi)\cos\alpha = 93.7$V. So the ripple-free load current is $I_l = 11.71$A. Use this current as a source to replace the load in the simulation. (2) PSpice forces conduction at $t = 0$, but gives proper results thereafter.) (Ans: **(a)** 1.261, **(b)** 0.793, **(c)** $\hat{I}_1 = 14.91$A, $\hat{I}_3 = 4.97$A, $\hat{I}_5 = 2.98$A, $\hat{I}_7 = 2.13$A, $\hat{I}_9 = 1.657$A, $\hat{I}_{11} = 1.36$A, $\hat{I}_{13} = 1.148$A, **(d)** 44.5%, **(e)** 0.758.)

Drill Exercise D8.3.2
Consider the circuit diagram of a single-phase converter in Fig. 8.4. The circuit specifications are as follows.

$$V_s = 100\text{V(rms) at 50Hz},\ R_s = 1\text{m}\Omega,\ L_s = 5\text{mH},\ \text{no diode},\ R_l = 5\Omega,$$
$$C = 0,\ L_l = 30\text{mH},\ \alpha = 60°.$$

Do a PSpice simulation using the thyristor model named TH_GAT. Plot traces of the input current i_s, the output current i_l, the thyristor currents i_{TH1}, i_{TH3} and the load voltage v_l. Determine **(a)** the form factor FF of the load current and **(b)** the total harmonic distortion THD of the input current. (Ans: **(a)** 1.116, **(b)** 6.53%.)

There are Extra Drill Exercises on the WEBsite.

8.4 BIPHASE CONTROLLED RECTIFICATION

We were introduced to the biphase uncontrolled rectifier in Section 7.4. A circuit diagram of the controlled biphase rectifier is shown in Fig. 8.6a. This is a single-phase, full-wave converter, whose output is similar to that of the bridge rectifier (Section 8.3). The biphase ac-dc converter has only two thyristors but has a transformer with two secondary windings. A simplification of the circuit to ease simulation is illustrated in Fig. 8.6b.

(a)

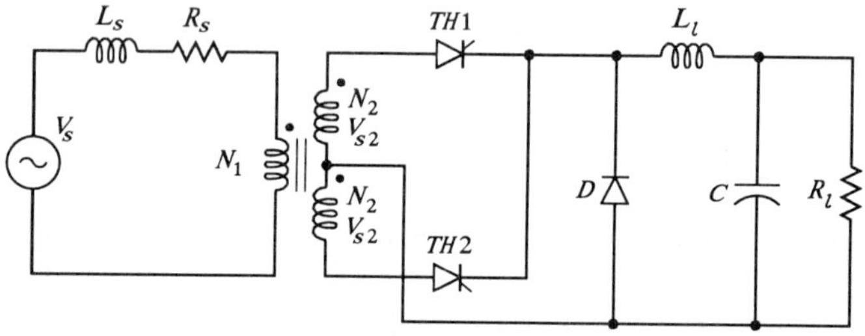

(b)

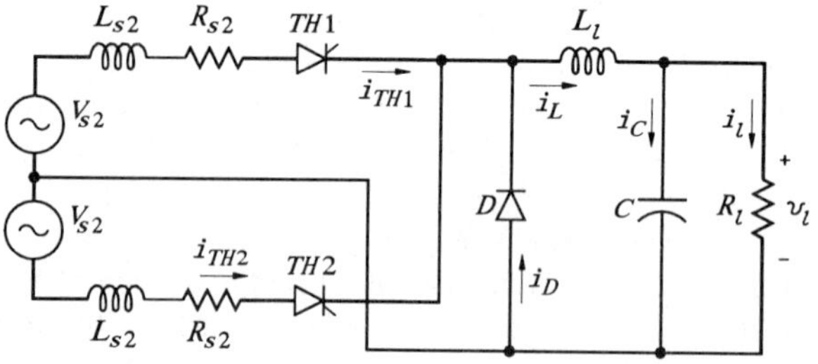

Fig. 8.6 Biphase controlled converter. (a) Circuit diagram, (b) simplified form.

To perform PSpice simulations of biphase converters does not require more techniques than have been used previously. The ac supply may have source impedance (R_s L_s). Models of the switches can be near-ideal since we are interested in system performance at this stage. Consequently, subcircuits TH_GAT and D_IDEAL can be used to simulate the thyristors and the diode respectively. Filtering action is represented by L_l and C in Fig. 8.6b. The freewheeling diode may be included if the load has inductance. This section can be considered to be one of simulation practice.

Drill Exercise D8.4.1
Consider the circuit diagram of a biphase converter in Fig. 8.6b. The circuit parameters have the following specifications.

$$V_{s2} = 100\text{V(rms) at 50Hz},\ R_{s2} = 0,\ L_{s2} = 0,\ \text{no diode},\ R_l = 10\Omega,$$
$$C = 0,\ L_l = 0,\ \alpha = 30°.$$

From a PSpice simulation, plot traces of i_{TH1}, i_{TH2} and i_l. Determine **(a)** the average value of the load current, **(b)** the rms value of the load current and **(c)** the average value of the load power.
(Ans: **(a)** 8.3627A (8.4A calculated), **(b)** 9.835A, **(c)** 967.3W.)

There are Extra Drill Exercises on the WEBsite.

8.5 THREE-PHASE HALF-WAVE CONTROLLED RECTIFICATION

The circuit diagram of a three-phase, half-wave converter is shown in Fig. 8.7. There may be filtering, natural or designed, in terms of L_l and C. Also, there may be a freewheeling diode for an inductive load. The three-phase supply is usually fed through a three-phase transformer. However, we can dispense with the transformer and use equivalent voltage supplies with source impedance R_s and L_s.

The important part of a simulation of a three-phase controlled system is to set correctly the phase angles of the voltage sources and the correct trigger angle α of the thyristors. It is conventional to have $\alpha = 0$ measured from $\omega t = \pi/6$ radian for phase a. Consequently, for phase b and phase c, the trigger angles are 120° and 240° phase displaced respectively from phase a.

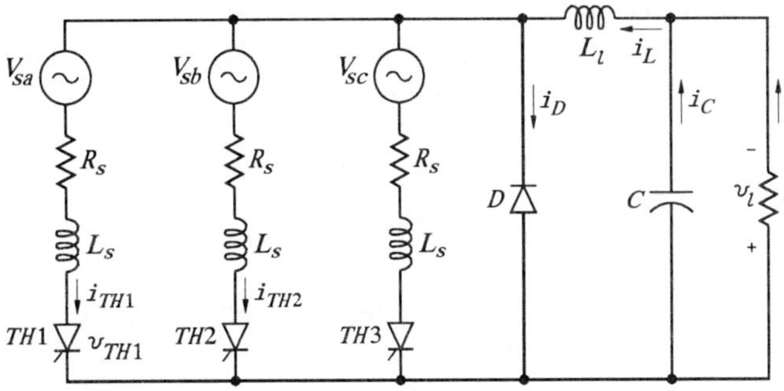

Fig. 8.7 A three-phase, half-wave, controlled rectifier.

The thyristor can be modelled by the subcircuit named TH_GAT which is found in DRIVER.LIB. The thyristor model and the gate driver are also shown in Fig. 8.2. The following Drill Exercise allows us to review the powerful tool of analogue behavioural modelling as well as parameter values and subcircuits.

Drill Exercise D8.5.1

Consider a three-phase, half-wave, controlled rectifier whose circuit diagram is shown in Fig. 8.7. The circuit specifications are as follows.

$$V_s = 120\text{V(rms, phase) at 60Hz},\ R_s = 0,\ L_s = 0,\ \text{no diode},\ R_l = 10\Omega,$$
$$C = 0,\ L_l = 0,\ \alpha = 35°.$$

Write a PSpice circuit file to include all the specific data and control statements. Then generalize the circuit file to have parameter statements and subcircuits. Plot from the simulation, using either circuit file, v_l, i_{TH1}, v_{TH1}, and the three-phase voltages of the source v_{sa}, v_{sb} and v_{sc}. Determine **(a)** the average value of the load voltage, **(b)** the average power consumed by the load, **(c)** the average thyristor current, **(d)** the total rms value of the input line current, **(e)** the fundamental rms value of the input line current, **(f)** the displacement power factor DPF at the input, **(g)** the total harmonic distortion THD at the input and **(h)** the input power factor PF. (Ans: **(a)** 115.4V, **(b)** 1648W, **(c)** 3.85A, **(d)** 7.41A, **(e)** 4.82A, **(f)** 0.94, **(g)** 83.1%, **(h)** 0.73.)

An Extra Drill Exercises is on the WEBsite.

8.6 HALF-CONTROLLED THREE-PHASE BRIDGE RECTIFICATION

The half-controlled, three-phase, bridge rectifier is an extension of the three-phase, half-wave rectifier of the previous section. See Fig. 8.8. The three diodes make full use of the negative half waves of the supply voltage. Consequently, the ripple component of the output voltage is reduced and the average load voltage is increased. The mean value of the output voltage is given by

$$V_{lav} = \frac{3\sqrt{3/2}}{\pi} V_{sa}(1 + \cos\alpha) \tag{8.6.1}$$

where α is the phase-*a* trigger angle, measured from $\omega t = \pi/6$ radian, and $v_{sa} = \sqrt{2}V_{sa}\sin\omega t$, where V_{sa} is the rms value. In this section we can model the thyristor with TH_GAT and the diode with D_IDEAL. These models are described in DRIVER.LIB and POWER.LIB respectively. In PSpice simulations, file OPTIONS is included in the circuit file for reasonably stable convergence.

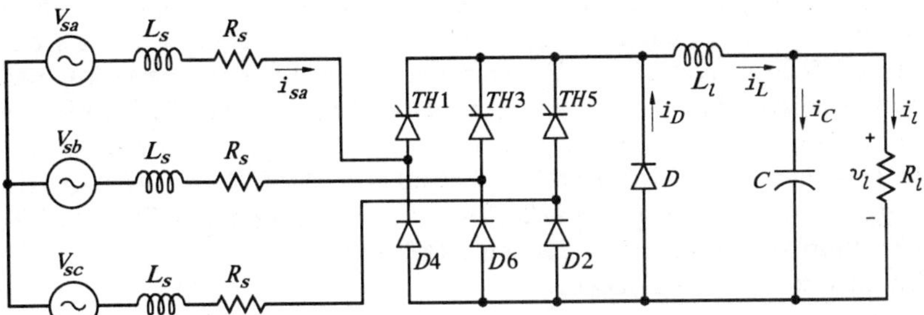

Fig. 8.8 Half-controlled, three-phase, bridge rectifier.

Drill Exercise 8.6.1

Consider the half-controlled, three-phase, bridge rectifier whose circuit diagram is shown in Fig. 8.8. The circuit specifications are as follows.

$$V_s = 120\text{V(rms, phase) at 60Hz}, \ R_s = 0, \ L_s = 0, \ \text{no diode}, \ R_l = 10\Omega,$$
$$C = 0, \ L_l = 0, \ \alpha = 35°.$$

Write a PSpice circuit file to include all the specific data and control statements. Then, generalize the circuit file to have parameter statements and subcircuits. From a simulation, plot traces of v_{sa}, v_{sb}, v_{sc}, v_{TH1}, i_{sa} and v_l. Determine **(a)** the average value of the load voltage, **(b)** the average power consumed by the load, **(c)** the total rms value of the input line current, **(d)** the fundamental rms value of the input line current, **(e)** the displacement power factor DPF at the input, **(f)** the total harmonic distortion THD at the input and **(g)** the input power factor PF.
(Ans: **(a)** 255.7V, **(b)** 6.74kW, **(c)** 21.3A, **(d)** 19.2A, **(e)** 0.97, **(f)** 43.6%, **(g)** 0.876.)

There are Extra Drill Exercises on the WEBsite.

8.7 THREE-PHASE FULLY-CONTROLLED RECTIFICATION

For high levels of power modulation from an ac supply to a dc load, the three-phase, fully-controlled, bridge rectifier is used. The fully-controlled, bridge rectifier has six thyristors as shown in the circuit diagram in Fig. 8.9a on page 233.

Taking the phase-a source as reference, the sequence of switching is thyristor $TH1$ at $\omega t = \pi/6 + \alpha$ radians, thyristor $TH3$ at $\omega t = \pi/6 + 2\pi/3 + \alpha$ radians and thyristor $TH5$ at $\omega t = \pi/6 + 4\pi/3 + \alpha$ radians. Note that α is measured from $\omega t = \pi/6$ radian. The other three thyristors are triggered in relation to each individual thyristor in the same leg. Thyristor $TH4$ is triggered at π radians later than $TH1$, thyristor $TH6$ is triggered at π radians later than $TH3$, and thyristor $TH2$ is triggered π radians later than $TH5$. Thus, from an inspection of Fig. 8.9b, the order of turn on from $\omega t = 0$ is $TH1$ on at $(\pi/6 + \alpha)$ radians.

$TH2$ on at $\pi/3 + (\pi/6 + \alpha)$ radians.	
$TH3$ on at $2\pi/3 + (\pi/6 + \alpha)$ radians.	$TH1$ reverse biased and turns off.
$TH4$ on at $\pi + (\pi/6 + \alpha)$ radians.	$TH2$ reverse biased and turns off.
$TH5$ on at $4\pi/3 + (\pi/6 + \alpha)$ radians.	$TH3$ reverse biased and turns off.
$TH6$ on at $5\pi/3 + (\pi/6 + \alpha)$ radians.	$TH4$ reverse biased and turns off.
$TH1$ on at $2\pi + (\pi/6 + \alpha)$ radians.	$TH5$ reverse biased and turns off.
$TH2$ on at $2\pi + \pi/3 + (\pi/6 + \alpha)$ radians.	$TH6$ reverse biased and turns off, etc.

An analysis yields the expression for the average voltage V_{lav}

$$V_{lav} \approx \left(3\sqrt{6}/\pi\right) V_s \cos\alpha . \tag{8.7.1}$$

This bridge rectifier is not self-starting. A path for current has to be provided by two thyristors in the on-state at any time. Only one thyristor is turned on initially, if the previous sequence of switching is followed. One solution is to shift the reference of the sources through an angle of $\pi/6 + \alpha$ radians (in advance). In this way initial conditions can be given to the gate circuits of $TH1$ and $TH6$ so that these two thyristors are turned on initially. The normal delay for the sequence of thyristor triggering can be followed after this initiation (starting) of the converter.

A simpler way to start the converter is to maintain the conventional phase reference for the source voltages. Using Fig. 8.9b it can be seen that gate pulses at $\omega t = 0$ to thyristors $TH5$ and $TH6$ will provide a path for load current. The gate pulses are created in PSpice by having initial conditions for a relevant branch voltage, a capacitor voltage or an inductor current, using .NODESET, .IC or IC.

Another method of starting can be determined from an inspection of Fig. 8.9b. It is seen that there are two thyristors on at any one time. The sequence is $TH6$, $TH1$, then $TH1$, $TH2$, then $TH2$, $TH3$, then $TH3$, $TH4$, then $TH4$, $TH5$, then $TH5$, $TH6$, then cyclically repeating. If the converter has its first gate pulse to thyristor $TH1$ at $\omega t = \pi/6 + \alpha$ radians, no current is conducted, because no other thyristor is on. However, if both thyristors $TH1$ and $TH2$ received gate pulses $\pi/3$ radians later, both thyristors would turn on and the converter would start. The sequence of firing thereafter would be $TH3$, $TH4$, $TH5$, $TH6$, $TH1$, $TH2$, etc., in $\pi/3$ radian intervals. In summary, all thyristors except $TH1$ have a single, cyclic gate signal. Thyristor $TH1$ receives two gate signals every cycle; the second signal is only useful for starting the converter, the first signal is for normal operation.

After starting, the second gate signal is ineffective because thyristor *TH*1 is always on when thyristor *TH*2 has its gate signal applied. This method is followed here.

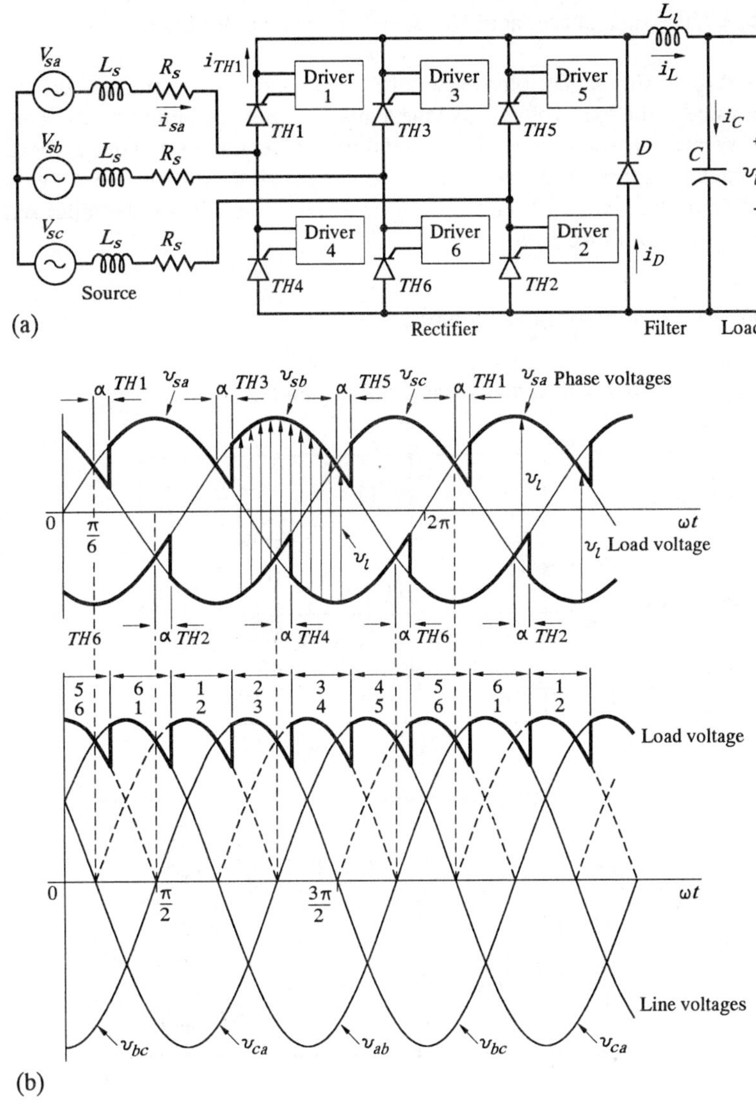

Fig. 8.9 A fully-controlled, three-phase rectifier. (a) Circuit diagram, (b) ideal waveforms.

EXAMPLE 8.7.1

Consider the circuit diagram in Fig. 8.9a, and the following specifications.

$V_s = 120\text{V}$(rms, phase) at 60Hz, $R_s = 0$, $L_s = 0$, no diode, $R_l = 10\Omega$, $C = 0$, $L_l = 0$, $\alpha = 35°$.

From a PSpice simulation, plot traces of i_{sa}, i_{TH1} and γ. Determine **(a)** the average value of the load voltage, **(b)** the average power consumed by the load, **(c)** the average thyristor current, **(d)** the total rms value of the input line current, **(e)** the fundamental rms value of the input line current, **(f)** the displacement power factor DPF at the input, **(g)** the total harmonic distortion THD at the input and **(h)** the input power factor PF.

Solution

There are four main steps in this solution.

STEP 1 Figure 8.10 shows a PSpice configuration.

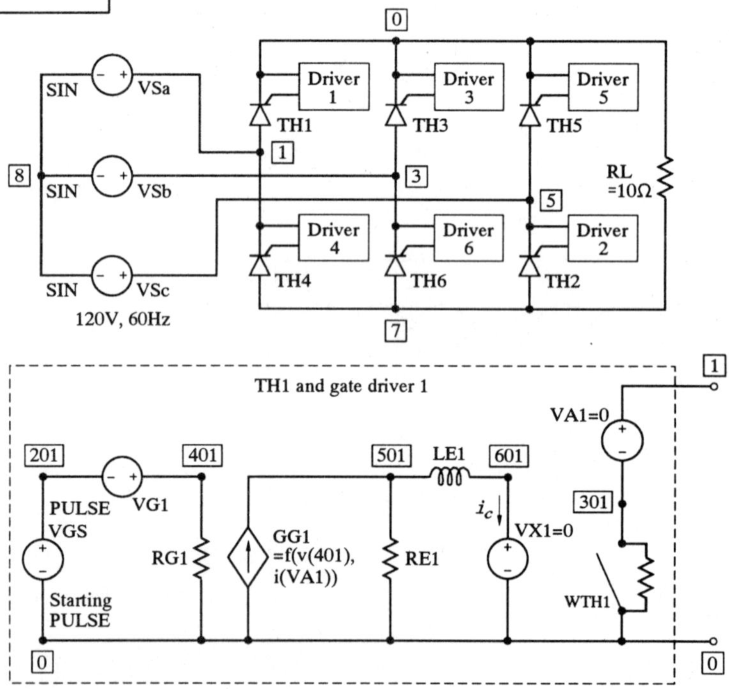

Fig. 8.10 Fully-controlled, three-phase, bridge configuration.

All the drivers are similar to the one shown in Fig. 8.10. However, driver 1 is the only one with the starting pulse VGS.

STEP 2 From the configuration in Fig. 8.10 we can write a circuit file (EX8_7_1.CIR). The gate pulses of the six thyristors are delayed from $t=0$ as follows. Since only thyristor $TH1$ has a starting pulse, this converter is satisfactory only for continuous load current. For discontinuous current all gate signals have a double pulse.

Delay for $TH1$, DEL1 $=(30+\alpha)/(360f)=(30+35)/(360\times 60)=3.0092$ms.
Delay for $TH2$, DEL2 $=$ DEL1 $+60/(360f)=$ DEL1 $+2.77778=5.7870$ms.
Delay for $TH1$ (starting), DELS $=$ DEL2 $=5.7870$ms.
Delay for $TH3$, DEL3 $=$ DEL2 $+60/(360f)=8.5648$ms.
Delay for $TH4$, DEL4 $=$ DEL3 $+60/(360f)=11.3426$ms.
Delay for $TH5$, DEL5 $=$ DEL4 $+60/(360f)=14.1204$ms.
Delay for $TH6$, if $\alpha<30°$ DEL6 $=$ DEL5 $+60/(360f)$.
Delay for $TH6$, if $\alpha\geq 30°$ DEL6 $=$ DEL5 $+60/(360f)-T=0.2315$ms
or DEL6 $=(\alpha-30)/(360f)$

»»»»» EX8_7_1.CIR »»»»»»»»»»»»»»»»»»»»»»»»»»»»»»»»»»»

```
FULLY-CONTROLLED THREE-PHASE RECTIFIER
* To determine the output and input performance of the converter.
.INC   OPTIONS;  See Sec. 7.5.1. A general options package for convergence.

* PARAMETERS
.PARAM  VRMS=120V   VMAX={1.414*VRMS};       Phase voltage.
.PARAM  FREQ=60Hz   PERIOD={1/FREQ};         Supply frequency.
.PARAM  V_P=2V;                   Gate-voltage pulse for each thyristor.
.PARAM  ALFA=35deg  DEL1={(30+ALFA)*PERIOD/360};    For TH1.
.PARAM  DEL2={DEL1+60*PERIOD/360} DELS={DEL2};      For starting.
.PARAM  DEL3={DEL2+60*PERIOD/360};  Time delay for TH3 gate pulse.
.PARAM  DEL4=(DEL3+60*PERIOD/360}
.PARAM  DEL5={DEL4+60*PERIOD/360}
*.PARAM DEL6={DEL5+60*PERIOD/360};              If ALFA < 30deg.
.PARAM  DEL6={(ALFA-30)*PERIOD/360};            If ALFA ≥ 30deg.
.PARAM  TON=1ms;                          Gate turn-on pulse width.
.PARAM  TR=1us   TF={TR};                 Gate pulse rise and fall times.
.PARAM  RLOAD=10ohms;                     The load resistance.

* SOURCES
VSa  1  8   SIN(0 {VMAX} {FREQ});                Phase-a voltage.
VSb  3  8   SIN(0 {VMAX} {FREQ} 0 0 -120)
VSc  5  8   SIN(0 {VMAX} {FREQ} 0 0 -240);       Phase-c voltage.
```

```
* DEVICES
WTH1   301   0   VX1   TH_IDEAL;      Current-controlled switch for TH1.
.MODEL   TH_IDEAL      ISWITCH(RON=1E-3   ROFF=1E6   ION=0.1
IOFF=0)
* Gate pulse to turn TH1 on.
VG1   401 201 PULSE(0 {V_P} {DEL1} {TR} {TF} {TON} {PERIOD})
* Start pulse.
VGS   201 0    PULSE(0 {V_P} {DELS} {TR} {TF} {TON} {PERIOD})
RG1   401 0    50;                    Not really necessary for simulation.
* Control current for WTH1 switch.
GG1   0    501 TABLE {v(401)+i(VA1)}=(1E-3,0) (105E-3,0.11)
RE1   501 0    1
LE1   501 601 0.1uH;                  Delay for stability.
VX1   601 0    0;                     Current control for the switch.
VA1   1    301 0;                     TH1 current sensor.

* SUBCIRCUITS, for regular thyristors and drivers.
.SUBCKT TH_GAT   200   100   PARAMS: DELAY=10ms;  Anode 200.
VG     400  100  PULSE(0 {V_P} {DELAY} {TR} {TF} {TON} {PERIOD})
RG     400  100  50
GG     100  500  TABLE {v(400,100)+i(VA)}=(1E-3,0) (105E-3,0.11)
RE     500  100  1
LE     500  600  0.1uH
VX     600  100  0;         Control current for switching the thyristor.
VA     200  300  0;         Thyristor current sensor.
WTH    300  100  VX  TH_IDEAL;               Model of the thyristor.
.MODEL TH_IDEAL  ISWITCH(RON=1E-3 ROFF=1E6 ION=0.1 IOFF=0)
.ENDS  TH_GAT
XTH2 7  5 TH_GAT  PARAMS:  DELAY={DEL2};     Call for TH2.
XTH3 3  0 TH_GAT  PARAMS:  DELAY={DEL3}
XTH4 7  1 TH_GAT  PARAMS:  DELAY={DEL4}
XTH5 5  0 TH_GAT  PARAMS:  DELAY={DEL5}
XTH6 7  3 TH_GAT  PARAMS:  DELAY={DEL6}

* LOAD
RL     0  7 {RLOAD}
.TRAN     0.05ms    33.3334ms  16.6666ms
.FOUR     60Hz      13         i(VSa);          Fourier analysis.
.PROBE    v(0,7),   i(VSa),    i(WTH1)
.END
```

»»»

STEP 3 Running the program with EX8_7_1.CIR gives rise to the results being written in EX8_7_1.DAT and EX8_7_1.OUT.

STEP 4 Using the PROBE, traces of i_{sa} (−i(VSa)), i_{TH1} (i(WTH1)) and v_l (v(0,7)) can be plotted. See Fig. 8.11. The waveform of the load voltage is similar to that sketched in Fig. 8.9b.

Part (a) of Solution. From the trace v_l, the average value of the load voltage is $V_{lav} = 229.4$V (theoretical value 229.9V).

Part (b) of Solution. From the trace v_l, the rms value of the load voltage is $V_{lrms} = 235.3$V. Average power $P = V_{lrms}^2/R_l = 5.537$kW.

Part (c) of Solution. From the trace of i_{TH1}, $I_{TH1av} = 7.65$A.

Part (d) of Solution. From the trace of i_{sa}, $I_{srms} = 19.38$A.

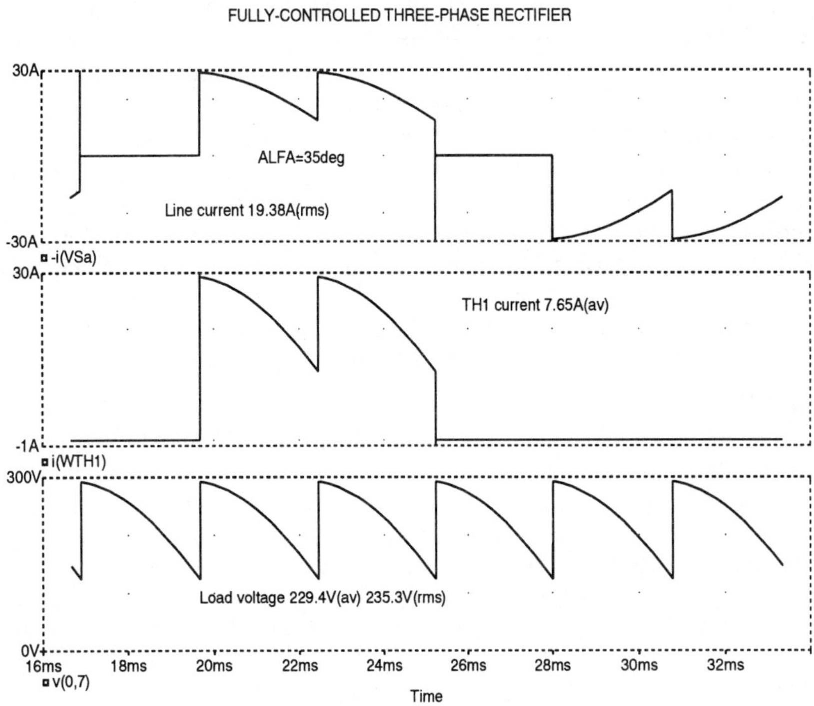

Fig. 8.11

Part (e) of Solution. From the output file EX8_7_1.OUT, the fundamental value of the input line current is $I_{s1\,\mathrm{rms}} = 17.93\mathrm{A}$.

Part (f) of Solution. From the output file EX8_7_1.OUT, $\theta_1 = -31.3°$. Hence, the displacement power factor is $\mathrm{DPF} = \cos\theta_1 = \cos 31.3° = 0.854$.

Part (g) of Solution. From the Fourier analysis, whose results are written in the output file, the total harmonic distortion is THD = 33.79%.

Part (h) of Solution. The power factor PF is given by
$\mathrm{PF} = P/(V_s I_{s\,\mathrm{rms}}) = (I_{s\,1\,\mathrm{rms}}/I_{s\,\mathrm{rms}})\cos\theta_1 = 0.790$.

Note: If the load current is discontinuous all the thyristors must be modelled like thyristor $TH1$ with two gate pulses.

END OF EXAMPLE 8.7.1

Drill Exercise D8.7.1

Consider the circuit diagram in Fig. 8.9a. This fully-controlled, three-phase converter has the following specifications.

$V_s = 120\mathrm{V}$(rms, phase) at 60Hz, $R_s = 0$, $L_s = 0$, no diode, $R_l = 10\Omega$, $C = 0$, $L_l = 5\mathrm{mH}$, $\alpha = 35°$.

From a PSpice simulation, plot traces of i_l, i_{TH1}, i_{sa} and v_l. Determine **(a)** the average value of the load voltage, **(b)** the average power absorbed by the load, **(c)** the load current excursion (ripple), **(d)** the total rms value of the input line current, **(e)** the fundamental rms value of the input line current, **(f)** the displacement power factor DPF at the input, **(g)** the total harmonic distortion THD at the input and **(h)** the input power factor PF. Note. $L_l = 10\mathrm{mH}$ produces a nonconvergence problem. (Ans: **(a)** 229.4V, **(b)** 5.353kW, **(c)** 9.33A, **(d)** 19.19A, **(e)** 17.97A, **(f)** 0.826, **(g)** 29.98%, **(h)** 0.773.)

Drill Exercise D8.7.2

A three-phase, fully-controlled rectifier is illustrated in Fig. 8.9. The circuit specifications are as follows.

$V_s = 120\mathrm{V}$(rms, phase) at 60Hz, $R_s = 0$, $L_s = 0$, no diode, $R_l = 10\Omega$, $C = 0$, $L_l = 5\mathrm{mH}$, $\alpha = 90°$.

The delay angle of firing α is such that there is discontinuous current operation. Adapt the circuit file of EXAMPLE 8.7.1 so that all thyristors receive a double pulse with a 60° shift to allow current conduction. From a PSpice simulation plot traces of i_l, i_{sa} and v_l. Determine **(a)** the average value of the load voltage, **(b)** the average power absorbed by the load and **(c)** the rms value of the input line current. (Ans: **(a)** 33.54V, **(b)** 207.8W, **(c)** 3.723A.) See TH_GAT6 in DRIVER.LIB. There are Extra Drill Exercises on the WEBsite.

Chapter 9

Switch-mode Inverters

9.1 INTRODUCTION

A converter, which modulates power from a dc source to an ac load, is called an inverter. Low power requirements are satisfied by a single-phase inverter, but medium to high power requires a three-phase inverter. The switches in the inverter help create pulses of voltage across the load. For low power requirements these voltages with rectangular waveforms are usually satisfactory. However, as the power requirement increases, it becomes more important to try to create an ac waveform that is close to being a sinewave. The types of inverter† studied in this chapter are centre-tapped source inverters, single-phase bridge inverters and three-phase bridge inverters.

9.2 CENTRE-TAPPED SOURCE INVERTER

Figure 9.1a, depicted on the next page, shows the circuit diagram of a centre-tapped source inverter††. The *RL* load bridges the tapping of two sources and two switches. The switches may be BJTs, MOSFETs, or IGBTs, all unidirectional switches that can be switched on and off at any instant. Since a load may be inductive, diodes are connected across the switches to provide a path for current if a switch is turned off.

† The basics of inverter circuits are described in Section 2.4 of PESS, from which a number of examples, exercises and problems are taken.

†† Otherwise known as a half-bridge inverter.

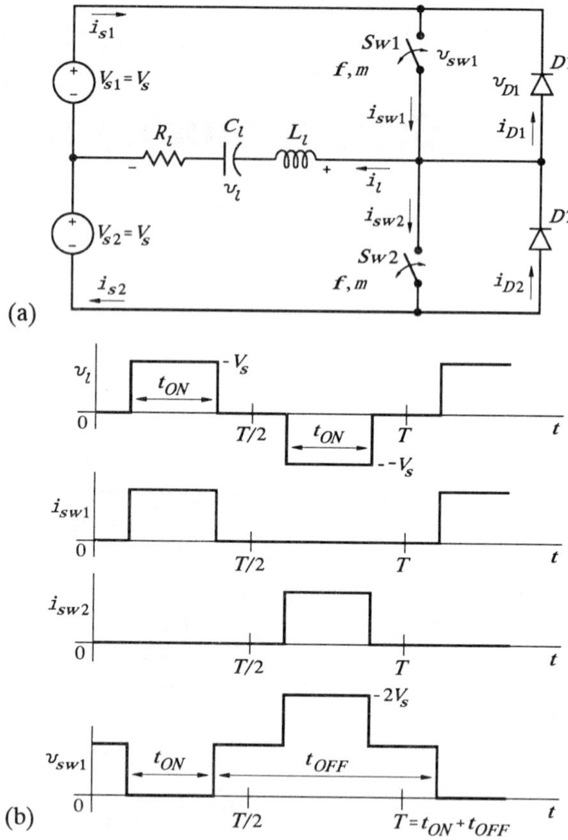

Fig. 9.1 A centre-tapped source inverter.
(a) Circuit diagram, (b) waveforms for a resistive load.

The operation of the inverter is accomplished by alternately turning on and off the two switches *Sw*1 and *Sw*2. Switch *Sw*1 on provides the voltage V_s across the load. Switch *Sw*2 on changes the load voltage to $-V_s$. This switching sequence results in the alternating current and voltage waveforms shown in Fig. 9.1b for a resistive load. In many inverters it is required to have adjustable voltage and adjustable frequency. Adjustable voltage is achieved by changing the on-time t_{ON} of the switches in the cycle. Adjustable frequency is controlled by the switching-cycle period *T*.

The inverter can be considered to be two choppers operating out of phase. The duty cycle is $m = t_{ON}/T$, but $m \leq 0.5$. The ideal rms value of the load voltage v_l is $V_{l\mathrm{rms}} = \sqrt{2m}\, V_s$.

EXAMPLE 9.2.1 (PESS EXAMPLE 2.7)

A centre-tapped source inverter is depicted in Fig. 9.1a. The circuit specifications are as follows.

$V_{s1} = V_{s2} = 200\mathrm{V}$, $L_l = 0$, $R_l = 2\Omega$, no diodes, *Sw*1 and *Sw*2 are IGBTs, $m = 0.4$, $f = 1/T = 400\mathrm{Hz}$. Single-pulse output per half cycle. Gate voltage 15V.

From a PSpice simulation plot traces of i_{s1}, i_{s2} and v_l. Determine **(a)** the average power absorbed by the load, **(b)** the total harmonic distortion THD of the output-voltage waveform, **(c)** the harmonic factor HF of the third harmonic of the output voltage and **(d)** the average current delivered by each source.

Solution

Let us choose an IGBT to switch at the 400-V, 100-A, 400-Hz level. The quasi-ideal model of the IGBT is described in Section 4.7. The drivers, that enable the switches to turn on and off at a frequency of 400Hz with a duty cycle $m = 0.4$, can be taken from Section 5.4.

There are four steps in the solution.

STEP 1 From the circuit diagram in Fig. 9.1a and from the given data, the PSpice configuration of the inverter circuit is drawn in Fig. 9.2.

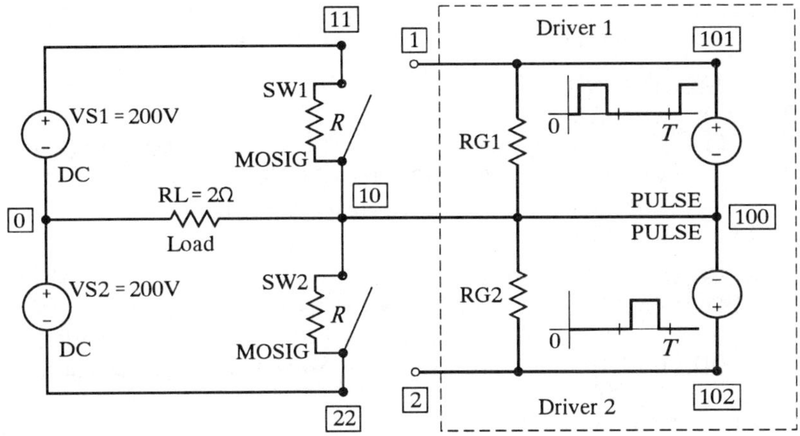

Fig. 9.2 PSpice configuration of the inverter and drivers.

STEP 2	From the PSpice configuration, from the description of an IGBT in Section 4.7 and from the description of the driver in EXAMPLE 5.4.1, the circuit file is written in EX9_2_1.CIR.

» » » » » EX9_2_1.CIR »

```
CENTRE-TAPPED SOURCE INVERTER
* To determine the waveforms, the harmonics and the output power.

* OPTIONS, a general package for numerical stability.  See Section 7.5.1.
. INC    OPTIONS
* The OPTIONS file must be in the PSpice subdirectory of the circuit file.

* PARAMETERS
. PARAM     VS=200V ;                                Value of source voltage.
. PARAM     RLOAD=2ohms
. PARAM     FREQ=400Hz        PERIOD={1/FREQ};  Output frequency.
. PARAM     DUTY=0.4;                                Duty cycle m.
. PARAM     TR=10ns  TF={TR};              Rise and fall times of gate pulse.
. PARAM     TON={DUTY*PERIOD-2*TR};        Pulsewidth of gate voltage.
. PARAM     DEL1={PERIOD/4-TON/2-TR};      Delay for driver 1 pulse.
. PARAM     DEL2={3*PERIOD/4-TON/2-TR};    Delay for driver 2 pulse.
. PARAM     VP=15V;                        Magnitude of gate voltage.

* SOURCES
VS1    11    0     DC    {VS}
VS2    0     22    DC    {VS}

* LOAD
RL      10    0     {RLOAD}

* SUBCIRCUIT for IGBT model.
. SUBCKT    IG_IDEAL  11    10  1
SW     11     10    1     10    MOSIG;      Voltage-controlled switch (IGBT).
. MODEL    MOSIG    VSWITCH(RON=1E-3 ROFF=1E7
+                                VON=10 VOFF=1E-3)
. ENDS     IG_IDEAL

* SUBCIRCUIT for IGBT driver.
. SUBCKT    IG_DRV   101    100    PARAMS: DELAY=0
VG     101  100  PULSE(0 {VP} {DELAY} {TR} {TF} {TON} {PERIOD})
RG     101  100  1E6;                       Not really necessary for simulation.
. ENDS     IG_DRV
* CALL subcircuits into main circuit.
XSW1 11    10    1      IG_IDEAL;     Switch Sw1.  Drain(11), S(10), G(1).
XSW2 22    10    2      IG_IDEAL;     Switch Sw2.
```

```
Xdriver1   1   10   IG_DRV  PARAMS:   DELAY={DEL1};   Driver 1.
Xdriver2   2   10   IG_DRV  PARAMS:   DELAY={DEL2};   Driver 2.

* ANALYSIS
.FOUR    400Hz    15       v(10);    Fourier analysis of load voltage.
.TRAN    20us     2.5ms;             One cycle.
.PROBE   i(VS1),  i(VS2),  v(10);    Source currents and load voltage.
.END
```

» »

STEP 3 A PSpice simulation can be run with the circuit file EX9_2_1.CIR. The results of the simulation are stored in EX9_2_1.OUT and EX9_2_1.DAT.

STEP 4 Using PROBE plot the traces of the source currents i_{s1} and i_{s2} and the load voltage v_l. See Fig. 9.3.

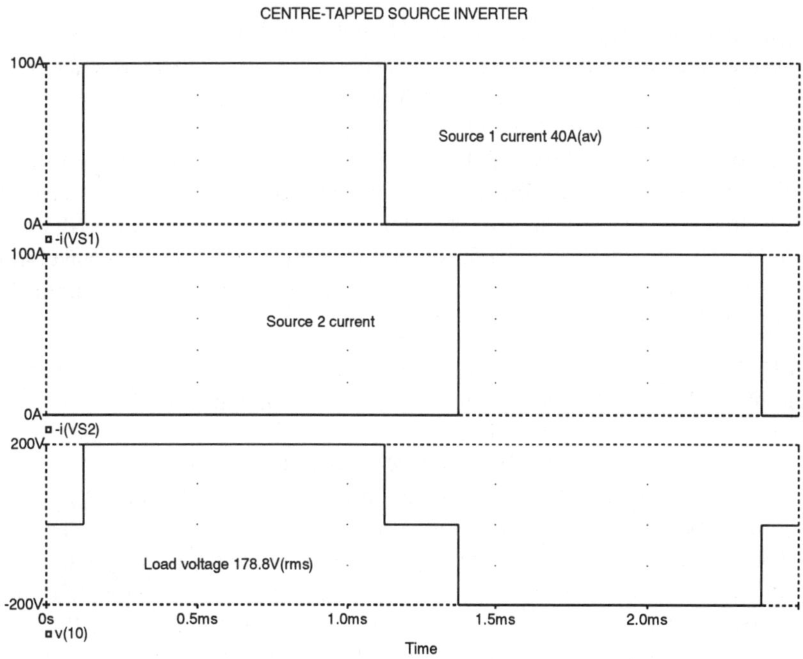

Fig. 9.3

Part (a) of Solution. From PROBE, the periodic rms value of the load voltage is $V_{l\mathrm{rms}} = 178.8\mathrm{V}$ (theoretically 178.9V). The average power absorbed by the load is $P = V_{l\mathrm{rms}}^2/R_l = 178.8^2/2 \approx 16\mathrm{kW}$.

Part (b) of Solution. From the output file EX9_2_1.OUT, the total harmonic distortion is THD = 27.19% (theoretically 32%).

Part (c) of Solution. For the third harmonic, $\mathrm{HF} = \hat{V}_{l3}/\hat{V}_{l1}$.

From the output file HF = 49.88/242.1 = 0.206 (theoretically 0.21).

Part (d) of Solution. From PROBE, the periodic average of the input current i_{s1} is $I_{s1av} \approx 40\mathrm{A}$. The total average power input is

$P = V_{s1}I_{s1av} + V_{s2}I_{s2av} = 2V_{s1}I_{s1av} \approx 2 \times 200 \times 40 = 16\mathrm{kW}$. Efficiency is 100%.

END OF EXAMPLE 9.2.1

Drill Exercise D9.2.1

Consider the circuit diagram of the converter in Fig. 9.1a. This is sometimes called a half-bridge inverter. The circuit has the following specifications.

$V_s = 200\mathrm{V}$, $L_l = 0$, $C_l = 0$, $R_l = 2\Omega$, IGBT switches,

$m = 0.3$, $f = 400\mathrm{Hz}$. Single-pulse output per half cycle. Gate voltage 15V.

Follow the pattern of EXAMPLE 9.2.1 to do a PSpice simulation. Plot traces of i_{s1}, i_{s2} and v_{sw1}. Determine **(a)** the total average power delivered by the two sources, **(b)** the total harmonic distortion THD of the output-current waveform and **(c)** the harmonic factor HF of the third harmonic of the output current i_l.
(Ans: **(a)** 11.99kW, **(b)** 33.11%, **(c)** 0.127.)

Drill Exercise D9.2.2 (PESS PROBLEM 2.25)

The circuit diagram of a centre-tapped source inverter is shown in Fig. 9.1a. The circuit has the following specifications.

$V_s = 300\mathrm{V}$, $L_l = 0$, $C_l = 0$, $R_l = 18\Omega$, IGBT switches,

$m = 0.3$, $f = 50\mathrm{Hz}$. Single-pulse output per half cycle. Gate voltage 15V.

Do a PSpice simulation and plot traces of the switch current i_{sw1} and the load voltage v_l. Determine **(a)** the average power delivered to the load, **(b)** the total harmonic distortion THD of the load voltage, **(c)** the harmonic factor HF of the fifth harmonic of the load voltage and **(d)** the current rating of the IGBT.
(Ans: **(a)** 3kW (theoretically 3kW), **(b)** 32.97% (theoretically 36%),
(c) 0.247 (theoretically 0.25), **(d)** >8.3A (theoretically >8.3A).)

9.2.1 UNIFORM MULTIPLE-PULSE INVERTER

We will examine multiple-pulse switching from a simulation point of view. We will not follow the theory to reduce or eliminate particular harmonics. Here, an arbitrary constraint is that, no matter how many pulses per half cycle there are, the half-wave average of the output voltage of the inverter remains the same. That means, for rectangular waves, the duty cycle is the same, the rms value of the voltage is the same and the average power absorbed by the load is the same for any number of pulses, if the load is resistive.

In Chapter 5, Section 5.4.3, a multiple-pulse driver is described. See Fig. 5.21. We can adapt and refine this driver (subcircuit INV_PWM_DRV) to suit a centre-tapped source inverter like that shown in Fig. 9.1. See INV_MLT_PLS in DRIVER.LIB. Use this in Drill Exercise D9.2.3.

Figure 9.4 depicts the waveform of the inverter output voltage with three pulses per half cycle. This waveform can be achieved by using similar waveforms for the gate signals to the switches of the inverter. Generalization for any number of pulses per half cycle is possible.

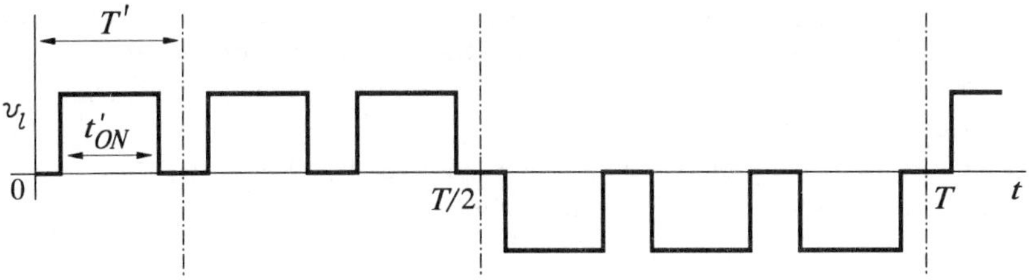

$$\text{Duty cycle } m = \frac{t_{ON}}{T} = \frac{3t'_{ON}}{6T'} = \frac{t'_{ON}}{2T'}$$

Fig. 9.4 Uniform multiple-pulse inverter output.

Drill Exercise D9.2.3

Consider the centre-tapped source inverter whose circuit diagram is shown in Fig. 9.1a. The circuit specifications are as follows.

$V_s = 200\text{V}$, $L_l = 0$, $R_l = 2\Omega$, no diodes, *Sw*1 and *Sw*2 are IGBTs, $m = 0.4$, $f = 400\text{Hz}$. Three-pulse output per half cycle. Gate voltage 15V.
From a PSpice simulation plot traces of the gate voltages, i_{s1} and v_l.
Determine **(a)** the rms value of the load voltage, **(b)** the average power absorbed by the load, **(c)** the total harmonic distortion THD of the output-voltage waveform

(d) the harmonic factor of the third harmonic of the load voltage, **(e)** the average value of the source current and **(f)** the total average power delivered by the sources. Use a configuration similar to EXAMPLE 9.2.1. (Ans: **(a)** 178.8V, **(b)** 15.98kW, **(c)** 62.6%, **(d)** 0.38, **(e)** 39.98A, **(f)** 15.99kW.)

9.2.2 SPWM INVERTER

The voltage magnitude of an inverter output can be controlled by pulse-width modulation (PWM). The shape of the inverter output waveform (ideally a sinewave) can be controlled by modulating the PWM. One type of modulation is called sinewave pulse-width modulation (SPWM). The aim of creating this modulating waveform is to reduce or eliminate certain of the harmonic content. The gating descriptions are found in Section 5.4.4, EXAMPLE 5.4.2 and in DRIVER .LIB (subcircuit named MOD_INV_PWM_DRV).

A common technique to obtain SPWM control of an inverter output waveform is illustrated in Fig. 9.5. The general principle of operation is to compare a reference sinewave signal v_R with a triangular carrier signal v_C of higher frequency. At any instant, if $v_R > v_C$, a comparator will have a gating signal at its output. The output of the inverter reflects this gating signal. As the number of carrier waves per period of the sinewave reference signal increases, so the harmonic content is transferred to higher frequencies. The amplitude of the reference voltage with respect to the carrier peak voltage alters the overall pulse width. In turn, this controls the rms value of the inverter output voltage.

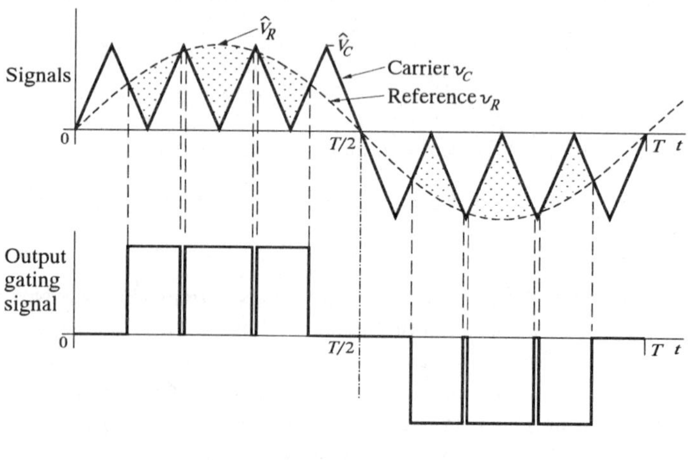

Fig. 9.5 SPWM gating.

Drill Exercise D9.2.4

Consider the centre-tapped source inverter whose circuit diagram is shown in Fig. 9.2.1. The circuit has the following specifications.

$V_s = 100$V, $L_l = 0$, $C_l = 0$, $R_l = 10\Omega$, *Sw*1 and *Sw*2 are IGBTs, control by SPWM, $f = 50$Hz. Gate voltage 15V.

The SPWM gate circuit is to be implemented according to the circuit diagram in Fig. 5.23. The modulation index ($M = \hat{V}_R/\hat{V}_C$) controls the output voltage magnitude. Let $M = 0.95$ and let the frequency of the sawtooth carrier wave be 400Hz. Do a PSpice simulation by combining the power circuit from EXAMPLE 9.2.1 and the driver circuit named MOD_INV_PWM_DRV from DRIVER.LIB. Plot traces of the gate voltages and the load voltage. Determine **(a)** the rms value of the load voltage, **(b)** the average power absorbed by the load, **(c)** the total harmonic distortion THD of the load voltage and **(d)** the fundamental, second, third, fifth and seventh rms harmonic values of the load voltage. (Note: An *RC* circuit must be added to the gate circuit to delay the gate voltage rise for convergence.) (Ans: **(a)** 74.0V, **(b)** 547.6W, **(c)** 43.55%, **(d)** 64.67V, 0, 7.88V, 12.78V, 13.49V.)

9.2.3 SPWM INVERTER DIGITAL DRIVER

What was done in Section 9.2.2 to produce an SPWM inverter with an analogue driver can be done equally well with a digital 555 driver. See the WEBsite for a detailed extension of this Section 9.2.3 with EXAMPLE W9.2.1.

9.2.4 CURRENT-REGULATED INVERTER

There is a PWM technique that has various names such as delta modulation, tolerance-band current control, adaptive current control and hysteresis-type current control. This technique aims to control the inverter output current very closely to a sinewave. Consider the half-bridge inverter illustrated in Fig. 9.1. It is desired to have a load current similar to a sinewave reference current, shown in Fig. 9.6a on the next page. The actual current is allowed a tolerance band, as shown by the broken lines either side of the ideal sinewave. The two small preset error limits are called hysteresis limits.

With Fig. 9.6 as reference, at some time t the actual current is given by the point P. The current is below the reference value, the error ($i_{ref} - i_l$) is positive, so it is required to apply a gate signal to switch *Sw*1 (Fig. 9.1a). With switch *Sw*1 on, the load current will tend to increase, and the positive value of the current error tends to reduce to zero. The switch *Sw*1 is kept on by the gate signal, the current error becomes negative until it reaches the preset limit at the point N. At this point N we must turn off switch *Sw*1 (remove its gate signal) and turn on switch *Sw*2 (apply its gate signal). Following these actions, the load current will tend to reduce so that the current error becomes less negative. Switch *Sw*2 is allowed to remain on until the reduction of load current reaches the predefined positive error $(i_{ref} - i_l)$ at the point P′. Here, the cycle repeats; switch *Sw*2 is turned off and switch *Sw*1 is turned on again. This strategy provides a load current which is a sinewave with a small ripple.

The control strategy is to compare the actual load current with a sinewave reference signal. The current error signal $(i_{ref} - i_l)$ is almost triangular. We have to convert this triangular wave to square waves that can be used as gate signals to turn on the inverter switches at the triangle "peaks" and "troughs". The gate signal must be maintained while the triangular wave traverses from peak to trough and *vice versa*. This can be accomplished with a bistable device such as a Schmitt trigger, since a Schmitt trigger can be modelled in analogue form in PSpice by a comparator with hysteresis (using positive feedback at the noninverting input).

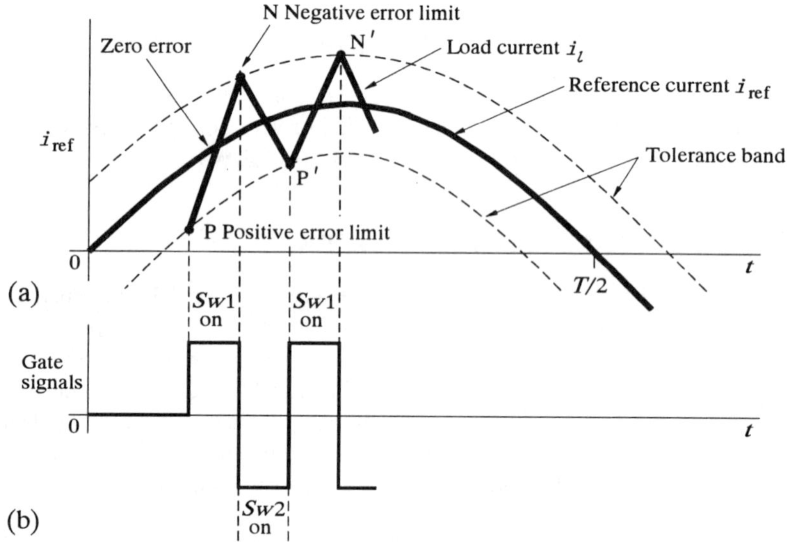

Fig. 9.6 A strategy for an inverter sinewave.
(a) Reference signal and load current, (b) gate signals to implement strategy.

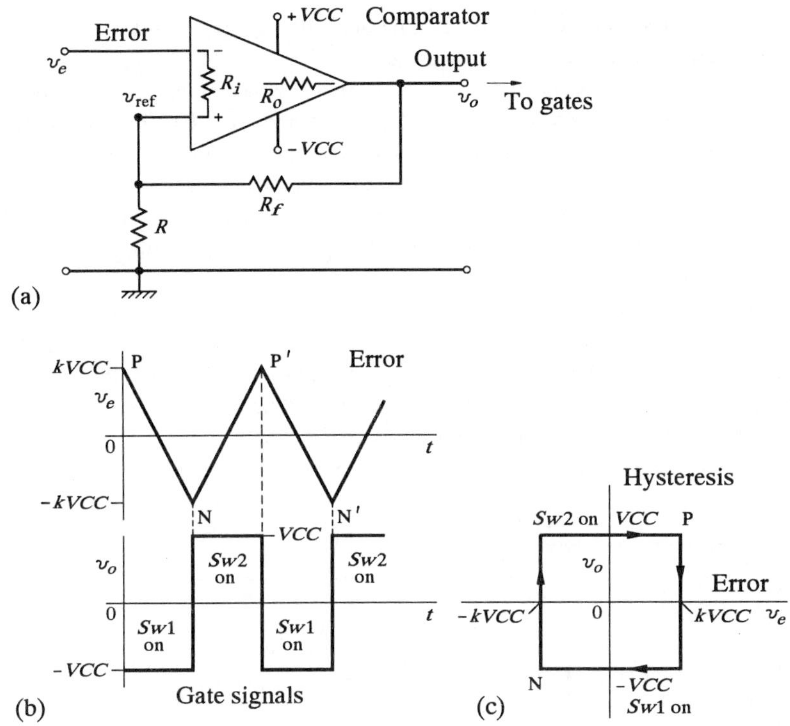

Fig. 9.7 PWM with preset tolerance.
(a) Comparator, (b) input and output signals, (c) hysteresis effect.

Figure 9.7a shows the circuit diagram of a comparator that is provided with hysteresis. The input signal v_e (proportional to the current error signal) is applied to the inverting input. At the noninverting input the reference voltage v_{ref} is given by the voltage divider principle to be

$$v_{ref} \approx v_o R/(R+R_f) = kv_o . \tag{9.2.1}$$

The output v_o is ideally one of the saturation levels $\pm VCC$. Thus, v_{ref} has one of two levels $\pm kVCC$.

The comparator works as follows. If the error signal v_e is greater than the reference signal v_{ref}, the output v_o goes negative. In this application the negative output $-VCC$ is to be a gate signal that will lead to a change (less positive) of the error signal v_e. When the error signal v_e is negative and less than the reference signal v_{ref} (at present $-kVCC$) the comparator output shifts states to $v_o = +VCC$ and the reference signal becomes $v_{ref} = kVCC$. The reference signal remains at $+kVCC$ until the error signal v_e is greater than v_{ref}, at which point the cycle repeats. While the comparator output is $v_o = +VCC$, the output is to be a gate signal that will lead to a change (less negative) in the error signal. Figure 9.7b shows how the error signal as a function of time is transformed by the comparator to a square-wave output. The negative output voltage v_o becomes the gate signal to keep switch $Sw1$ of the inverter in the on-state. The positive output voltage v_o is used to maintain switch $Sw2$ on.

If we use the results in Fig. 9.7b, showing how we have created a square wave from error limits, we can plot the comparator output voltage v_o against the comparator input voltage v_e. See Fig. 9.7c. This demonstrates the required hysteresis effects of the comparator if positive feedback is employed. The error limits are set by the value chosen for k $(k = R(R + R_f))$.

The control circuit for the inverter is depicted in Fig. 9.8.

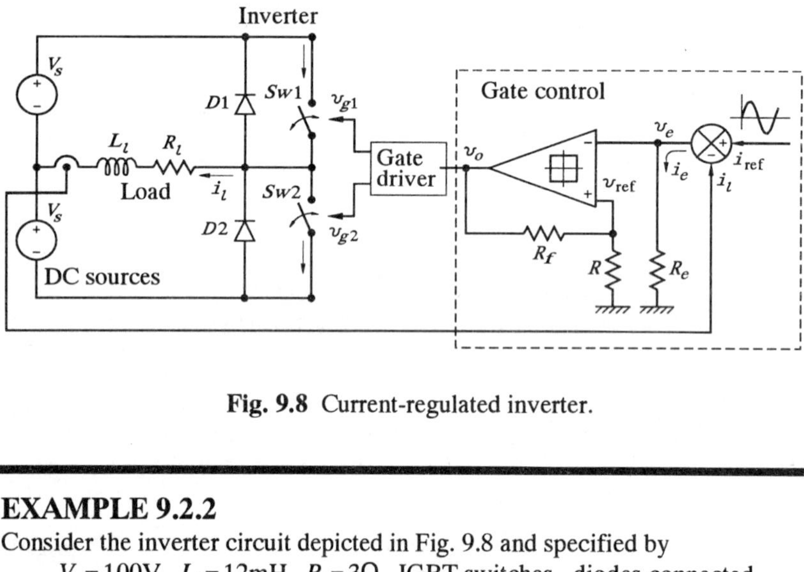

Fig. 9.8 Current-regulated inverter.

EXAMPLE 9.2.2

Consider the inverter circuit depicted in Fig. 9.8 and specified by

$V_s = 100$V, $L_l = 12$mH, $R_l = 3\Omega$, IGBT switches, diodes connected, inverter frequency 50Hz. Gate voltage 20V. Current-regulated PWM control to have a load current $i_l = 12\sin 314t$ A with a tolerance of about $\pm 10\%$.

Do a PSpice simulation and plot traces of a source current, the load current and reference current, and the load voltage. Determine **(a)** the average power delivered to the load, **(b)** the total and fundamental rms values of the load current and **(c)** the total harmonic distortion THD of the load-current waveform.

Solution

The solution can be achieved in four steps by implementing the description of the current-regulated inverter (in this subsection).

STEP 1 From the circuit diagram in Fig. 9.8 we can draw a PSpice configuration. This is shown in Fig. 9.9. *RC* elements are added to delay the transition of the gate voltage for the purpose of aiding convergence.

STEP 2 The load reference current amplitude is $\hat{I}_{\text{ref}} = 12\text{A}$. At 10% error, the tolerance band is $\pm 0.1 \times 12 = \pm 1.2\text{A}$ error. That is, $\hat{I}_e = 1.2\text{A}$.

Let $R_f = 9\text{k}\Omega$ and $R = 1\text{k}\Omega$. The hysteresis band is represented by

$$\hat{V}_e = kVCC = (R/(R+R_f))\,VCC = (1/(1+9)) \times 20 = 2.0\text{V}.$$

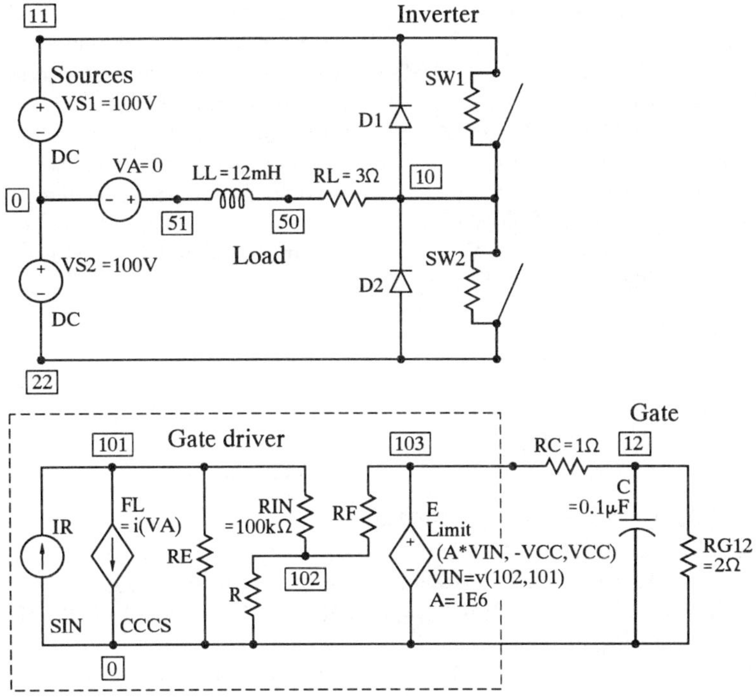

Fig. 9.9 PSpice configuration.

Hence, $R_e = \hat{V}_e/\hat{I}_e = 2/1.2 = 1.667\Omega$.

This indicates that an error of 1.2A would mean a voltage $\hat{V}_e = 2V$ which causes the output of the comparator (E) to change polarity. This causes one switch of the inverter to turn off and the complementary switch to turn on.

»»»»» EX9_2_2.CIR »»

```
CURRENT-REGULATED INVERTER (WITH HYSTERESIS)
* To determine the efficacy of tolerance-band PWM for RL load.
.INC   OPTIONS;              Include OPTIONS file for convergence stability.
* PARAMETERS
.PARAM   VS=100V;                        Source voltage.
.PARAM   FREQ=50Hz;                      Inverter output frequency.
.PARAM   RLOAD=3ohms;                    Load resistance.
.PARAM   LLOAD=12mH;                     Load inductance.
.PARAM   IRMAX=12A;                      Current reference amplitude.
.PARAM   VCC=20V;              Gate voltage magnitude, comparator source.
.PARAM   ERR=0.1;              Per unit maximum error tolerance.
.PARAM   IEMAX={ERR*IRMAX};              Maximum current error magnitude.
.PARAM   RNON=1kohm;                     Comparator positive feedback.
.PARAM   RFNON=9kohm;                    Comparator positive feedback.
.PARAM   KDIV={RNON/(RNON+RFNON)};               Voltage divider ratio k.
.PARAM   VEMAX={KDIV*VCC};               Comparator switching value.
.PARAM   REV={VEMAX/IEMAX}; Current error to voltage error conversion.
* SOURCES and LOAD
VS1   11  0   DC  {VS}
VS2   0   22  DC  {VS}
RL    10  50  {RLOAD}
LL    50  51  {LLOAD}
VA    51  0   0;                         Load-current sensor.
* SUBCIRCUITS for IGBT and DIODE.
.SUBCKT IG_IDEAL  11  10  12  100; Drain, source and gate pair nodes.
SW    11  10  12   100   MOSIG; Voltage-controlled switch.
.MODEL MOSIG VSWITCH(RON=1E-3 ROFF=1E6 VON=10 VOFF=1E-3)
.ENDS   IG_IDEAL
.SUBCKT D_IDEAL  10  11;          Anode, cathode nodes of diode.
D 10  11  DIO
.MODEL DIO   D(RS=1m  CJO=0.1pF  N=0.001)
.ENDS  D_IDEAL
```

```
* CALLS for the switches Sw1, Sw2 and diodes D1, D2.
XSW1   11   10   0    12   IG_IDEAL
XSW2   10   22   12   0    IG_IDEAL
XD1    10   11   D_IDEAL
XD2    22   10   D_IDEAL

* GATE DRIVER
IR     0    101  SIN(0 {IRMAX} {FREQ});        Reference current.
FL     101  0    VA  1;                        CCCS, load current.
RE     101  0    {REV};                        Path for error current.
RIN    101  102  100kohms;                     Comparator input resistance.
R      102  0    {RNON};                       Positive feedback resistor divider.
RF     103  102  {RFNON};                      Positive feedback resistor divider.
E      103  0    VALUE={LIMIT(1E6*v(102,101), -VCC,VCC)}
* E is the comparator with hysteresis.
RC     103  12   1
C      12   0    0.1uF    IC=-16.7V;           Delays gate voltage transition.
RG     12   0    2;                            Nominal gate resistance.

* ANALYSIS
.TRAN      40us    40ms    20ms    10us    UIC;  TMAX=10μs for slew rate.
.FOUR      50Hz    15      i(RL)   v(10)
.PROBE     i(IR),  i(RL),  i(VS1), v(10)
.END
```

» »

STEP 3 A PSpice simulation can be run with the circuit file EX9_2_2.CIR.

STEP 4 PROBE can be used to plot traces of a source current, the load current and the reference current, and the load voltage. See Fig. 9.10 on the next page.

Part (a) of Solution. From PROBE, the average current of each source is 1.09A and the rms value of the load current is 8.5A. The average power P delivered to the load is

$$P \approx 2V_s I_{sav} = I_{l\,\mathrm{rms}}^2 R_l = 8.5^2 \times 3 = 216.8\,\mathrm{W}.$$

Part (b) of Solution. From PROBE, the rms value of the load current is $I_{l\mathrm{rms}} = 8.525\mathrm{A}$. From the results of the Fourier analysis, found in the output file EX9_2_2.OUT, the fundamental rms value of the load current is $I_{l1\mathrm{rms}} = 8.50\mathrm{A}$.

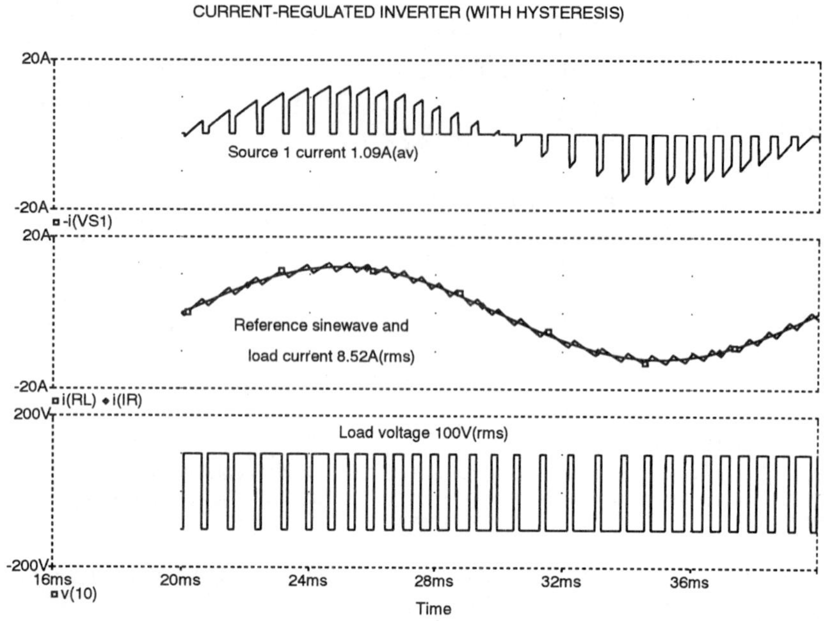

Fig. 9.10

Part (c) of Solution. From the results of the Fourier analysis, found in the output file, the total harmonic distortion THD of the load current is only 0.326%. However, the load voltage distortion is 13.3%.

END OF EXAMPLE 9.2.2

Drill Exercise D9.2.5

A current-regulated inverter is illustrated in Fig. 9.8 and is specified by

$V_s = 100\text{V}$, $L_l = 6\text{mH}$, $R_l = 3\Omega$, IGBT switches, diodes connected, inverter frequency $f = 50\text{Hz}$. Gate voltage 20V.

Use current-regulated PWM control to obtain a load current $i_l = 8\sin 314t\,\text{A}$ with a tolerance band of about $\pm 5\%$.

Follow the pattern of EXAMPLE 9.2.2 and do a PSpice simulation. Plot traces of a source current and the inverter load current. Determine **(a)** the average power delivered to the load, **(b)** the rms value of the load current and **(c)** the total harmonic distortion THD of the load-current waveform.

(Note the switching frequency.) (Ans: **(a)** 94.7W, **(b)** 5.62A, **(c)** 0.65%.)

There are Extra Drill Exercises for this Section on the WEBsite.

9.3 SINGLE-PHASE BRIDGE INVERTER

The circuit diagram of a single-phase, bridge inverter is shown in Fig. 9.11. Pairs of switches *Sw*1, *Sw*2 and *Sw*3, *Sw*4 are switched in sequence in order to have alternating voltage across the load. The diodes provide a path for the inductive current when the switches are turned off. The switches can be any of the unidirectional devices such as MOSFETs, BJTs, IGBTs, thyristors, GTOs or MCTs.

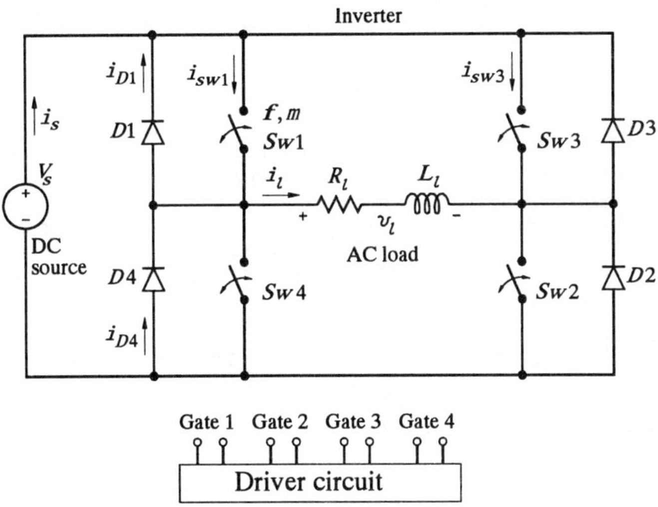

Fig. 9.11 Single-phase, bridge inverter.

EXAMPLE 9.3.1

A single-phase, bridge inverter is illustrated in Fig. 9.11, and specified by $V_s = 100\text{V}$, $L_l = 0$, $R_l = 5\Omega$, IGBT switches, no diodes. Gate voltage 15V. $m = 0.4$ ($m = t_{ON}/T$), $f = 1/T = 50\text{Hz}$. Single-pulse output per half cycle ($T/2$). From a PSpice simulation plot i_s and v_l. Determine **(a)** the average power absorbed by the load and **(b)** the total harmonic distortion of the output-voltage waveform. **(c)** Plot the frequency spectrum of the output-voltage waveform.

Solution

We will follow the pattern of the solution in EXAMPLE 9.2.1. The gate drivers for the four switches are taken from Section 5.4. The inverter is treated as two choppers (*Sw*1 and *Sw*2 switched simultaneously for one chopper, and *Sw*3 and *Sw*4 switched simultaneously for the other). The load is to have one pulse per half cycle of the required frequency and each pulse to the load has an alternating polarity. There are four steps in the solution.

STEP 1 From the circuit diagram in Fig. 9.11, from the given specifica-tions, and from EXAMPLE 9.2.1, the PSpice configuration of the single-phase, bridge inverter is shown in Fig. 9.12.

STEP 2 From the PSpice configuration and from the circuit file EX9_2_1.CIR, the circuit file for this example is written and named EX9_3_1.CIR.

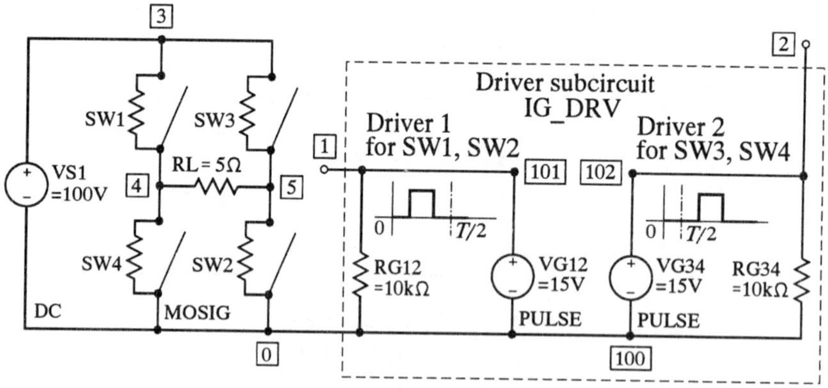

Fig. 9.12 PSpice configuration for a single-phase, bridge inverter.

```
»»»»» EX9_3_1.CIR »»»»»»»»»»»»»»»»»»»»»»»»»»»»»»»»»»»»»»
SINGLE-PHASE BRIDGE INVERTER (SINGLE PULSE)
* To determine average power and harmonic content.
. INC    OPTIONS;                The file OPTIONS is a .OPTIONS package.

* PARAMETERS
. PARAM   VS=100V;                          The power source voltage.
. PARAM   RLOAD=5ohms;                      The load resistance.
. PARAM   FREQ=50Hz   PERIOD={1/FREQ};  Output frequency.
. PARAM   DUTY=0.4;                         Duty cycle m of output voltage.
```

```
. PARAM   TR=20ns   TF={TR};                 Rise and fall times of gate pulse.
. PARAM   TON={DUTY*PERIOD-2*TR};            Pulsewidth of gate signal.
. PARAM   DEL1={PERIOD/4-TON/2-TR};          Delay for driver 1 pulse.
. PARAM   DEL2={3*PERIOD/4-TON/2-TR};        Delay for driver 2 pulse.
. PARAM   VP=15V;                            Magnitude of gate voltage.

* SOURCE
VS1   3   0   DC  {VS}

* LOAD
RL    4   5   {RLOAD}

* SUBCIRCUIT for IGBT model.
. SUBCKT  IG_IDEAL  13  14  15  16;   Drain(13).Source(14).Gate(15,16).
SW    13  14  15  16  MOSIG;          Voltage-controlledswitch(IGBT).
.MODEL MOSIG VSWITCH(RON=1E-3 ROFF=1E6 VON=10 VOFF=1E-3)
. ENDS    IG_IDEAL
* CALLS for subcircuit into main circuit.
XSW1 3   4   1   0   IG_IDEAL;           Switch Sw1.
XSW2 5   0   1   0   IG_IDEAL;           Switch Sw2.
XSW3 3   5   2   0   IG_IDEAL;           Switch Sw3.
XSW4 4   0   2   0   IG_IDEAL;           Switch Sw4.

* SUBCIRCUIT for IGBT driver.
. SUBCKT  IG_DRV 101 100  PARAMS: DELAY=0
VG   101   100   PULSE(0 {VP} {DELAY} {TR} {TF} {TON} {PERIOD})
RG   101   100   1E4
. ENDS    IG_DRV
* CALLS for subcircuit into main circuit.
Xdriver1  1 0  IG_DRV  PARAMS:  DELAY={DEL1};  Gate for Sw1, Sw2.
Xdriver2  2 0  IG_DRV  PARAMS:  DELAY={DEL2};  Gate for Sw3, Sw4.

* ANALYSIS
. FOUR    50Hz  15    v(4,5);        Fourier analysis of load voltage.
. TRAN    40us  20ms  UIC;           One cycle.
. PROBE   i(VS1),     v(4,5);        Source current and load voltage.
. END
```

» »

STEP 3 A PSpice simulation can be run with the circuit file EX9_3_1.CIR.

STEP 4 Plot the load voltage v(4,5) and the source current -i(VS1). Create a frequency spectrum of the load-voltage waveform. See Figs. 9.13 and 9.14 on the following page.

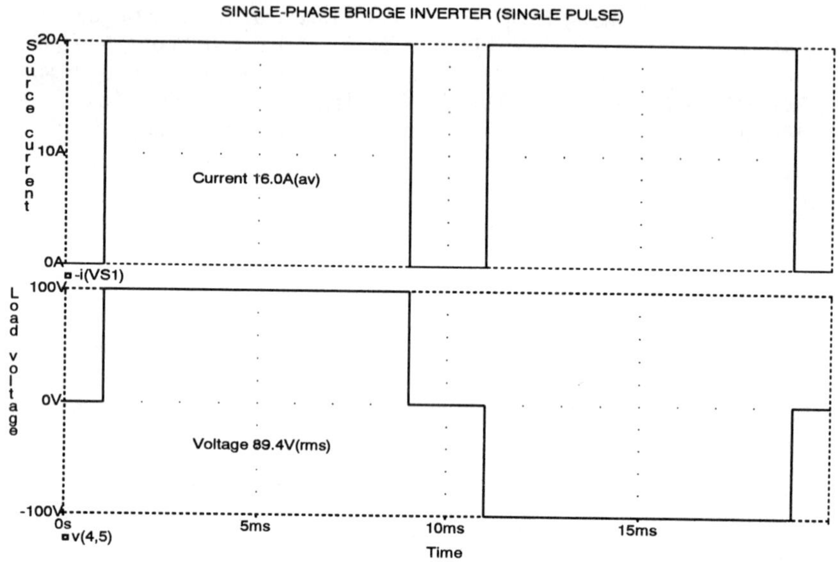

Fig. 9.13

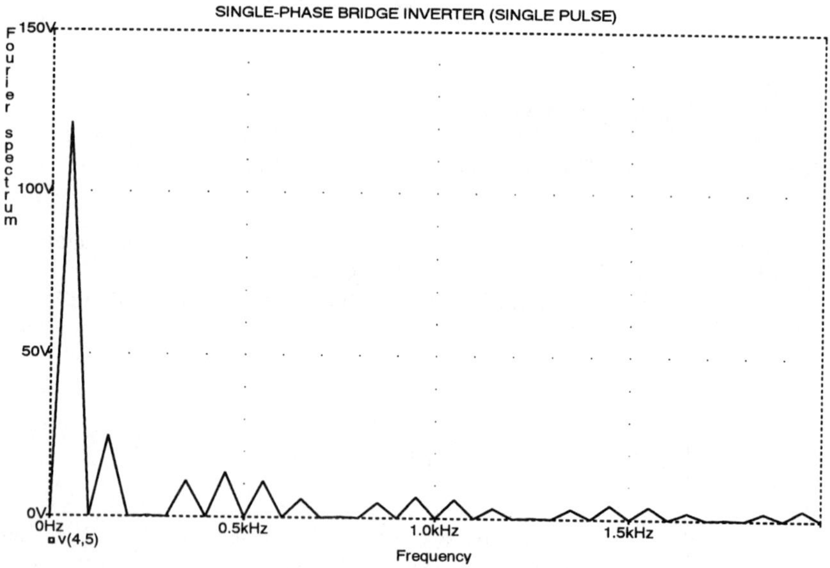

Fig. 9.14

Part (a) of Solution. From the PROBE traces, the rms value of the load voltage is $V_{lrms} \approx 89.4$V and the average value of the source current is $I_{sav} = 16.0$A. The average power absorbed by the load is

$$P = V_{lrms}^2/R_l = 89.4^2/5 \approx 1.6\text{kW}.$$

Check. The average power delivered by the source is

$$P = V_s I_{sav} = 100 \times 16 = 1.6\text{kW}.$$

Part (b) of Solution. From the output file EX9_3_1.OUT, the total harmonic distortion of the output-voltage waveform is THD = 27.05%.

END OF EXAMPLE 9.3.1

Drill Exercise D9.3.1

A single-phase, bridge inverter is depicted in Fig. 9.11, and specified by $V_s = 200$V, $L_l = 0$, $R_l = 2\Omega$, IGBT switches, no diodes, $m = 0.4$, $f = 400$Hz. Single-pulse output per half cycle. Gate voltage 15V. From a PSpice simulation, using EXAMPLE 9.3.1 as a pattern, plot traces of the source current and the load voltage. Determine **(a)** the rms value of the load voltage, **(b)** the total harmonic distortion THD of the output-voltage waveform, **(c)** the average value of the source current and **(d)** the average power absorbed by the load and delivered by the supply.
(Ans: **(a)** 178.7V, **(b)** 27.19%, **(c)** 79.93A, **(d)** 15.97kW, 15.98kW.)

Drill Exercise D9.3.2

A single-phase, bridge inverter, illustrated in Fig. 9.11, is specified by $V_s = 100$V, $L_l = 0$, $R_l = 1\Omega$, IGBT switches, $m = 0.4$, $f = 100$Hz. Single-pulse output per half cycle. Gate voltage 15V.
The gate-driver output signals V_g are depicted in Fig. 9.15. Do a PSpice simulation and plot traces of the source current, the gate and load voltages. Determine **(a)** the average power delivered by the source and **(b)** the total harmonic distortion of the load-voltage waveform. (Ans: **(a)** 7.97kW, **(b)** 26.98%.)

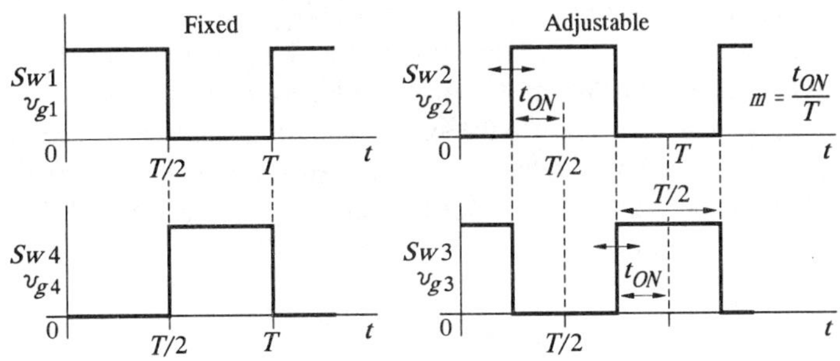

Fig. 9.15

9.3.1 UNIFORM MULTIPLE-PULSE INVERTER

Multiple pulses in each half cycle of an inverter output give control over both the harmonic content and the rms value of the voltage across a load. See Fig. 9.4. Both the duty cycle and the number of pulses are variables and both are controlled by the gate signals to the inverter switches. We can adapt and expand the gate drive circuits that were described in Chapter 5 and in Section 9.2.1 for the single-phase, bridge inverter, depicted in Fig. 9.11. View the subcircuit MPLS_TRI_INV in DRIVER.LIB.

An EXAMPLE W9.3.1 appears on the WEBsite in Section 9.3.1.

Drill Exercise D9.3.3

Consider the single-phase, bridge inverter, that is illustrated in Fig. 9.11. The circuit has the following specifications.

$V_s = 100\text{V}$, $L_l = 0$, $R_l = 5\Omega$, IGBT switches,

$m = 0.4$, $f = 50\text{Hz}$. Three-pulse output per half cycle. Gate voltage 15V.

Use the gate driver MPLS_TRI_INV, described in DRIVER.LIB, and do a PSpice simulation. Plot traces of the source current and the load current over an interval of two cycles for steady conditions. Determine **(a)** the steady-state initial conditions, **(b)** the average power absorbed by the load and **(c)** the total harmonic distortion THD of the load voltage.

(Ans: **(a)** 0, **(b)** 1.56kW, **(c)** 64.77%.)

EXAMPLE 9.3.2

This example is concerned with particular harmonic elimination. Consider the single-phase, bridge inverter, illustrated in Fig. 9.11, and specified by

$V_s = 100\text{V}$, $L_l = 0$, $R_l = 5\Omega$, IGBT switches, $f = 100\text{Hz}$.

Three, unequal pulses per half cycle at the inverter output as shown in Fig. 9.16. Here, $\alpha_1 = 17.8°$ and $\alpha_2 = 38.0°$. Gate voltage 15V.

Simulate a gate driver to satisfy the above specifications accounting for the symmetry of waveforms. Do a PSpice circuit simulation and plot traces of the dc source current and the inverter output voltage over two cycles. Determine **(a)** the average power delivered to the load, **(b)** the fundamental rms value of the output voltage, **(c)** the rms values of the third, fifth and seventh harmonic components of the output voltage and **(d)** the total harmonic distortion THD of the output-voltage waveform.

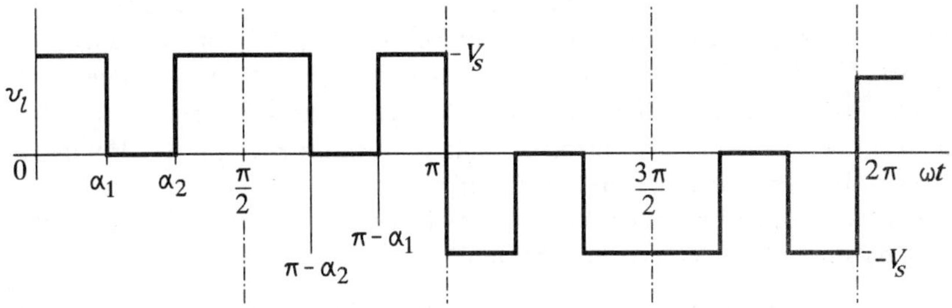

Fig. 9.16 Unipolar waveform for harmonic elimination.

Solution

A harmonic analysis† of the waveform in Fig. 9.16 provides the amplitude of the nth harmonic $\hat{V}_{ln}$ of the output voltage

$$\hat{V}_{ln} = \frac{4V_s}{n\pi}(1 - \cos n\alpha_1 + \cos n\alpha_2)\,. \tag{9.3.1}$$

If the third and fifth harmonic components are to be zero,

$$0 = 1 - \cos 3\alpha_1 - \cos 3\alpha_2 \tag{9.3.2}$$

$$\text{and } 0 = 1 - \cos 5\alpha_1 - \cos 5\alpha_2\,. \tag{9.3.3}$$

The values of α_1 and α_2 in these transcendental equations can be obtained from equation solvers like MATLAB™, MATHCAD™, MAPLE™, etc. The solution is $\alpha_1 = 17.8°$ and $\alpha_2 = 38°$, as required in this example.

There are five steps to achieve a solution.

STEP 1 The driver waveform we want has two notches per half cycle every cycle. In PSpice versions 5.4 and later there is a periodic PWL source that satisfies our requirement. A subset†† of the general PSpice form of the repetitive piecewise-linear voltage source is

```
V  N+  N-  PWL  REPEAT  FOREVER
+  (T1,V1,  T2,V2 ··· TN,VN ··· )  ENDREPEAT
```

The REPEAT FOREVER indicates that the piecewise-linear waveform described by corner points TN,VN (time, and magnitude) is to be repeated for as long as the analysis lasts. ENDREPEAT ends the description within the parentheses. Time TN must be literal, but VN can be a parameter.

† See Fisher "Power Electronics" p388.

†† MicroSim's Circuit Analysis Reference Manual, Analog Devices, Independent Voltage Source V, PWL.

The general form of PWL is versatile. For example, a waveform can be repeated for a number of cycles, then the waveform can be changed for another number of cycles. There can be REPEAT FOR <n> ··· ENDREPEAT nests. There can be scaling factors for time and voltage magnitude. Parameters VN can be altered for a sweep with a .STEP statement and data for the corner points TN,VN can be read by PSpice from a prepared ASCII file.
The subset of the PWL statement above will be used in the circuit file to simulate gate pulses with waveforms like that in Fig. 9.16.

STEP 2 From the given specifications, from the circuit diagram in Fig. 9.11 and from STEP 1 above, we can draw a PSpice configuration of the inverter system, as in Fig. 9.17.

STEP 3 From the configuration schematic in Fig. 9.17 we can write a circuit file named EX9_3_2.CIR. We must calculate the times for α_1 and α_2.

$f = 100\text{Hz}$, $T = 10\text{ms}$, $\alpha_1 = 17.8°$, $\alpha_2 = 38.0°$.

Hence, $t_1 = \alpha_1/\omega = \alpha°_1 T/360 = \alpha°_1 \times 27.7778\mu s = 0.49444\text{ms}$.

$t_2 = \alpha_2 \times 27.7778\mu s = 1.05555\text{ms}$.

$\omega t_3 = \pi - \alpha_2$ and $\omega t_4 = \pi - \alpha_1$. So, $t_3 = 3.94443\text{ms}$ and $t_4 = 4.50554\text{ms}$.

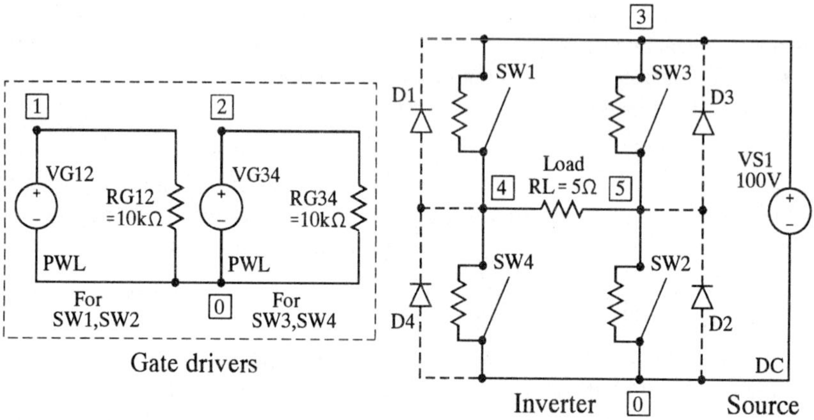

Fig. 9.17 Inverter configuration for nonuniform PWL.

```
»»»»» EX9_3_2.CIR »»»»»»»»»»»»»»»»»»»»»»»»»»»»»»»»»»»»»»»»»
HARMONIC CANCELLATION IN A SINGLE-PHASE INVERTER
. INC   OPTIONS;                An OPTIONS file to aid convergence.
* PARAMETERS
. PARAM    VS=100V;                          Source voltage.
. PARAM    RLOAD=5ohms;                      Load resistance.
. PARAM    VP=15V;                           Magnitude of gate voltage.
* SOURCE and LOAD
VS1    3    0    DC   {VS};                  Source voltage.
RL     4    5    {RLOAD};                    Load resistance.
RG12   1    0    1E4;                        Gate resistance of Sw1, Sw2.
RG34   2    0    1E4;                        Gate resistance of Sw3, Sw4.
* SUBCIRCUITS for IGBT and diode models.
. SUBCKT   IG_IDEAL   17   18   19   20; Drain(17). Source(18). Gate(19,20).
SW     17   18   19  20  MOSIG;   IGBT as a voltage-controlled switch.
. MODEL MOSIG VSWITCH(RON=1E-3 ROFF=1E6 VON=10 VOFF=1E-3)
. ENDS     IG_IDEAL
. SUBCKT   D_IDEAL   21   22;                Anode(21).
DIO    21   22   DIODE
. MODEL    DIODE      D(RS=1m  CJO=0.1pF  N=0.001)
. ENDS     D_IDEAL
* CALLS for IGBTs and diodes.
XSW1   3   4   1   0   IG_IDEAL;             Switch Sw1.
XSW2   5   0   1   0   IG_IDEAL;             Switch Sw2.
XSW3   3   5   2   0   IG_IDEAL;             Switch Sw3.
XSW4   4   0   2   0   IG_IDEAL;             Switch Sw4.
XD1    4   3   D_IDEAL;                      Diode D1.
XD2    0   5   D_IDEAL;                      Diode D2.
XD3    5   3   D_IDEAL;                      Diode D3.
XD4    0   4   D_IDEAL;                      Diode D4.
* DRIVERS for switches.
VG12   1   0   PWL    REPEAT  FOREVER
+  (0, 0   20ns, {VP}  0.49444ms, {VP}   0.49446ms, 0
+  1.05555ms, 0   1.05557ms, {VP}   3.94443ms, {VP}
+  3.94445ms, 0   4.50554ms, 0   4.50556ms, {VP}
+  4.99998ms, {VP}  5.0ms, 0   10.0ms)  ENDREPEAT
VG34   2   0   PWL    REPEAT FOREVER
+  (0, 0   5ms, 0   5.00002ms, {VP}   5.49444ms, {VP}
+  5.49446ms, 0   6.05555ms, 0   6.05557ms, {VP}
+  8.94443ms, {VP}  8.94445ms, 0   9.50554ms, 0
+  9.50556ms, {VP}  9.99998ms, {VP}   10.0ms, 0)   ENDREPEAT
```

```
* ANALYSIS
. TRAN     20us     20ms    UIC
. FOUR     100Hz    15      v(4,5) ;     Fourier analysis of load voltage.
. PROBE    i(VS1),  v(4,5) ;             Source current and load voltage.
. END
```

» »

STEP 4 A PSpice simulation can be run with the circuit file EX9_3_2.CIR.

STEP 5 Using PROBE, traces can be plotted for the source current i_s and the load voltage v_l. See Fig. 9.18.

Part (a) of Solution. From PROBE, the rms value of the load voltage is $V_{l\mathrm{rms}} = 88.03\mathrm{V}$. Therefore, the average power P delivered to the load is $P = V_{l\,\mathrm{rms}}^2/R_l = 88.03^2/5 = 1.55\,\mathrm{kW}$.

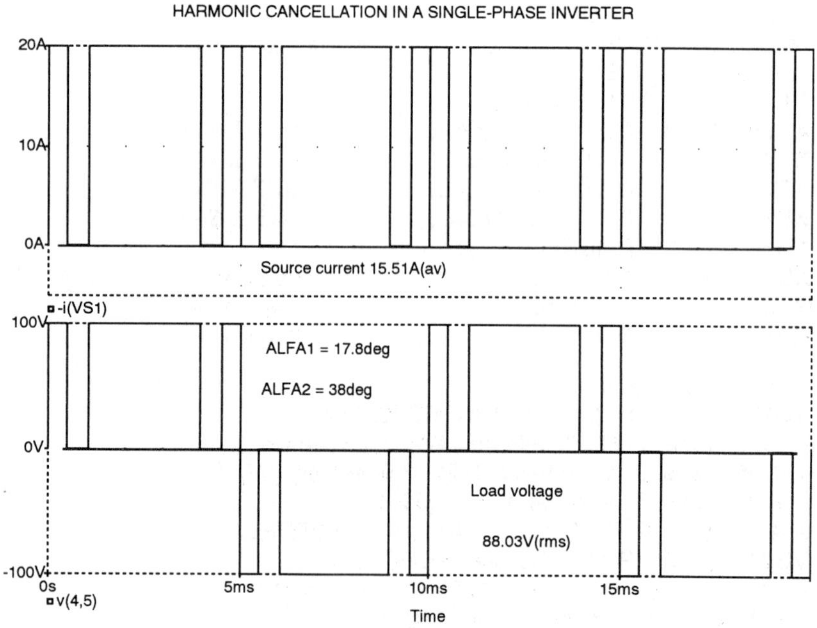

Fig. 9.18

Part (b) of Solution. From the output file EX9_3_2.OUT for the Fourier analysis, the fundamental rms value of the load voltage is

$$V_{l1\,\mathrm{rms}} = \hat{V}_{l1}/\sqrt{2} = 106.5/\sqrt{2} = 75.32\,\mathrm{V}.$$

Part (c) of Solution. Also from the same output file, the output-voltage harmonic components are

$$V_{l3\,\mathrm{rms}} = \hat{V}_{l3}/\sqrt{2} = 1.08\times10^{-2}/\sqrt{2} \approx 0.$$

$$V_{l5\,\mathrm{rms}} = \hat{V}_{l5}/\sqrt{2} = 5.0\times10^{-1}/\sqrt{2} \approx 0.$$

$$V_{l7\,\mathrm{rms}} = \hat{V}_{l7}/\sqrt{2} = 26.49/\sqrt{2} = 18.73\,\mathrm{V}.$$

Part (d) of Solution. From the output file for the output voltage the total harmonic distortion is THD = 53.97%.

END OF EXAMPLE 9.3.2

Drill Exercise D9.3.4
Consider the circuit diagram of a single-phase, bridge inverter in Fig. 9.11. The circuit has the following specifications.

$V_s = 48\mathrm{V}$, $L_l = 0$, $R_l = 2\Omega$, IGBT switches. Gate voltage 15V.

Consider two output waveforms, (i) a rectangular waveform with the duty cycle $m = 0.5$ (no notches) and (ii) a waveform with two notches per half cycle as shown in Fig. 9.16 with $\alpha_1 = 29.8°$ and $\alpha_2 = 57.6°$. Plot the frequency spectra. Make a comparison of **(a)** the total harmonic distortion THD of the two output waveforms, **(b)** the amplitudes of the fundamental components of the output voltages, **(c)** the amplitudes of the third harmonic of the output voltage, **(d)** the amplitudes of the fifth harmonic of the output voltage and **(e)** the total rms values of the output voltage. (Ans: **(a)** 44.98%, 86.28%, **(b)** 61.11V, 40.87V (controlled by α), **(c)** 20.37V, 0, **(d)** 12.22V, 26.52V (high), **(e)** 48V, 39.9V.)

9.3.2 SINEWAVE PULSE-WIDTH MODULATION SPWM

Here, we will use a triangular waveform compared with a sinewave, as in Fig. 9.5 to produce SPWM modulation in an inverter. Remember that the modulation index $M = V_{\mathrm{ref\,max}}/V_{c\,\mathrm{max}}$ of the reference and carrier signals determines the rms value of the inverter output voltage. Also, the number of carrier pulses per cycle of the reference signal determines the harmonic content of the inverter output voltage.

EXAMPLE 9.3.3

The SPWM gating technique is to be employed with the single-phase, bridge inverter that is illustrated in Fig. 9.11. The circuit has the following specifications.

$V_s = 100\text{V}$, $L_l = 0$, $R_l = 5\Omega$, IGBT switches, $f = 50\text{Hz}$.
Arbitrary modulation index of 0.833. Gate voltage 15V.
SPWM with six pulses per half cycle of the output waveform.

Follow the patterns of EXAMPLES 5.4.2 and W9.3.1 (on the WEBsite) to do a PSpice simulation. Plot traces of the source current and the load voltage. Determine **(a)** the average power absorbed by the load, **(b)** the total harmonic distortion THD of the output voltage and **(c)** the harmonic factors ($\text{HF} = V_{ln}/V_{l1}$) of the third and fifth harmonic components of the load voltage.

Solution

There are four steps to achieve a solution.

STEP 1 From Fig. 9.11, from Fig. 5.24, and from the data, we can create a PSpice configuration. See Fig. 9.19.

STEP 2 A circuit file can be composed using the configuration in Fig. 9.19. The gate-driver circuit is put in the form of a subcircuit, so that it can be written to DRIVER.LIB.

»»»»» EX9_3_3.CIR »»»»»»»»»»»»»»»»»»»»»»»»»»»»»»»»»»»»»»

```
SPWM SINGLE-PHASE INVERTER WITH MODULATING PULSES
* To determine power and harmonic distortion.
.INC  OPTIONS;            An OPTIONS file for convergence stability.
* PARAMETERS
.PARAM  VS=100V;                  Source voltage.
.PARAM  RLOAD=5ohms;              Load resistance.
.PARAM  FREQ=50Hz PERIOD={1/FREQ}; Inverter output frequency.
.PARAM  TR=20ns TF={TR};          Pulse rise and fall times.
.PARAM  TON={PERIOD/2-2*TR};      On-time of frequency generator.
.PARAM  VP=15V;                   Magnitude of gate voltage to the switches.
.PARAM  NUM=6;                    Number of inverter pulses per half cycle.
.PARAM  PER={0.5*PERIOD/NUM};     Period of pulses and carrier signals.
.PARAM  VC=12V;                   Magnitude of each carrier pulse.
.PARAM  VREF={0.9*VC};            Comparator reference voltage amplitude.
.PARAM  DELC={TR/2};              Delay of carrier pulse each cycle.
.PARAM  TRC={PER/2-TR};           Rise time of carrier pulse.
.PARAM  TFC={TRC};                Fall time of carrier pulse.
.PARAM  WID={TR};                 Carrier pulse on-time.
.PARAM  DEL={PERIOD/2};           Delay for frequency generator pulse.
```

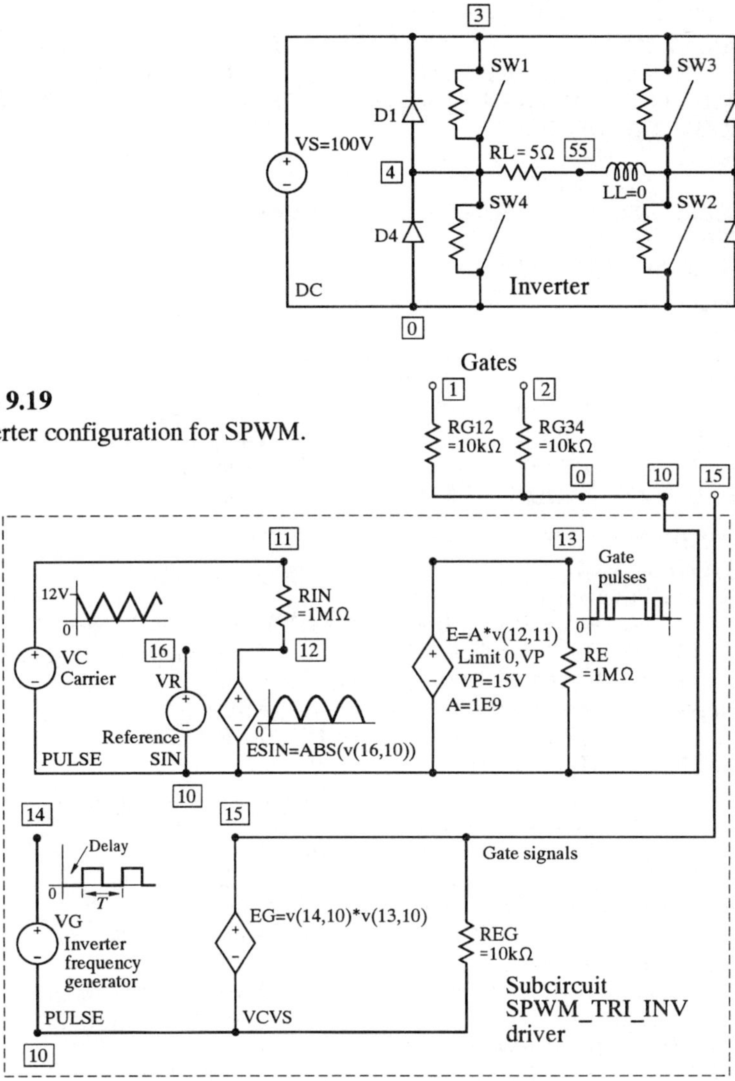

Fig. 9.19
Inverter configuration for SPWM.

```
* SOURCE and LOADS
VS1   3  0  DC  {VS};       The dc power source.
RL    4  5  {RLOAD}
RG12  1  0  1E4;            Gate resistance of Sw1 and Sw2.
RG34  2  0  1E4;            Gate resistance of Sw3 and Sw4.
```

```
* SUBCIRCUITS for IGBTs and diodes.
. SUBCKT IG_IDEAL    17  18    19 20;  Drain(17).Source(18).Gate(19,20).
SW     17  18  19  20  MOSIG;              Voltage-controlled switch.
. MODEL MOSIG VSWITCH(RON=1E-3 ROFF=1E6 VON=10 VOFF=1E-3)
. ENDS     IG_IDEAL
. SUBCKT D_IDEAL     21  22;               Anode(21), cathode(22).
DIO 21     22  DIODE
. MODEL    DIODE        D(RS=1m CJO=0.1pF N=0.001)
. ENDS     D_IDEAL
* CALLS for IGBTs and diodes.
XSW1       3   4   1   0   IG_IDEAL;       Switch Sw1.
XSW2       5   0   1   0   IG_IDEAL;       Switch Sw2.
XSW3       3   5   2   0   IG_IDEAL;       Switch Sw3.
XSW4       4   0   2   0   IG_IDEAL;       Switch Sw4.
XD1   4    3   D_IDEAL;                    Diode D1.
XD2   0    5   D_IDEAL;                    Diode D2.
XD3   5    3   D_IDEAL;                    Diode D3.
XD4   0    4   D_IDEAL;                    Diode D4.

* SUBCIRCUIT for IGBT driver.
. SUBCKT SPWM_TRI_INV  15  10  PARAMS: DELAY=0
* The carrier triangular-wave source.
VC     11  10  PULSE({VC} 0 {DELC} {TRC} {TFC} {WID} {PER})
VR     16  10  SIN(0 {VREF} {FREQ});     Comparator ref. voltage source.
ESIN  12  10  VALUE={ABS(v(16,10))}
* Comparator output, voltage is clipped between zero and VP.
E      13  10  VALUE={LIMIT(v(12,11)*1E9, 0, VP)}
RE     13  10  1E6
* Frequency generator for the inverter.  Generates frequency f (FREQ).
VG     14  10  PULSE(0 1 {DELAY} {TR} {TF} {TON} {PERIOD})
* The timing of this half-period pulse is controlled by DELAY.
* DELAY=0 for gates 1,2 and DELAY=T/2 (that is, DEL) for gates 3,4.
* Gate-pulse generator for all switches, Sw1, Sw2, Sw3 and Sw4.
EG     15  10  VALUE={v(14,10)*v(13,10)};  Pulses in a half cycle.
. ENDS     SPWM_TRI_INV
* SPWM (Sine PWM), TRI (TRIangular carrier wave), INV (INVerter).
* CALLS for gate drivers, one for Sw1, Sw2, and one for Sw3, Sw4.
Xdriver1   1   0   SPWM_TRI_INV
Xdriver2   2   0   SPWM_TRI_INV  PARAMS: DELAY={DEL}

* ANALYSIS
. TRAN     40us     40ms    UIC
. FOUR     50Hz     15      v(4,5)
. PROBE    i(VS1),  v(4,5)
. END
```

»»»

STEP 3 A PSpice simulation can be run with the circuit file EX9_3_3.CIR.

STEP 4 Traces of the source current (−i(VS1)) and the invert load voltage (v(4,5)) are in Fig. 9.20. The results show a high value of slew rate for the comparator. This can be reduced by setting a low value of step size TMAX in the .TRAN statement.

Part (a) of Solution. The average power P absorbed by the load is $P = V^2_{l\mathrm{rms}}/R_l$. From PROBE $P = 75.81^2/5 \approx 1150$W.

Part (b) of Solution. From the file EX9_3_3.OUT, the total harmonic distortion is THD = 48.86%.

Part (c) of Solution. From the results of the Fourier analysis recorded in EX9_3_3.OUT, $HF_3 = 6.519/87.96 = 0.074$ and $HF_5 = 1.803/87.96 = 0.020$.

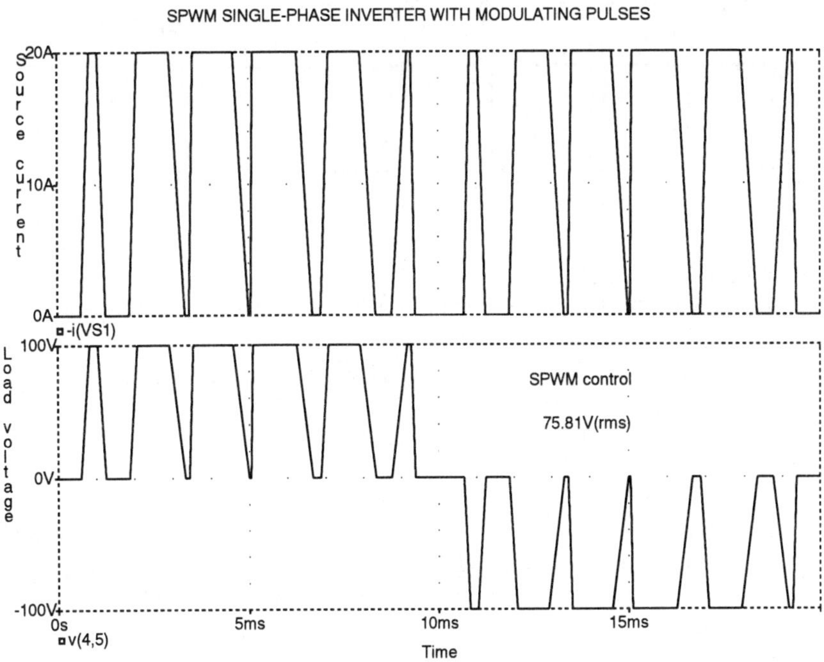

Fig. 9.20

END OF EXAMPLE 9.3.3

Drill Exercise D9.3.5

A single-phase, bridge inverter, depicted in Fig. 9.11, is specified by

$V_s = 48$V, $L_l = 0$, $R_l = 2\Omega$, IGBT switches, $f = 100$Hz. Gate voltage 15V.

SPWM modulation index $M = 0.75$. Seven-pulse output per half cycle.

Follow the pattern in EXAMPLE 9.3.3 and do a PSpice simulation using the SPWM switching strategy with that illustrated in Fig. 9.21. The sinewave reference is compared with the carrier signal. If the reference signal's value is greater than that of the carrier, gate pulses are applied to switches *Sw*1, *Sw*2. Otherwise, gate pulses are applied to switches *Sw*3, *Sw*4. In this way, there is always one switch on in each leg of the bridge (bipolar operation). Plot traces of the source current and the load voltage. Determine **(a)** the average power delivered by the source and **(b)** the total harmonic distortion of the inverter-output waveform. (Ans: **(a)** 1152W, **(b)** 119.1%.)

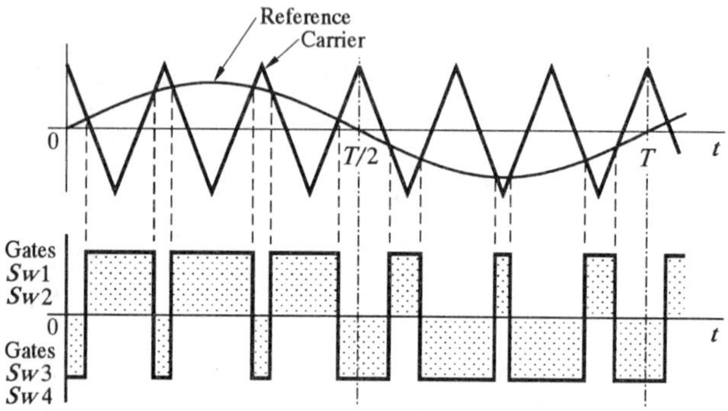

Fig. 9.21

Drill Exercise D9.3.6

A single-phase, bridge inverter is illustrated in Fig. 9.11. The circuit has the following specification.

$V_s = 100$V, $L_l = 4$mH, $R_l = 4\Omega$, IGBT switches, diodes connected,
$f = 100$Hz. Gate voltage 15V.

Use the gate strategy suggested in Fig. 9.22 to obtain an inverter output that has a unipolar nature each half cycle. The modulation index is 75%. The number of carrier cycles is two per half cycle of the reference signals. This is similar to the bipolar operation in Fig. 9.21 except that there are two reference sinewaves (180° phase difference). If $v_{ref14} > v_c$, *Sw*1 is gated on; otherwise, *Sw*4 is on.If $v_{ref32} > v_c$ *Sw*3 is gated on; otherwise, *Sw*2 is on. Do a PSpice simulation and plot traces of the output voltage and load current for steady conditions. Determine **(a)** the initial conditions for steady state, **(b)** the peak and rms values of the load current, **(c)** the average power absorbed by the load and **(d)** the total harmonic distortion THD of the load-current waveform.

(Ans: **(a)** IC = -9.05A, **(b)** 19.0A, 11.38A, **(c)** 517.6W, **(d)** 17.06%.)

Fig. 9.22

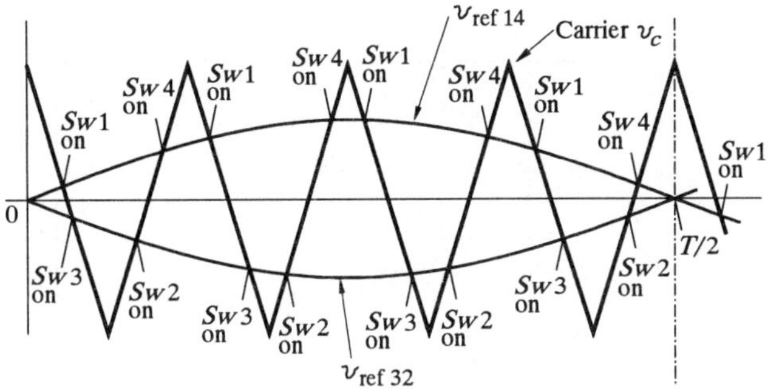

9.3.3 CURRENT-SOURCE INVERTER

Current-source inverters (CSI) are used to drive induction motors. A description and an EXAMPLE can be viewed on the WEBsite. See EXAMPLE W9.3.2 in Section 9.3.3.

9.3.4 CYCLOCONVERTER

A cycloconverter is an ac-ac converter, converting an ac supply of one frequency to an ac source of another frequency at the load, with or without voltage adjustment. See the WEBsite for details of EXAMPLE W9.3.3.

There are Extra Drill Exercises for Section 9.3 on the WEBsite.

9.4 THREE-PHASE SWITCH-MODE INVERTERS

9.4.1 SIX-STEP MODE INVERTER

For ac loads that require high power, three-phase inverters are used. The power circuit and gate drivers of three-phase, switch-mode inverters are extensions of the single-phase, bridge inverters. A three-phase inverter could be constructed from three single-phase, bridge inverters. However, it is usual for a three-phase inverter to have only six switches.

Figure 9.23, depicted on the next page, shows the circuit diagram of the power circuit. The supply is a dc source. The output is an *RL* load. The inverter has three half bridges. Each half bridge consists of a leg (arm or pole) with two controlled switches and two anti-parallel diodes.

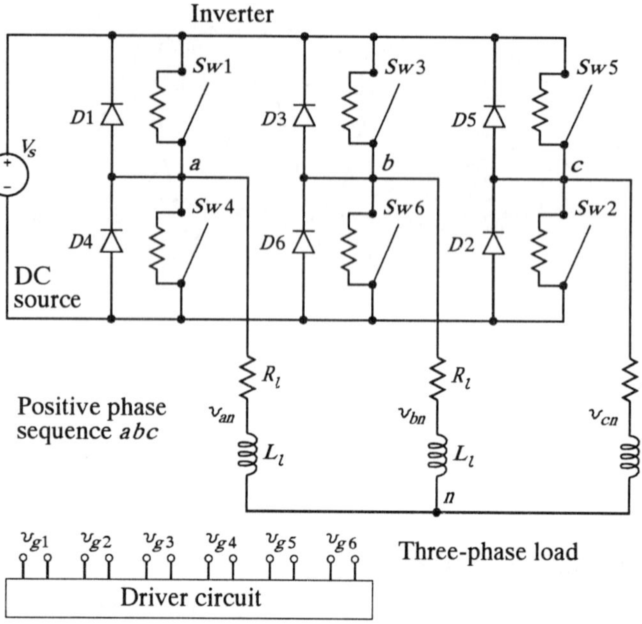

Fig. 9.23 A three-phase inverter.

For a particular sequence of inverter switching, called a six-pulse, 180° mode, the ideal output waveforms and switching intervals are shown in Fig. 9.24. The 180° mode signifies that each switch is on for one half of the output cycle. In each cycle there are six steps (60° per step), or patterns of switching. In this case the pattern is three switches on for each 60° step. The sequence is *Sw*1,2,3, *Sw*2,3,4, *Sw*3,4,5, *Sw*4,5,6, *Sw*5,6,1, *Sw*6,1,2, repeating with *Sw*1,2,3. In order to generate gate signals in PSpice it is best to follow the *intervals of switching* from Fig. 9.24.

EXAMPLE 9.4.1 (PESS EXAMPLE 2.13)

Consider the three-phase inverter depicted in Fig. 9.23. The circuit has the following specifications.

V_s=600V, L_l=0, R_l=20Ω per phase, six-pulse, 180° mode inverter, f=100Hz. Do a PSpice simulation and plot traces of the three output phase voltages and a line voltage (v_{ab}). Determine **(a)** the rms value of the phase voltage across the load, **(b)** the average power delivered to the three-phase load, **(c)** the rms value

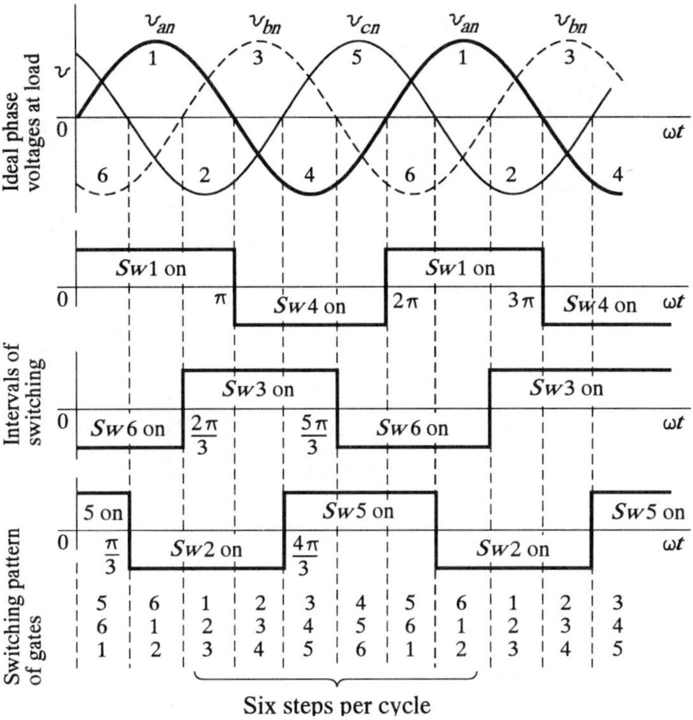

Fig. 9.24 Inverter waveforms, six-step 180° mode.

of the line voltage across the load, **(d)** the amplitude of the third harmonic in the line-voltage waveform of the load and **(e)** the total harmonic distortion THD of the output line-voltage waveform.

(Ans: **(a)** 282.8V, **(b)** 12kW, **(c)** 490V, **(d)** 6.4V, **(e)** 27.1%.)

Solution

Two steps of the solution are described.

STEP 1 From the data of the example and from the circuit diagram in Fig. 9.23 it is possible to set up a PSpice configuration. See Fig. 9.25 on the next page. Because the state of one switch is opposite the state of the second switch in the same leg, both switches can share the same bipolar gate signals. The positive gate signal can be for the upper switch and the negative gate signal can be for the lower switch. In this case, the intervals of switching in Fig. 9.24 can represent the gate signals.

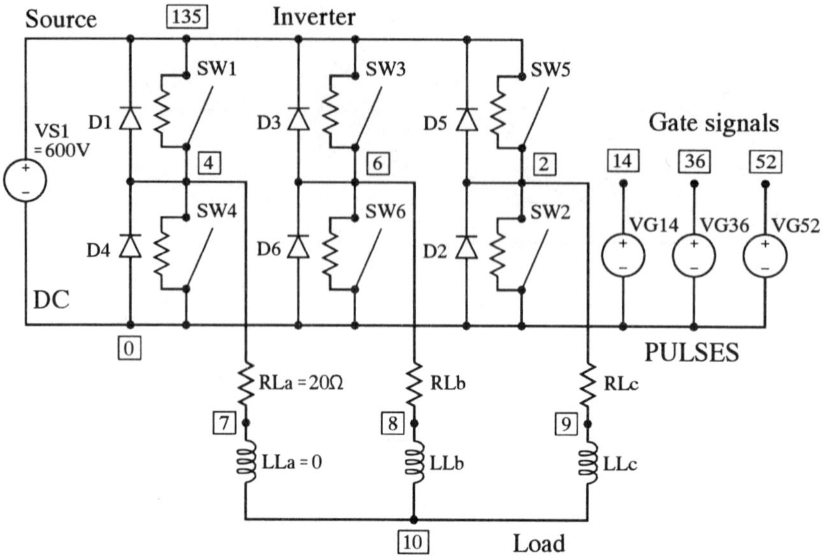

Fig. 9.25 PSpice configuration of a three-phase inverter.

STEP 2 From the data of his example and from the PSpice configuration in Fig. 9.25, we can write a circuit file to represent the inverter system. Run the simulation.

```
»»»»» EX9_4_1.CIR »»»»»»»»»»»»»»»»»»»»»»»»»»»»»»»»»»»»»»»»

THREE-PHASE INVERTER  (6-STEP 180-DEG MODE)
.INC  OPTIONS;                     An OPTIONS file to aid convergence.
* PARAMETERS
.PARAM  VS=600V;                              DC source voltage.
.PARAM  RLOAD=20ohms;              Load resistance per phase.
.PARAM  FREQ=100Hz  PERIOD={1/FREQ};  Output frequency.
.PARAM  VP=15V;                               Gate-voltage magnitude.
.PARAM  TR=40ns  TF={TR};          Rise and fall of gate pulses.
.PARAM  PW={PERIOD/2-TF};                     Gate pulse widths.
.PARAM  DEL14={PERIOD/2-TR};                  Gates 1,4 delay (0 deg).
.PARAM  DEL36={PERIOD/3-TF};                  Gates 6,3 delay (120 deg).
.PARAM  DEL52={PERIOD/6-TF};                  Gates 5,2 delay (-120 deg).
```

```
* SOURCE and LOAD
VS1  135  0     {VS}
RLa  4    10    {RLOAD}
RLb  6    10    {RLOAD}
RLc  2    10    {RLOAD}

* IGBT power switches.
. SUBCKT IG_IDEAL 17  18  19 20;  Drain(17). Source(18). Gate (19,20).
SW  17   18   19  20  IGBT;             Voltage-controlled switch.
. MODEL   IGBT VSWITCH(RON=1E-3 ROFF=1E6 VON=10 VOFF=1E-3)
. ENDS    IG_IDEAL
* CALLS for the subcircuits of the controlled switches.
XSW1      135  4   14  0   IG_IDEAL;     Switch Sw1 and its gate v(14).
XSW2      2    0   0   52  IG_IDEAL;     Switch Sw2 and its gate -v(52).
XSW3      135  6   0   36  IG_IDEAL;     Switch Sw3 and its gate -v(36).
XSW4      4    0   0   14  IG_IDEAL;     Switch Sw4 and its gate -v(14).
XSW5      135  2   52  0   IG_IDEAL;     Switch Sw5 and its gate v(52).
XSW6      6    0   36  0   IG_IDEAL;     Switch Sw6 and its gate v(36).

* DIODES
. SUBCKT D_IDEAL  21  22;  Anode (21).
DIO        21   22  DIODE
. MODEL   DIODE     D(RS=1m  CJO=0.1pF  N=0.001)
. ENDS    D_IDEAL
* CALLS for the subcircuits of the diodes.
XD1       4     135      D_IDEAL
XD2       0     2        D_IDEAL
XD3       6     135      D_IDEAL
XD4       0     4        D_IDEAL
XD5       2     135      D_IDEAL
XD6       0     6        D_IDEAL

* GATE DRIVER
VG14 14  0  PULSE({VP} {-VP} {DEL14} {TR} {TF} {PW} {PERIOD})
VG36 36  0  PULSE({VP} {-VP} {DEL36} {TR} {TF} {PW} {PERIOD})
VG52 52  0  PULSE({VP} {-VP} {DEL52} {TR} {TF} {PW} {PERIOD})

* ANALYSIS
. TRAN    40us      20ms      UIC;           Two cycles.
. FOUR    100Hz     15        v(4,6);        Fourier analysis.
. PROBE   v(4,10),  v(6,10),  v(2,10);       Phase voltages.
. END
```

»»

END OF EXAMPLE 9.4.1

Drill Exercise D9.4.1 (PESS PROBLEM 2.33)
Consider the three-phase, bridge inverter illustrated in Fig. 9.23. The circuit has the following specifications.

$V_s = 600\text{V},\ L_l = 0,\ R_l = 20\Omega,\ f = 100\text{Hz}$. Gate voltage 15V.
Inverter mode of operation 180°, six steps per cycle.

Do a PSpice simulation and plot traces of the dc source current and the load current. Determine **(a)** the instantaneous value of the load current as a Fourier series, **(b)** the harmonic factors HF for the seventh and ninth harmonics of the load-current waveform, **(c)** the average current in the power semiconductor switches and **(d)** the average power delivered to the load.
(Ans: **(a)** $i_a(t) = 19.14\sin(628.3t) + 0.16\sin(3\times628.3t + 1.58)$
$+ 3.77\sin(5\times628.3t + 0.03) + 2.78\sin(7\times628.3t + 0.05)$
$+ 0.16\sin(9\times628.3t + 1.6) + \ldots,$
(b) 0.15, 0.008, **(c)** 6.667A, **(d)** 12kW.)

Drill Exercise D9.4.2 (PESS PROBLEM 2.36)
Consider the three-phase, bridge inverter depicted in Fig. 9.23. The circuit has the following specifications.

$V_s = 600\text{V},\ L_l = 0,\ R_l = 20\Omega$ per phase, $f = 100\text{Hz}$,
six-pulse, 120°-mode inverter (see Fig. 9.26).

Do a PSpice simulation and plot traces of the three output phase voltages and a line voltage (v_{ab}). Determine **(a)** the rms value of the phase voltage across the load, **(b)** the average power delivered to the three-phase load, **(c)** the rms value of the line voltage across the load, **(d)** the amplitudes of the first, third and fifth harmonic in the line-voltage waveform of the load and **(e)** the total harmonic distortion THD of the output line-voltage waveform. (Ans: **(a)** 244.9V, **(b)** 8.999kW, **(c)** 424.2V, **(d)** 570.4V, 4.55V, 117.1V, **(e)** 27.6%.)

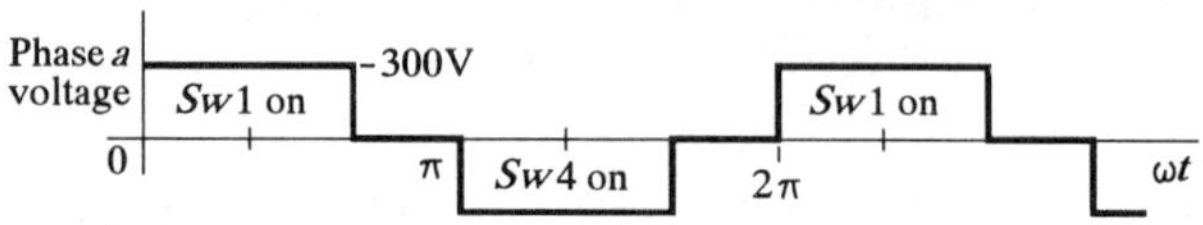

Fig. 9.26

9.4.2 SINEWAVE PULSE-WIDTH MODULATION SPWM

In Section 9.3.2 we described an SPWM technique of comparing a sinewave reference voltage with a triangular carrier wave.

Consider the control signals depicted in Fig. 9.27. The triangular carrier wave is compared with three individual sinewaves which are phase displaced by 120°. The comparison of the carrier wave with each sinewave provides gate

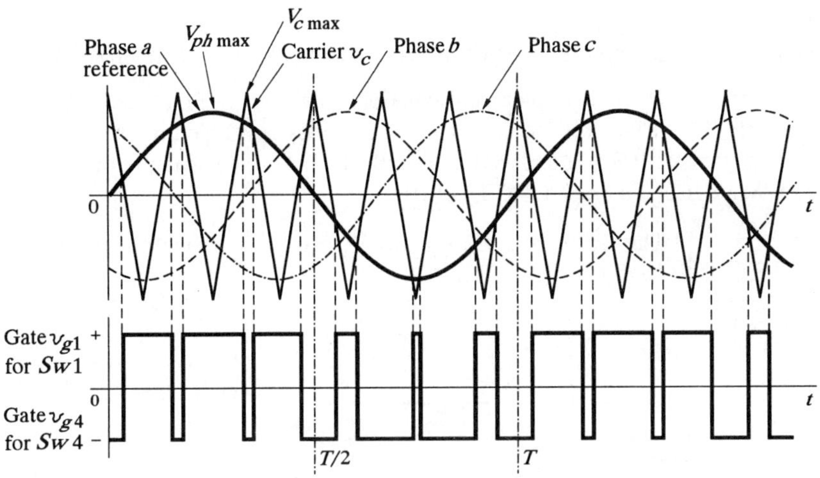

Fig. 9.27 SPWM control signals.

signals for switches in the leg of the respective half bridge. Figure 9.27 illustrates gate signals for switches *Sw*1 and *Sw*2 in the leg of phase a (see Fig. 9.23). The other switches receive a similar group of gate signals. An important characteristic of this technique is that one or other of the switches in a leg is always on but two switches in a leg are never on together. The frequency of the sinewave reference signals govern the inverter frequency and phase. The modulation index $M = V_{c\max}/V_{ph\max}$ and the number of carrier pulses per reference cycle determine the rms value of the inverter output.

EXAMPLE 9.4.2

A three-phase, bridge inverter is illustrate in Fig. 9.23. The control strategy for the SPWM gating is shown in Fig. 9.27. Specifications for the circuit are

$V_s = 600\text{V},\ L_l = 0,\ R_l = 20\Omega$ per phase, $f = 100\text{Hz}.\ M = 0.75.$

Five carrier pulses per half cycle of the inverter period.

Do a PSpice simulation and plot traces of the dc source current i_s, a gate voltage for phase a, the line voltage v_{ab} and the phase voltage v_{an}. Determine **(a)** the rms values of the phase and line voltages of the load, **(b)** the average source current, **(c)** the average power delivered to the load, **(d)** the amplitudes of the first, third, fifth and seventh harmonics in a line voltage of the load and **(e)** the total harmonic distortion THD of the output line-voltage waveform.

Solution

Following the pattern of EXAMPLE 9.4.1 there are four steps in the solution.

STEP 1 Draw a PSpice configuration, as shown in Fig. 9.28. The gate driver for the three half bridges was developed in Drill Exercise D9.3.5 and is described in the subcircuit named SPWM_FULL_TRI (see DRIVER.LIB).

STEP 2 The circuit file EX9_4_2.CIR can be written from the PSpice configuration in Fig. 9.28.

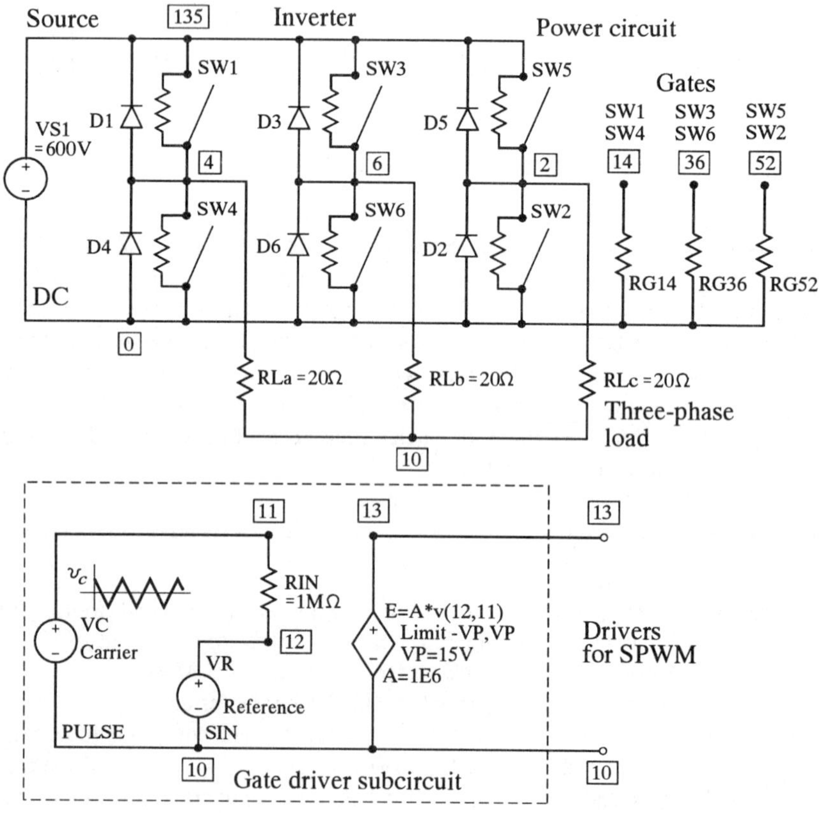

Fig. 9.28 PSpice configuration for the SPWM inverter.

```
»»»»» EX9_4_2.CIR »»»»»»»»»»»»»»»»»»»»»»»»»»»»»»»»»»»»»»»»
SPWM THREE-PHASE INVERTER
* To determine the output responses for an R load.
. INC    OPTIONS;                        An OPTIONS file to aid convergence.
* PARAMETERS
. PARAM   VS=600V;                       DC source voltage.
. PARAM   RLOAD=20ohms;                  Load resistance per phase.
. PARAM   FREQ=100Hz  PERIOD={1/FREQ}; Inverter output frequency.
* SOURCE and THREE-PHASE LOAD
VS1  135  0    {VS}
RLa  4    10   {RLOAD}
RLb  6    10   {RLOAD}
RLc  2    10   {RLOAD}
* IGBT power switches.
. SUBCKT IG_IDEAL 17   18   19   20; Drain(17). Source(18). Gate (19,20).
SW   17   18   19   20   IGBT;        Voltage-controlled switch.
. MODEL IGBT VSWITCH(RON=1E-3 ROFF=1E6 VON=10 VOFF=1E-3)
. ENDS     IG_IDEAL
* CALLS for the subcircuits of the controlled switches.
XSW1   135   4    14   0    IG_IDEAL;   Switch Sw1 and its gate v(14).
XSW2   2     0    0    52   IG_IDEAL;   Switch Sw2 and its gate -v(52).
XSW3   135   6    36   0    IG_IDEAL;   Switch Sw3 and its gate v(36).
XSW4   4     0    0    14   IG_IDEAL;   Switch Sw4 and its gate -v(14).
XSW5   135   2    52   0    IG_IDEAL;   Switch Sw5 and its gate v(52).
XSW6   6     0    0    36   IG_IDEAL;   Switch Sw6 and its gate -v(36).
* DIODES
. SUBCKT D_IDEAL  21   22;              Anode (21), cathode (22).
DIO      21      22   DIODE
. MODEL   DIODE    D(RS=1m  CJO=0.1pF  N=0.001)
. ENDS    D_IDEAL
* CALLS for the subcircuits of the diodes.
XD1     4      135      D_IDEAL
XD2     0      2        D_IDEAL
XD3     6      135      D_IDEAL
XD4     0      4        D_IDEAL
XD5     2      135      D_IDEAL
XD6     0      6        D_IDEAL
*********************************************
```

```
* GATE DRIVER from D9_3_5.CIR (adapted for three phases).
.PARAM  VP=15V;               Magnitude of the gate voltage to the switches.
.PARAM  NUM=5;                Number of inverter pulses per half cycle.
.PARAM  PER={0.5*PERIOD/NUM}; Period of pulses and carrier signal.
.PARAM  VC=12V;                     Magnitude of each carrier pulse.
.PARAM  MOD=0.75;                   Modulation index M=Vref/Vc.
.PARAM  VREF={MOD*VC}
* Reference voltage amplitude VREF at the comparator input.
.PARAM  WID=40ns;                   Carrier-pulse on-time.
.PARAM  DELC={WID/2};               Delay of the carrier pulse each cycle.
.PARAM  TRC={PER/2-WID} TFC={TRC};  Pulse rise and fall times.
RG14    14  0   1E4;                Gate resistances of Sw1 and Sw4.
RG36    36  0   1E4;                Gate resistances of Sw3 and Sw6.
RG52    52  0   1E4;                Gate resistances of Sw5 and Sw2.
* SUBCIRCUIT for the IGBT driver. (For two switches in a leg.)
.SUBCKT SPWM_FULL_TRI3  13  10    PARAMS: DELAY=0
* The carrier triangular-wave source.
VC   11   10   PULSE({VC} {-VC} {DELC} {TRC} {TFC} {WID} {PER})
VR   12   10   SIN(0 {VREF} {FREQ} 0 0 {DELAY})
* Reference source VR for the comparator.
* Comparator output, voltage clipped between -VP and VP.
E     13   10   VALUE={LIMIT((v(12,11)*1E6), -VP,VP)}
.ENDS SPWM_FULL_TRI3
* SPWM (Sine PWM), TRI3 (TRIangular carrier wave for three phases),
* FULL (positive and negative triangles).
* CALL for gate drivers.
Xdriver14  14   0   SPWM_FULL_TRI3;         For Sw1 and Sw4.
Xdriver36  36   0   SPWM_FULL_TRI3  PARAMS: DELAY=-120; Sw3,6.
Xdriver52  52   0   SPWM_FULL_TRI3  PARAMS: DELAY=-240; Sw5,2.
* * * * * * * * * * * * * * * * * * * * * * * * * * * * * * * * * * * * * * * * * * *
* ANALYSIS
.TRAN     20us    20ms    0     5us    UIC
* TMAX=5us to increase the numerical slew rate of the comparator.
.FOUR     100Hz  15        v(4,6);                Line voltage.
.PROBE   v(14),  i(VS1),  v(4,10),    v(4,6)
.END
```

» »

STEP 3 A PSpice simulation can be run with the circuit file EX9_4_2.CIR.

STEP 4 Plot traces of the source current, a gate signal, a line voltage and a phase voltage. Figure 9.29 shows these traces.

Part (a) of Solution. Using the PROBE facilities,
$V_{an\,\mathrm{rms}} = 222.2\mathrm{V}$ and $V_{ab\,\mathrm{rms}} = 385.7\mathrm{V}$.

Parts (b) and (c) of Solution. From PROBE, the average value of the source current is $I_{s\,av} = 12.4\mathrm{A}$. Hence, the average power P delivered to the three-phase load is $P = V_s I_{s\,av} = 600 \times 12.4 = 7.44\mathrm{kW}$.
(Check: $P = 3\,V^2_{an\,\mathrm{rms}}/R_l = 7.41\mathrm{kW}$.)

Part (d) of Solution. From the file EX9_4_2.OUT
$\hat{V}_{ab1} = 389.3\mathrm{V}$, $\hat{V}_{ab3} = 2.2\mathrm{V}$, $\hat{V}_{ab5} = 1.4\mathrm{V}$, $\hat{V}_{ab7} = 0.8\mathrm{V}$.

Part (e) of Solution. From the file EX9_4_2.OUT, the total harmonic distortion THD of the line voltage is THD = 36.45%. Note. The subcircuit named SPWM_FULL_TRI3 is written in the file DRIVER.LIB.

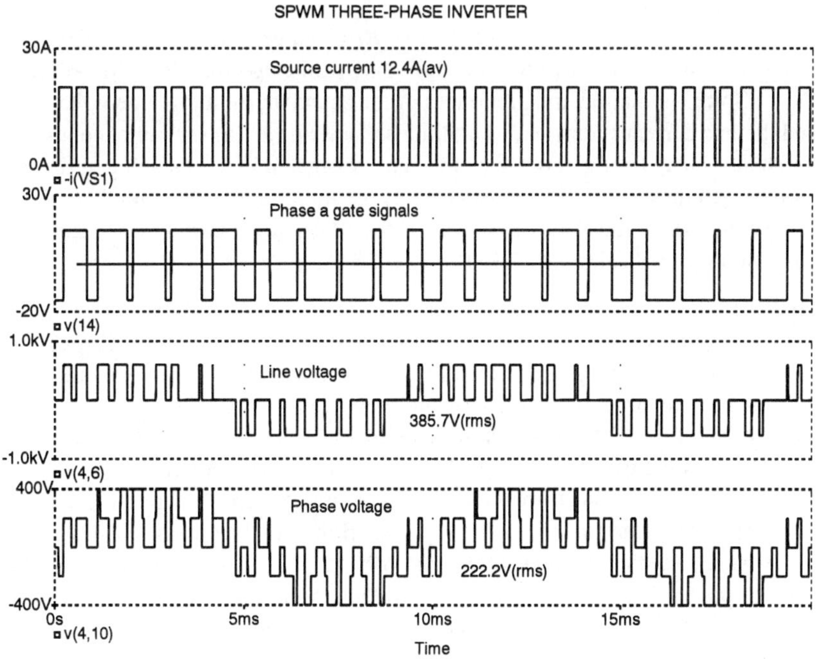

Fig. 9.29

END OF EXAMPLE 9.4.2

Drill Exercise D9.4.3

Consider the three-phase, bridge inverter depicted in Fig. 9.23. The control strategy for sinusoidal pulse-width modulation as described in EXAMPLE 9.4.2 is chosen. The circuit has the following specifications.

V_s=300V, L_l=0, R_l=10Ω per phase, f=60Hz, M=0.7. Gate voltage 14V. Four carrier pulses per half cycle of the inverter period.

Do a PSpice simulation and plot traces of the line voltage and the three phase voltages of the load. **(a)** Determine the average power absorbed by the three-phase load. **(b)** Show that the rms value of the line voltage is $\sqrt{3}$ times the phase voltage. **(c)** What is the total harmonic distortion THD of the line-voltage waveform? (Ans: **(a)** 3.457kW, **(b)** 107.35V, 186.0V, **(c)** 62.0%.)

Drill Exercise D9.4.4

Consider the three-phase bridge inverter depicted in Fig. 9.23. The circuit has the following specifications.

V_s=600V, L_l=6.67mH per phase, R_l=20Ω per phase, f=100Hz.
Gate voltage 15V. Inverter operation 120°, six-step mode (see Fig. 9.26).

Do a PSpice simulation and plot traces of the source current, the line voltage v_{ab} and the line current i_a for steady conditions. Determine **(a)** the initial conditions for steady state, **(b)** the rms value of the line voltage, **(c)** the rms value of the line current, **(d)** the average power delivered by the source and **(e)** the total harmonic distortion THD of the load-current waveform. (Ans: **(a)** 0, current discontinuous, **(b)** 436.4V, **(c)** 11.4A, **(d)** 7.734kW, **(e)** 25.6%.)

Drill Exercise D9.4.5

Consider the three-phase, bridge inverter depicted in Fig. 9.23. The control strategy for sinusoidal pulse-width modulation as described in EXAMPLE 9.4.2 is chosen. The circuit has the following specifications.

V_s=600V, L_l=6.67mH per phase, R_l=20Ω per phase, f=100Hz, M=0.75. Five carrier pulses per half cycle of the inverter period.

Do a PSpice simulation and plot traces of the source current i_s, a gate voltage, the line voltage v_{ab} and the line current i_a. Determine **(a)** the initial conditions for steady state, **(b)** the rms value of the line voltage and current of the load, **(c)** the average power delivered by the supply, **(d)** the amplitudes of the first, third, fifth and seventh harmonics of the line current of the load and **(e)** the total harmonic distortion THD of the line current of the load.

(Ans: **(a)** i_a=-3.19A, i_b=-6.9A, i_c=10.1A, **(b)** 384.6V, **(c)** 3.8kW, **(d)** 10.96A, 0.04A, 0.03A, 0.02A, **(e)** 16.98%.)

Chapter 10

Resonant Converters

10.1 INTRODUCTION

As the demand for higher performance of power electronics circuits increases, new power conversion techniques are being developed. A relatively new power conversion technique, broadly classified as *resonant converters*, has begun to emerge. The term *resonant* is used since many (but not all) of these topologies require some form of *LC* resonance. In general, resonant converters are used to generate nearly pure sinewaves, or they are used to provide zero-current or zero-voltage switching for lower switching losses.

The purpose of this chapter is to demonstrate the use of PSpice for simulating the operation of these relatively new converters. Three basic representative classifications have been chosen for simulation in this chapter, namely

1. Resonant-load dc-ac inverters.
2. Resonant-switch dc-dc converters.
3. Integral-cycle ac-ac converters.

10.2 RESONANT-LOAD INVERTERS

Resonant circuits can be employed to generate nearly pure sinewaves in power circuits. These techniques can be applied to inverter applications where low harmonic distortion is paramount. In inverters, the resonant circuit is often placed in the load section. Most loads already possess inductance. If capacitance is added, as illustrated in Fig. 10.1 on the next page, the result is a resonant circuit which can be set to oscillate at the desired output frequency. Two configurations for resonant-load inverters are series-load elements and parallel-load elements.

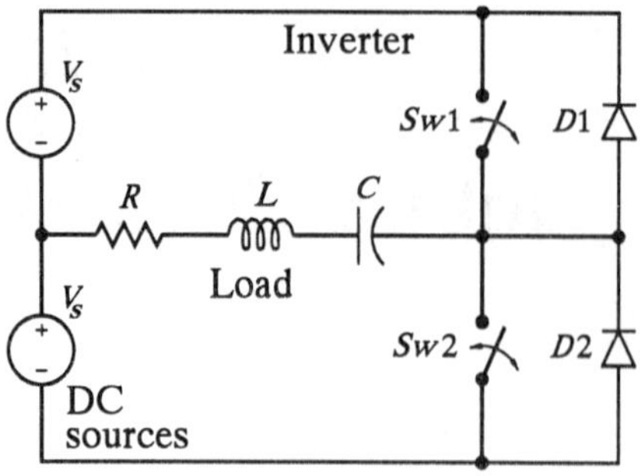

Fig. 10.1
Resonant half-bridge with series *RLC* load.

10.2.1 SERIES-LOAD RESONANT INVERTER

A basic inverter circuit is illustrated in Fig. 10.1. This circuit is called a series resonant inverter because the *LC* network and the load are all in series with each other. An intuitive appreciation for the operation of this circuit can be obtained by recalling that a series resonant circuit presents a low impedance at its natural frequency. A set of power switches is then introduced to enable the application of a square wave of voltage to this resonant combination of elements. If the resonant frequency is equal to the fundamental frequency of the square wave, the higher order harmonics will be filtered out and a nearly pure sinewave of current will be the response.

In Fig. 10.1 the square wave of voltage is produced by the half bridge consisting of the unidirectional switches $Sw1$, $Sw2$. When $Sw1$ is fired, the load conducts the first half cycle of the quasi-sinewave of current until the current reaches zero. At this point, the current through $Sw1$ also reaches zero and the semiconductor switch turns off naturally. Then, $Sw2$ is turned on and the load sees a supply voltage V_s of reverse polarity. The load experiences a negative quasi-sinewave of current until the current reaches zero and $Sw2$ turns off. At this point, one cycle has been completed and the process begins again. The angular frequency of the resulting quasi-sinewave is the frequency of the *RLC* combination and is given by

$$\omega = \left[1/(LC) - R^2/(4L^2)\right]^{1/2}. \tag{10.2.1}$$

Thyristors are employed in this example because of the convenience of natural turn-off. However, in high frequency inverter applications faster power semiconductors may need to be used with proper gating signals for turn-off.

EXAMPLE 10.2.1 (PESS EXAMPLE 2.8)
Consider the half-bridge inverter illustrated in Fig. 10.1. The dc source has a voltage $V_s = 200V$. The load has a resistance $R = 1ohm$, an inductance $L = 1mH$, and a capacitance of $25\mu F$. The frequency of the inverter is to be 1kHz. Simulate the operation of this inverter using PSpice. Plot traces of the load current and voltage. Determine **(a)** the harmonic distortion THD of the load current, **(b)** the peak voltage across the inductor and **(c)** the change in output if there is a 10% increase in the value of the load inductance L.

Solution
There are four steps to achieve a solution.

STEP 1 — The first step involves transforming the circuit diagram in Fig.10.1 into a PSpice configuration, as shown in Fig. 10.2.

STEP 2 — The second step is to use the PSpice configuration to create a circuit file that we will call EX10_2_1.CIR.

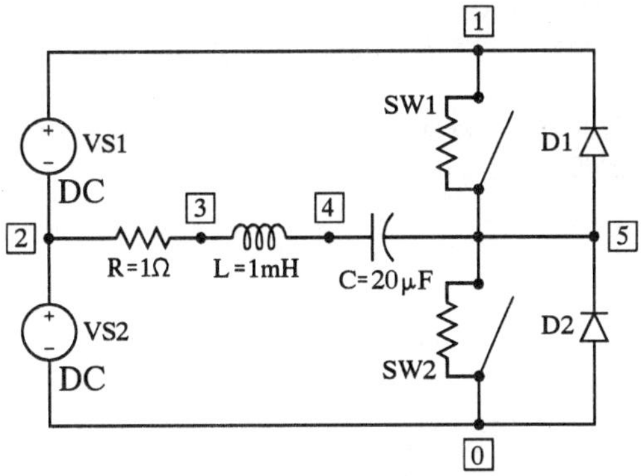

Fig. 10.2
PSpice configuration of resonant inverter.

»»»»» EX10_2_1.CIR »»»»»»»»»»»»»»»»»»»»»»»»»»»»»»»»»»»

```
RESONANT INVERTER (SINEWAVE BY SERIES RESONANCE)
* The POWER and DRIVER libraries are used for the thyristors and drivers.
. LIB   DRIVER.LIB
. LIB   POWER.LIB
. INC   OPTIONS;                         Convergence aid. See Section 7.5.1.
```

```
* PARAMETERS
.PARAM  FREQ=1kHz  PERIOD={1/FREQ};     Inverter frequency.
.PARAM  TON=50us;                       TON is the gate pulse width.
.PARAM  TR=1us       TF=1us   VP=2V;    Gate pulse parameters.

* SOURCES
VS1   1  2   200V;                      DC voltage sources.
VS2   2  0   200V

* CALLS for a thyristor subcircuit for Sw1 and Sw2.
XTHY1   1  5  TH_GAT  PARAMS: DELAY=0; Thyristor with gate drive.
XTHY2   5  0  TH_GAT  PARAMS: DELAY=500us; From DRIVER.LIB.

* CIRCUIT ELEMENTS
D1     5  1   D_IDEAL;                  Near ideal diode. From POWER.LIB.
D2     0  5   D_IDEAL
*.MODEL       D_IDEAL D(RS=1m CJO=0.1pF N=0.001)
R      3  2   1ohm
L      4  3   1mH
C      5  4   25uF

* ANALYSIS
.TRAN    20us   8ms   0   25us;         Time-step ceiling set to 25μs.
.PROBE
.FOUR    1kHz   i(RL),  v(5,2);         Fourier analysis.
.END
```

»»»

STEP 3 The third step is to run the PSpice program with the circuit file EX10_2_1.CIR.

STEP 4 The fourth step is to run PROBE and view the load current and load voltage. The results are shown on the PROBE plot in Fig. 10.3.

Part (a) of Solution. After running PSpice we can view the Fourier analysis results in the EX10_2_1.OUT text file. The pertinent information from this output file is shown below.

FOURIER COMPONENTS OF TRANSIENT RESPONSE i(R)
DC COMPONENT = -8.025890E-1

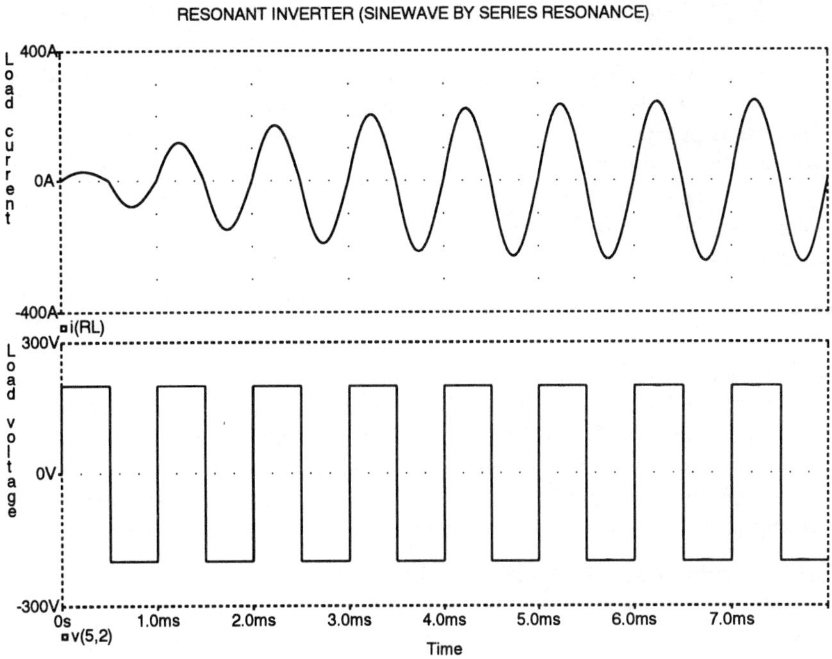

Fig. 10.3

HARMONIC NORMALIZED NO	FREQUENCY (HZ)	FOURIER COMPONENT
1	1.000E+03	2.441E+02
2	2.000E+03	5.392E-01
3	3.000E+03	4.847E+00
4	4.000E+03	1.086E-01
5	5.000E+03	1.643E+00
6	6.000E+03	4.701E-02
7	7.000E+03	8.464E-01
8	8.000E+03	2.636E-02
9	9.000E+03	5.172E-01

The total harmonic distortion THD = 2.14%. This is impressive when compared to the harmonic content of many square-wave inverters.

Part (b) of Solution. From PROBE, the peak inductor voltage is 1.76kV. Was this expected?

Part (c) of Solution. If the value of the inductance L is increased, the resonant frequency of the *RLC* circuit is decreased. This increases the period of oscillation. As a result, switch *Sw*1 is still conducting current from the first half cycle when the second switch is turned on. When this occurs both devices latch on and a large *shoot-through* current ensues. In practice this would result in destroying the switches, the dc supply, or both. These are occasions that the reader is grateful he/she is sitting at his/her computer and not at the test bench. PSpice is useful in isolating faults such as these before prototyping begins.
The resonance is sensitive to the element values. Try $C = 25.2\mu$F.

END OF EXAMPLE 10.2.1

Drill Exercise D10.2.1
Use the circuit from EXAMPLE 10.2.1 but change the values of L and C for an operating frequency $f = 100$Hz. L and C can be determined by either using eq. (10.2.1) or the reader may choose to *parameterize* the *LC* values and sweep till appropriate values are determined. Once appropriate values are determined use PSpice and PROBE to display the resulting load-current waveform.

Drill Exercise D10.2.2 (PESS EXAMPLE 2.9)
Consider the resonant inverter illustrated in Fig. 10.1, and specified by
$V_s = 200$V, $R = 1\Omega$, $L = 1$mH, $C = 25.0\mu$F, switching frequency $f = 1$kHz.
Follow the pattern of EXAMPLE 10.2.1 and do a PSpice simulation. For steady conditions plot traces of a source current, the load voltage and load current over an interval of two cycles. Determine **(a)** the initial conditions for steady state, **(b)** the average power delivered to the load, **(c)** the current ratings of the thyristors and the diodes and **(d)** the peak load current. (Ans: **(a)** IC = -1.616kV, IC = 5.207A, **(b)** 32.3kW, **(c)** 80.8A, negligible, **(d)** 254.9A.)

10.2.2 PARALLEL-LOAD RESONANT INVERTER

The parallel resonant inverter has the load resistance shunted by a parallel *LC* network. An example is illustrated in Fig. 10.4. It should be noted that this inverter is fed by a current source which is approximated by a large inductance behind a voltage source. In Fig. 10.4, an alternating square wave of current is produced by controlling the power switches in the single-phase bridge. The operation of this circuit can be realized if one recalls that a parallel *LC* network has a high impedance at its resonant frequency. Thus, the resonant circuit will attenuate all frequencies except its natural frequency. Therefore, if the switching is at the natural frequency of the resonant circuit then the fundamental frequency will appear across the load resistance while all harmonics will be reduced.

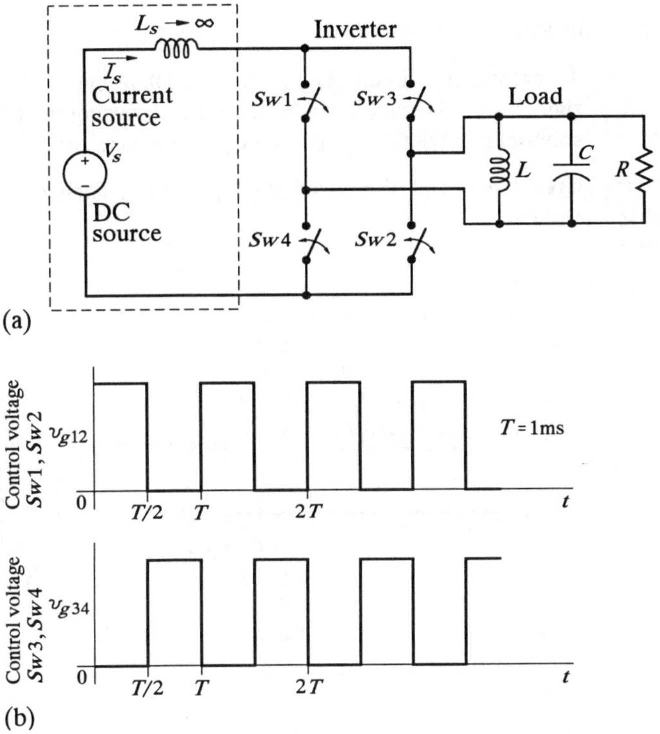

Fig. 10.4 Parallel-load resonant inverter.
(a) Circuit diagram, (b) gate waveforms.

EXAMPLE 10.2.2

Simulate the parallel-load resonant inverter illustrated in Fig. 10.4a. The switching frequency is 1kHz using the control voltages shown in Fig. 10.4b. The dc input voltage V_s is 100volts. Set $L_s = 1$H, $L = 1$mH, $C = 25\mu$F and $R = 1\Omega$. In order to avoid a lengthy start-up transient due to the source inductance, use the initial conditions $I_s = 6$A and the current in L equal to -24A. Use PROBE to **(a)** display the current in the load resistance and the load voltage. **(b)** Perform a Fourier analysis and use PROBE to produce a spectral plot of the load current between 0 and 10kHz. Use the SWITCH model in POWER.LIB for each of the switches. (THD = 4.1%.)

Solution

Two steps of the solution are described.

STEP 1 — Transform the circuit diagram in Fig. 10.4a into a PSpice configuration. The circuit can be simplified by using the INV_DRV subcircuit in DRIVER.LIB to produce the gate control voltages.

STEP 2 — Create a circuit file based on Fig. 10.5. Then do a PSpice simulation.

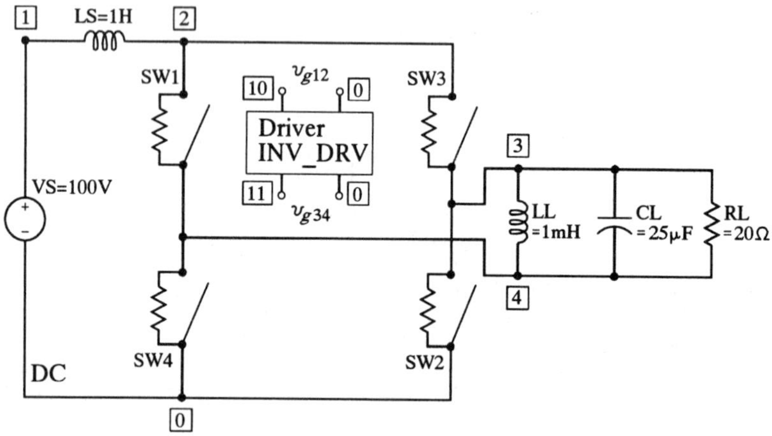

Fig. 10.5 PSpice configuration of a parallel resonant inverter.

»»»»» EX10_2_2.CIR »»»»»»»»»»»»»»»»»»»»»»»»»»»»»»»»»»»

```
PARALLEL RESONANT INVERTER (SINEWAVE BY RESONANCE)
* To demonstrate the operation of a parallel-load resonant inverter circuit.
* The POWER and DRIVER libraries are used.
.LIB   DRIVER.LIB
.LIB   POWER.LIB
.INC   OPTIONS;                               Convergence aid.
* PARAMETERS
.PARAM   FREQ=1kHz  PERIOD={1/FREQ}  VGATE=15V  TON=0.5ms
.PARAM   RLOAD=20ohms  LLOAD=1mH   CLOAD=25uF
* SOURCE
VS1   1   0   DC   100V;                       DC voltage source.
```

```
* CIRCUIT ELEMENTS
LS    1  2   1H IC=6A ;      Source inductance with initial conditions.
LL    3  4   {LLOAD}  IC=-24A;                  Resonating element.
CL    3  4   {CLOAD};                           Resonating element.
RL    3  4   {RLOAD};                           Load element.

* CALLS for driver subcircuit for the inverter.
Xdriver    10  11   0  INV_DRV;                 From DRIVER.LIB.
SW1  2   3   10   0  SWITCH;          Switch parameters in POWER.LIB.
SW2  4   0   10   0  SWITCH
SW3  2   4   11   0  SWITCH
SW4  3   0   11   0  SWITCH

* ANALYSIS
.TRAN    25us   5ms  1us  50us  UIC;   Time-step ceiling set to 50μs.
.FOUR    1kHz   15   i(RL);            Fourier analysis.
.PROBE
.END
```

»»

END OF EXAMPLE 10.2.2

Drill Exercise D10.2.3

Using the parallel resonant inverter circuit in EXAMPLE 10.2.2, perform a sweep of the value of C from 1μF up to 100μF in steps of 25μF. Plot the resulting collection of transient-voltage waveforms and comment on the effects of a poorly tuned resonant inverter.

Drill Exercise D10.2.4

A parallel-load resonant inverter, depicted in Fig. 10.4, is specified by V_s=48V, L_s=1H, R=20Ω, L=100mH, C=25.0μF, frequency f=100Hz. Follow the pattern of EXAMPLE 10.2.2 and do a PSpice simulation. Plot the traces of the source current, and the load resistor, inductor and capacitor currents for the steady state. Determine **(a)** the transient time to reach steady conditions, **(b)** the initial conditions for steady state, **(c)** the average power delivered to the load and **(d)** the total harmonic distortion THD of the load voltage. **(e)** Comment on the results.

(Ans: **(a)** >1s, **(b)** i_{LS}=2.722A, i_L = -1.196A,
(c) 130.7W, **(d)** 29.13%, **(e)** overdamped circuit.)

10.3 RESONANT-SWITCH CONVERTERS

The second class of resonant converters to be explored comes under the general heading of resonant-switch converters. In these topologies, *LC* resonance is utilized to shape the voltage and/or current associated with a power switch. A quick overview will show that most resonant-switch topologies in this class can be visualized as a conventional converter circuit with the addition of a resonant circuit. With a properly designed driver circuit the power switch can be made to operate with zero current or zero voltage during switching on or off. Consequently, these converters are often subclassed as zero-current switching (ZCS) converters or zero-voltage switching (ZVS) converters. PSpice is an ideal tool for analyzing and comparing the operation of different resonant-converter circuit options and variations. This is particularly true since hand analysis of resonant-switch topologies can be daunting.

Resonant-switch converters all operate in essentially the same way. A square pulse of current or voltage is delivered by the power switches to a resonant circuit. Energy then circulates in the resonant circuit and some or all of it is then tapped off to supply the load. During these time periods, resonant-switch converters can experience resonant as well as nonresonant operating intervals.

10.3.1 ZERO-CURRENT SWITCHING CONVERTERS

Zero-current switching (ZCS) is one variety of resonant-switch circuits that provides *lossless* switching. In theory, during a resonant cycle there are periodic instances when the current is zero. This zero-current instant within a power switch can be used to facilitate lossless switching during turn-off.

Figure 10.6 illustrates a *half-wave, parallel-loaded,* ZCS converter. This circuit topology should appear similar to the standard buck converter. See Section 6.2. It is a traditional buck converter with the addition of a resonant circuit placed near the switching device. This particular circuit is described as half-wave because the switch conducts for half of the resonant cycle period. It is parallel-loaded because the resonant circuit is placed in parallel with the output load.

The ZCS driver in Fig. 10.6 senses the switch current and controls the switch operation. Figure 10.7 illustrates the details of the zero-current switch control using PSpice analogue behavioural modelling. The switch SW is initially turned on by a small trigger pulse from the source Vclock. The initial pulse is left on long enough to allow current to rise in the resonant switch so that when the trigger pulse is removed the switch remains latched-on until the resonant cycle returns the switch current to zero†. The switching frequency in this case is fixed and the cycle begins again at the onset of another trigger pulse from Vclock.

For the ZCS converter, the period T of switching is determined by the clock frequency. The switch's on-time t_{ON} is determined by the resonant frequency $f_r = 1/T_r$. So, $t_{ON} \approx T_r/2$. The value of t_{ON} is fixed, so the chopper voltage is adjusted by controlling the clock frequency. (Usually we change t_{ON} and have T fixed for conventional choppers.)

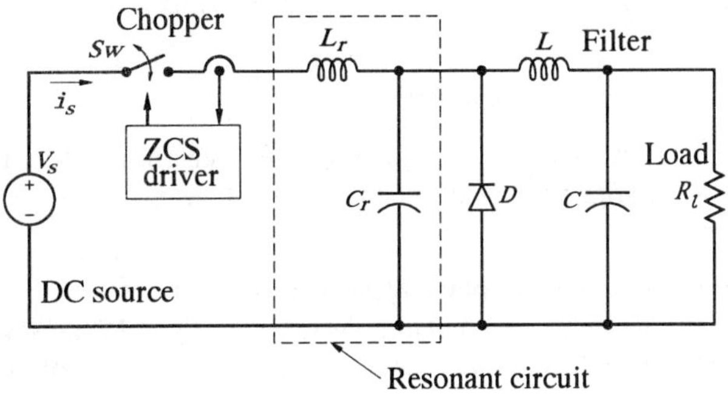

Fig 10.6
Resonant zero-current switching buck converter.

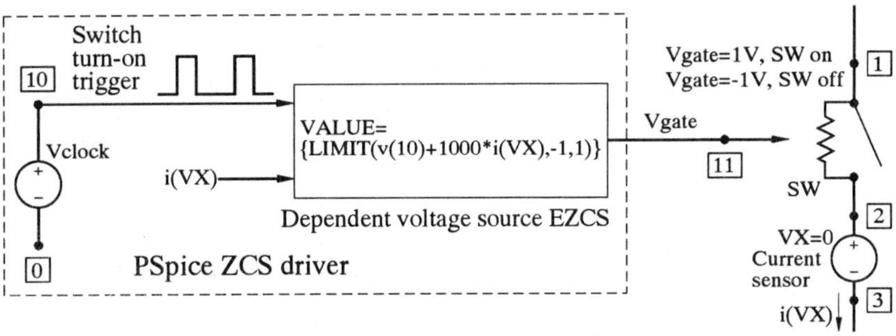

Fig. 10.7
Analogue behavioural model to achieve switch turn-off at zero current.

† This is similar to the control of the thyristor system model (Section 4.4).

EXAMPLE 10.3.1

Consider the half-wave, parallel-loaded, ZCS converter in Fig. 10.6. The dc source has a voltage of 15volts and the switching frequency is f=70kHz. Set L_r=500nH, C_r=1μF, L=25μH, C=10μF, R_l=1Ω. Use the driver illustrated in Fig. 10.7. Do a PSpice simulation. **(a)** Use PROBE to plot traces of the switch current and the load voltage for the first 100μs of operation. **(b)** Use PROBE to plot traces of the switch gate voltage and the instantaneous power being dissipated in the switch. Magnify one of these transient waveforms at the point when the circuit is approaching steady state. What can you conclude regarding switching losses?

Solution

Two steps of the solution are described.

STEP 1 Transform the circuit into PSpice elements and nodes. The resulting configuration is illustrated in Fig. 10.8.

STEP 2 The circuit file of the ZCS converter can be created using Fig. 10.8. The ZCS driver circuit, illustrated in Fig. 10.7, must be included. The UIC option is used to ensure proper zero start-up conditions for the various inductors and capacitors. In addition, a time-step ceiling is imposed to enhance the number of points on the output plots during transition periods. Run a PSpice simulation.

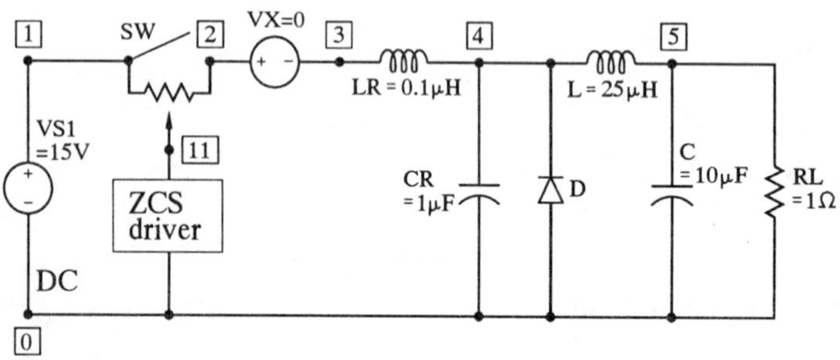

Fig. 10.8 PSpice configuration of the parallel-resonant inverter.

»»»»» EX10_3_1.CIR »»»»»»»»»»»»»»»»»»»»»»»»»»»»»»»»»»»»»

```
HALF-WAVE PARALLEL-LOADED ZCS RESONANT BUCK CONVERTER
* The POWER library is used for the switch model.
. LIB     POWER.LIB
. INC     OPTIONS;                                Convergence aid.
* PARAMETERS
. PARAM   FREQ=70kHz  PERIOD={1/FREQ};Chopper switching frequency.
* SOURCES
VS1    1  0  DC  15V;                             The DC source.
Vclock 10 0  PULSE(-1V 1V 0 2ns 2ns 1us {PERIOD}); Gating.
VX     2  3  0;                                   Current sensor.
* DEVICES
SW     1  2  11  0  SWITCH;     Voltage-controlled switch of the chopper.
* SWITCH parameters in POWER.LIB.
D 0    4  D_IDEAL
. MODEL  D_IDEAL  D(RS=1m CJO=0.1pF N=0.001)
* CIRCUIT ELEMENTS
LR     3  4  500nH;         LR, CR form a resonant network for SW turn-off.
CR     4  0  1uF
L      4  5  25uH;                                L, C form the output filter.
C      5  0  10uF
RL     5  0  1ohm;                                Output load resistance.
* ZCS DRIVER
EZCS 11  0  VALUE={LIMIT(v(10)+i(VX)*1000,-1,1)}
* ANALYSIS
* A time-step ceiling is used to increase the plot resolution.
. TRAN    200ns  100us  0  200ns  UIC;            UIC for proper start up.
. PROBE
. END
```

»»

END OF EXAMPLE 10.3.1

Drill Exercise D10.3.1

Consider the resonant zero-current switching (ZCS) buck converter depicted in Fig. 10.6 The circuit has the following specifications.

$$V_s = 24\text{V},\ L_r = 3\mu\text{H},\ C_r = 2\mu\text{F},\ L = 25\mu\text{H},\ C = 10\mu\text{F},\ R_l = 1\Omega,$$
chopper switching frequency $f = 50\text{kHz}$.

Follow the pattern of EXAMPLE 10.3.1 and do a PSpice simulation. Plot the traces of the source current, the gate voltage and the load current. Determine **(a)** the peak value of the source current, in steady state, **(b)** the switching duty cycle $m = t_{ON}/T = t_{ON}f$, **(c)** the average power delivered to the load and **(d)** the load current excursion. **(e)** How long is the transient condition?
(Ans: **(a)** 33.6A, **(b)** 0.6, **(c)** 277W, **(d)** 1.889A, **(e)** 100μs.)

Drill Exercise D10.3.2

Figure 10.9 represents the circuit diagram of a series-loaded ZCS buck converter. Follow the pattern of EXAMPLE 10.3.1 and do a PSpice simulation. Plot traces of the source current, switch current and load current. How do the results compare with EXAMPLE 10.3.1? Determine the power delivered to the load for steady conditions.

Note that the resonant capacitor must be charged initially.
(Ans: 43.4W.)

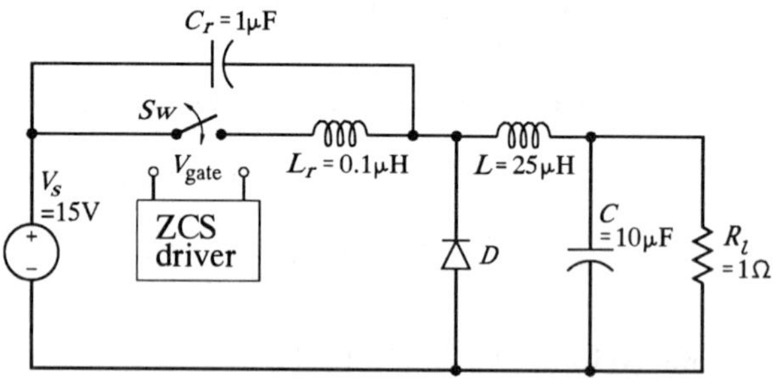

Fig. 10.9

EXAMPLE 10.3.2

Simulate the full-wave, series-loaded ZCS flyback converter shown in Fig. 10.10. The dc source has a voltage of 50volts and the switching frequency is $f = 200\text{kHz}$. Set $L_r = 400\text{nH}$, $C_r = 0.33\mu\text{F}$, $L = 20\mu\text{H}$, $C = 3\mu\text{F}$, $R_l = 20\Omega$. Use the ZCS driver from the previous example. Do a PSpice simulation and plot traces of the switch current and load voltage over the transient interval. Determine **(a)** the steady-state average load voltage, **(b)** the peak overshoot in the load voltage, **(c)** the switch's peak current, the peak diode D_s current, and the switch's peak voltage.

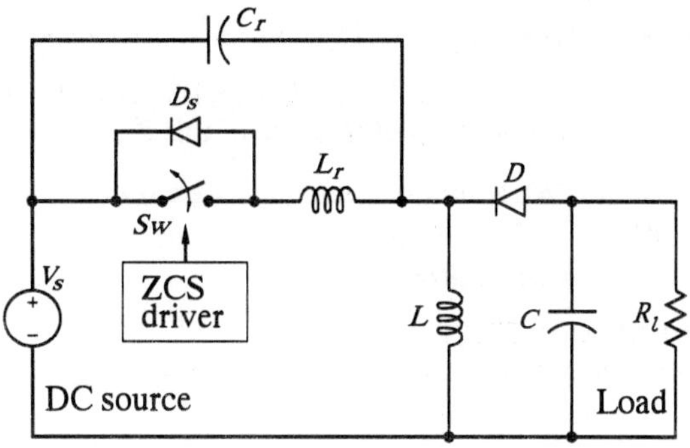

Fig. 10.10
Full-wave series-loaded ZCS flyback converter.

Solution

There are four steps to achieve the solution.

STEP 1 The first step is to draw the PSpice configuration. This is done in Fig. 10.11.

STEP 2 We are given the element values so we can proceed to write the circuit file EX10_3_2.CIR using Fig. 10.11.

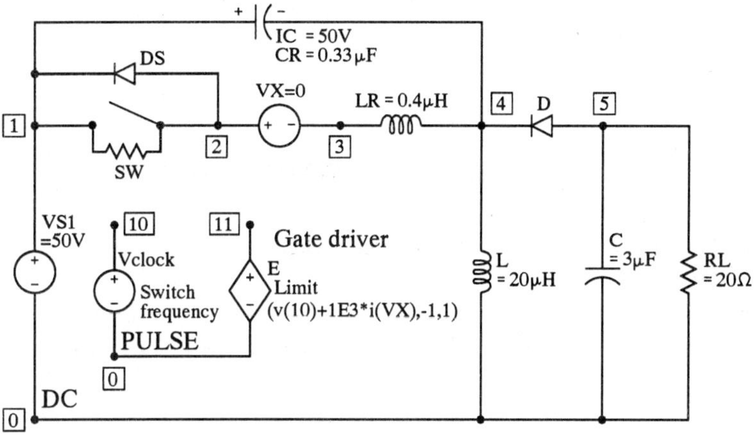

Fig. 10.11 PSpice configuration of series-loaded ZCS converter.

```
»»»»» EX10_3_2.CIR »»»»»»»»»»»»»»»»»»»»»»»»»»»»»»»»»»»»
FULL-WAVE SERIES-LOADED ZCS RESONANT FLYBACK CONVERTER
* To determine the performance of this ZCS converter.

* The POWER library components are used.
.LIB     POWER.LIB
.INC     OPTIONS;                         Convergence aid.

* PARAMETERS
.PARAM  FREQ=200kHz  PERIOD={1/FREQ};     Switching frequency.
.PARAM  VS=50V

* SOURCES
VS1      1   0   DC  {VS};                The dc source voltage.
Vclock   10  0   PULSE(-1V 1V 0 2ns 2ns 1us {PERIOD}); Gating.
VX       2   3   0;                       Current sensor.

* DEVICES
SW       1   2   11 0 SWITCH; The chopper switch. Model in POWER.LIB.
DS       2   1   D_IDEAL;                 Switch diode.
D        5   4   D_IDEAL
.MODEL D_IDEAL   D(RS=1m  CJO=0.1pF  N=0.001)

* CIRCUIT ELEMENTS
LR       3   4   400nH;                   For resonance.
CR       1   4   0.33uF   IC=50V;         For resonance.
L        4   0   20uH
C        5   0   3uF
RL       5   0   20ohms;                  Load resistance.

* ZCS DRIVER
EZCS     11  0   VALUE={LIMIT(v(10)+i(VX)*1000,-1,1)}

* ANALYSIS
.TRAN    200ns   150us   0   200ns   UIC
.PROBE
.END
»»»»»»»»»»»»»»»»»»»»»»»»»»»»»»»»»»»»»»»»»»»»»»»»»»»»»»
```

STEP 3 The third step is to run the PSpice program with the circuit file EX10_3_2.CIR.

STEP 4 The fourth step is to run PROBE with EX10_3_2.DAT to plot the traces. The results are displayed in Fig. 10.12.

Part (a) of Solution. From a PROBE trace, the average load voltage is 40V.

Part (b) of Solution. From the same PROBE trace, the peak overshoot of the load voltage is 72.24V.

Part (c) of Solution. From the PROBE traces in Fig. 10.12, the peak switch current is 115.9A, the peak current in the diode across the switch is 105.8A and the peak switch voltage is 224V.

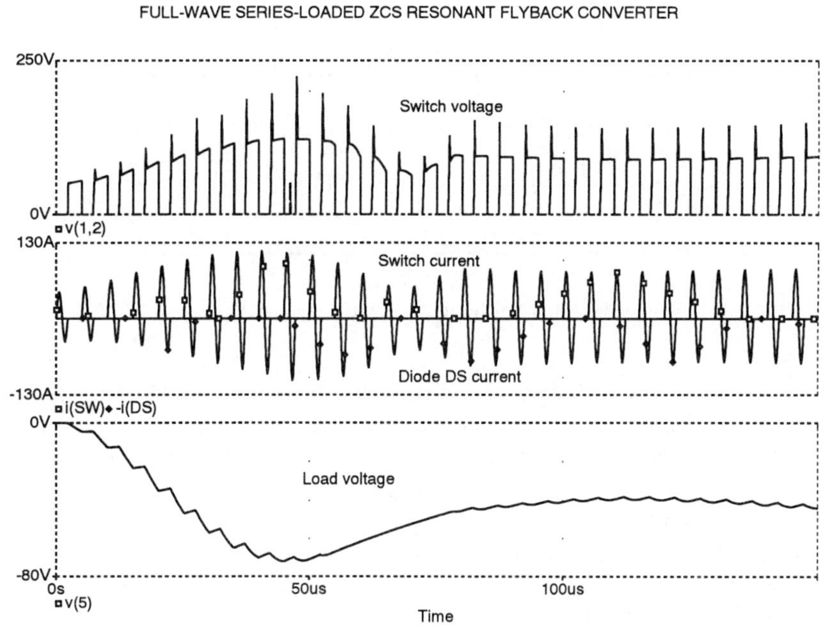

Fig. 10.12

END OF EXAMPLE 10.3.2

Drill Exercise D10.3.3
Repeat EXAMPLE 10.3.2 using an ideal BJT model in place of the ideal switch *Sw*. What other kinds of power switches might be used in this high frequency converter?

10.3.2 ZERO-VOLTAGE SWITCHING CONVERTERS

Zero-voltage switching (ZVS) converters are yet another class of resonant-switch converters which provide *lossless* switching. As the name implies, switching losses are eliminated by switching when the voltage across the switch is zero. A special ZVS driver is used and this is quite different from the ZCS driver described in the previous section. We will start by examining a zero-voltage switching buck converter as illustrated in Fig. 10.13. It is a traditional buck converter with the addition of a resonant circuit and a special driver.

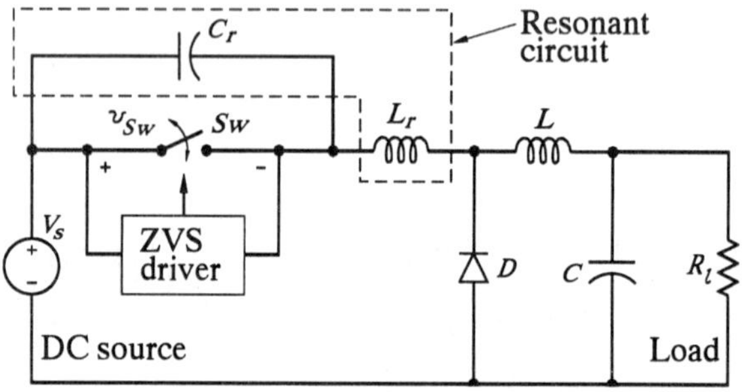

Fig. 10.13 ZVS buck-converter circuit diagram.

A ZVS driver in the form of an analogue behavioural model is illustrated in Fig. 10.14. In the previous section we observed that ZCS converters resonate during the switch's on-time where switch-off is set to correspond with zero-current instants. In the case of ZVS converters, the resonant cycle occurs during the switch's off-time ($t_{OFF} = (1-m)T$). During this off-time the voltage across the switch traverses zero to its peak and back down again to zero. At this point of zero voltage the switch is reactivated and lossless zero-voltage switching at turn-on is achieved. A trigger source denoted by Vclock produces a brief off-pulse to begin each resonant off-cycle. The driver ensures the switch is then latched off until the voltage across it returns to zero. When the switch voltage returns to zero, the switch is allowed to turn on once again and the cycle repeats. The switching frequency in this case is fixed and the cycle ends when another trigger pulse from Vclock arrives.

The average voltage V_{lav} across the chopper load is given by

$$V_{lav} = mV_s = (t_{ON}/T)V_s = ((T-t_{OFF})/T)V_s. \qquad (10.3.1)$$

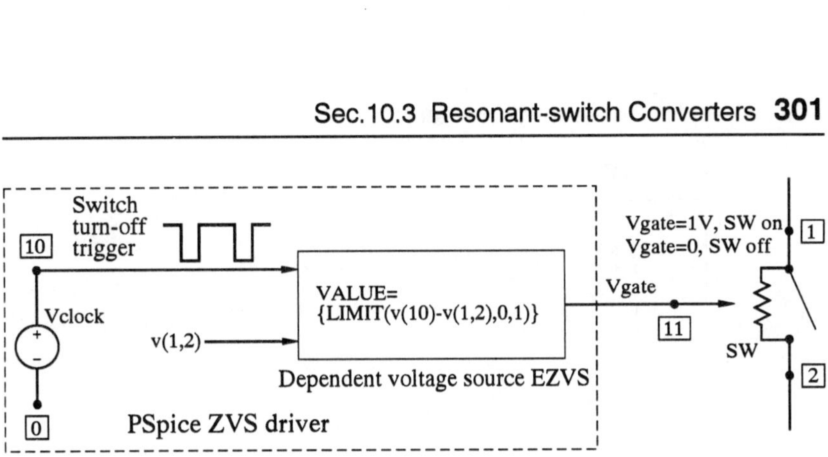

Fig. 10.14 ZVS driver configuration.

For the condition of zero-voltage switching, the switch's off-time t_{OFF} is determined by the resonant-circuit frequency $f_r = 1/T_r$; that is, $t_{OFF} = T_r/2$. This is fixed by the resonant parameter values to be $T_r \approx 2\pi\sqrt{L_r C_r}$. Consequently, the controlling variable for changing the load voltage is T, the period of the clock pulse (and the period of switching *Sw*).

EXAMPLE 10.3.3

Simulate the half-wave ZVS buck converter shown in Fig. 10.13. Use a fixed frequency of 100kHz for the ZVS driver in Fig. 10.14. Use component values $L_r = 1\mu$H, $C_r = 400$nF, $L = 10\mu$H, $C = 10\mu$F, $R_l = 1$ohm, $V_s = 12$V. Confirm the zero-voltage switching operation of this circuit by using PROBE to plot the trigger voltage Vclock, the switch control voltage Vgate and the actual switch voltage v_{Sw}.

Solution

There are four steps to achieve the solution.

STEP 1 Transform the circuit into PSpice elements and nodes. A configuration is illustrated in Fig. 10.15 on the next page.

STEP 2 The next step is to construct the PSpice circuit file. The ZVS-driver behavioural model, shown in Fig. 10.14, was created using a voltage-controlled voltage source E. Initial conditions are used to ensure proper circuit start-up. Since the switch voltage must be zero for the switch to be activated, the voltage across CR is set to zero initially.

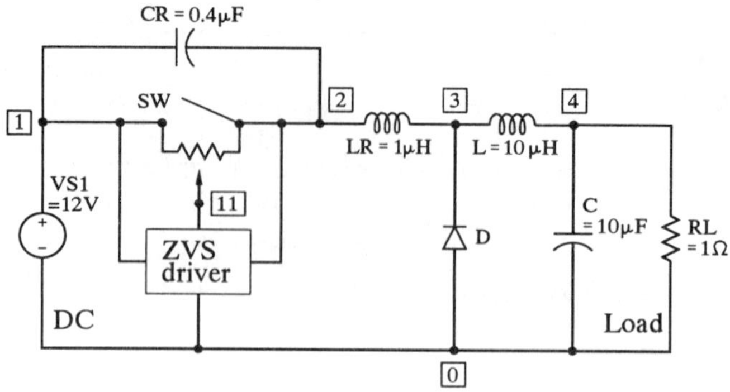

Fig. 10.15 PSpice configuration of ZVS resonant buck converter.

```
»»»»» EX10_3_3.CIR »»»»»»»»»»»»»»»»»»»»»»»»»»»»»»»»»»»
HALF-WAVE SERIES-LOADED ZVS RESONANT BUCK CONVERTER
* The POWER library is used for the switch's model parameters.
. LIB     POWER.LIB
. INC     OPTIONS;                              Convergence aid.
* PARAMETERS
. PARAM  FREQ=100kHz  PERIOD={1/FREQ};          Switching frequency.
. PARAM  VS=12V;                                DC source voltage.
. PARAM  RLOAD=1ohm;                            Load resistance.
. PARAM  LLOAD=10uH
. PARAM  CLOAD=10uF
* SOURCES
VS1    1  0  DC {VS};                           The dc source voltage.
Vclock 10 0  PULSE(0 2 0 2ns 2ns {PERIOD-1us} {PERIOD})
* Vclock sets chopper frequency.
* DEVICES
SW    1  2  11  0  SWITCH ;               The ZVS switch in POWER.LIB.
D     0  3  D_IDEAL ;                     The diode and model parameters.
. MODEL  D_IDEAL  D(RS=1m  CJO=0.1pF  N=0.001)
* CIRCUIT ELEMENTS
CR    1  2  400nF  IC=0;                  Resonance for switch off-time.
LR    2  3  1uH;                          Lr, Cr form resonant network.
L     3  4  {LOAD};                       L, C form output filter.
C     4  0  {CLOAD}
RL    4  0  {RLOAD};                      Output load resistance.
* ZVS DRIVER
EZVS  11 0  VALUE={LIMIT(v(10)-v(1,2),0,1)};   Switch gate signal.
```

```
* ANALYSIS
* A time-step ceiling is used to increase the plot resolution.
.TRAN   50ns   16us   0   50ns   UIC
.PROBE  v(11),  v(1,2),  v(10);              Gate, switch and clock voltages.
.END
```

» »

STEP 3 The third step is to run the PSpice program with the circuit file EX10_3_3.CIR.

STEP 4 The fourth step is to run PROBE and view the results which are displayed in Fig. 10.16. The switch's off-voltage v_{SW} is sinusoidal in nature and it is seen from the trace why this circuit is referred to as a half-cycle converter. When viewing the switch voltage it is apparent that the peak off-voltage is high. This can be a concern for devices whose blocking voltage is low (possibly causing avalanche conditions). The design engineer can use transient PROBE plots such as these to verify the peak voltage requirements of a power switch.

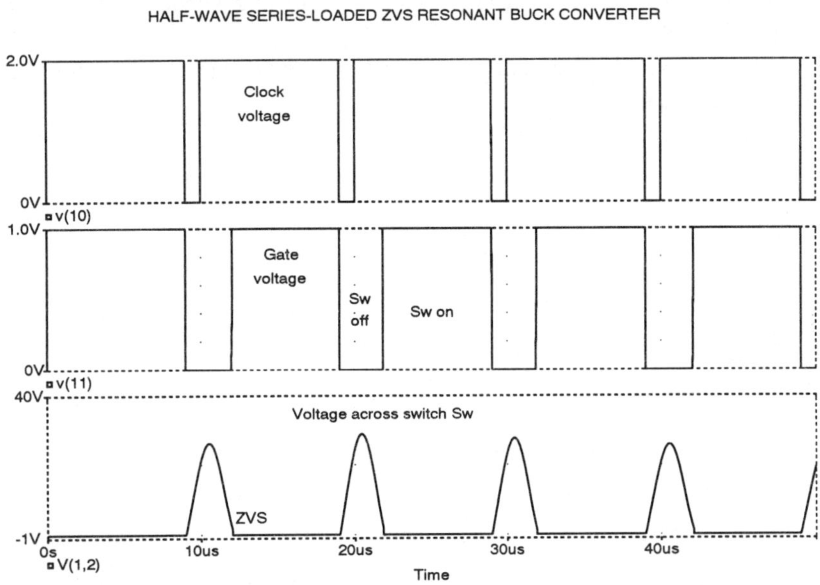

Fig. 10.16

END OF EXAMPLE 10.3.3

Drill Exercise D10.3.4

Consider the circuit used in EXAMPLE 10.3.3 and double the operating frequency. Determine new values for L_r, C_r, L and C. Use PROBE to display the load voltage.

10.4 INTEGRAL-CYCLE CONVERTERS

Integral-cycle converters can perform direct ac to ac conversion without intermediate ac-dc, dc-ac conversion and it can be accomplished with lossless switching. See PESS FIG. 1.5.

The basis of integral-cycle control is that a sinusoid can be approximated by combining and removing half cycles from a higher frequency ac source. Since only half or full cycles are ever used, the converter switches are turned on or off only during half-cycle boundaries at which the input voltage passes through zero. As a result, these converters achieve zero-voltage switching without the need for a resonant circuit.

The WEBsite gives an addendum to this Section and includes EXAMPLES and Drill Exercises.

Chapter 11

Advanced Topics

11.1 INTRODUCTION

In Chapters 1 through 10 we have given a guide to the use of PSpice for power electronics circuits. The analyses have been completely concerned with system responses. Because detailed device characteristics do not alter system responses significantly, we have been able to model devices by near-ideal characteristics (see Chapter 4).

There are situations for which it is desired to know the detailed characteristics of a device. Such situations may involve having semiconductor switches connected in series and parallel, or the need to determine the maximum frequency of operation of a particular switch from the device's average power dissipation.

In order to ascertain the performance of a device, PSpice gives us choices of models. If the production version of PSpice is used, there are libraries of particular devices available. The list contains the names of over nine thousand devices, whose detailed models are provided. The model may be in the form of a subcircuit or model parameters. In the evaluation version of PSpice there are a few representative models of devices described†. We can use these models.

There are other options to enable us to use detailed models of devices in power electronics circuits. We can use the available PSpice models and enhance them by using analogue behavioural models (dependent sources) and ideal circuit elements to represent more accurately the known characteristics. We could use macro models, that is, models built entirely from analogue elements†† (circuit elements and mathematical expressions).

A useful book for SPICE models of diodes and power transistors is "SPICE Practical Device Modeling" by Ron M. Kielkowski, (McGraw-Hill, Inc., 1995) (1SBN 0-07-911524-1)

† The list is provided at the beginning of EVAL.LIB. It is worth browsing through this file.

†† Some device manufacturers provide these PSpice models.

There is yet another facility provided by PSpice that allows us to characterize the behaviour of a device. PSpice calls it the PARTS option. PARTS is a processor. For example, in DIODE.LIB of the production version 6.0, there are over 340 diodes described. The method of description is the .MODEL statement, that gives the parameter values associated with the physics of the device. We might have designed our own diode or obtained a commercial diode not listed in the library. This is where PARTS comes to our help. As long as we have experimental results of the device's external characteristics, or data sheets from the manufacturer, we can obtain .MODEL parameter values of our diode. We provide the external data to PARTS, and PARTS does the transformation and provides us with parameter values that a PSpice simulation can use in the analysis†.

The PSpice OPTIMIZER (formerly PARAGON) is another facility of PSpice (Windows version). For a required optimum condition, PSpice can be run a number of times to determine a particular parameter value that will satisfy the condition. This is useful for any circuit design.

PERFORMANCE ANALYSIS is a facility within PROBE to allow us to plot variables for specified conditions. The specifications are set out as macros in special files that are used by PROBE to find particular data that are required for a particular plot.

We will complete this introductory guide to PSpice for power electronics by giving a brief explanation of PARTS, PSpice OPTIMIZER and PERFORMANCE ANALYSIS. Examples will be chosen to incorporate detailed device modelling.

11.2 THE JUNCTION DIODE MODEL

SPICE2 has a model of a diode based on the physics of the device and PSpice uses this model with added enhancements. The detailed modelling is well documented††. Essentially, a diode in a circuit, like that depicted in Fig. 11.1a, will have a dc *I-V* characteristic similar in form to that shown in Fig. 11.1b. The large-signal dc equivalent circuit for the static characteristic is illustrated in Fig. 11.1c. Over a large range of operation the equations governing the dc behaviour of the diode are

† See MicroSim PSpice A/D Circuit Analysis User's Guide, V6.2 for DOS, Chapter 10.

†† "Semiconductor device modelling with SPICE" by G. Massobrio and P. Antognetti, McGraw-Hill, Inc., 2nd edition, 1993.

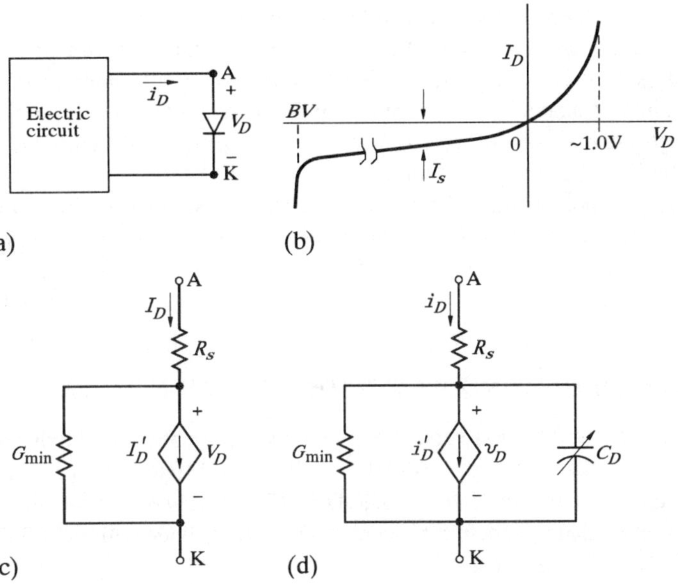

Fig. 11.1 The diode. (a) Circuit diagram, (b) dc *I-V* characteristic, (c) dc equivalent circuit, (d) transient equivalent circuit.

$$V_{AK} = R_s I_D + V_D \tag{11.2.1}$$

and

$$I_D = I_s\left(e^{\frac{qV_D}{nkT}} - 1\right) + G_{min} V_D \tag{11.2.2}$$

where
V_{AK} = voltage, anode to cathode,
R_s = resistance of the neutral regions of the *pn* device and the contacts,
I_D = equivalent current source, $I_D = f(V_D)$, according to eq. (11.2.2),
V_D = diode internal voltage drop,
I_s = reverse saturation current,
q = electron charge,
n = emission coefficient,
k = Boltzmann's constant,
T = junction temperature in kelvin and
G_{min} = SPICE minimum conductance (for convergence).
There are three other regions modelled for reverse bias conditions.

For the large-signal transient behaviour the equivalent circuit, depicted in Fig. 11.1d, allows C_D to model the charge storage in the depletion layer (C_J) and the charge storage due to minority carriers injected into the neutral regions (C_d).

There is a small-signal ac model of the diode. This model is derived from the large-signal model by linearizing about an operating point. This is transparent to us because PSpice does it for us.

Let us look at some of the different options that are available to determine the detailed performance of a junction diode. The options are 1. a library model, 2. a model with known parameters, 3. a hybrid model and 4. a model created by the PARTS option.

11.2.1 LIBRARY MODELS OF THE JUNCTION DIODE

If we use the production version of PSpice we can check the library reference manual for generic diode names such as 1N6098 or MBR8045 or SD51 to see if one of them is in the model library DIODE.LIB. If we use the evaluation version of PSpice we find that there is a model of a switching diode with the PSpice name D1N4148 in EVAL.LIB.

In the right circumstances, we can use one of these diodes in a circuit without knowing anything about the diode. In fact, it is probable that we have the manufacturer's data sheets on the diode, because we would have to choose a diode with required ratings of voltage, current, speed of switching, etc., compatible with a particular circuit application.

EXAMPLE 11.2.1

Examine the dc (static) large-signal model of a diode whose *area* factor is equivalent to 10 times the diode D1N4148 that is found in EVAL.LIB. From a PSpice simulation, plot the *I-V* characteristic of the diode. Determine **(a)** the value of the voltage drop across the diode for a diode current of 0.9A, **(b)** the value of the current conducted by the diode if the diode voltage drop is 1V and **(c)** the value of the reverse leakage current for a reverse-bias voltage of −1V.

Solution

There are four steps to achieve the solution.

STEP 1 The first step is to set up a PSpice configuration with an adjustable dc source and a diode under test (DUT). This is shown in Fig. 11.2.

Fig. 11.2
PSpice configuration for *I-V* test.

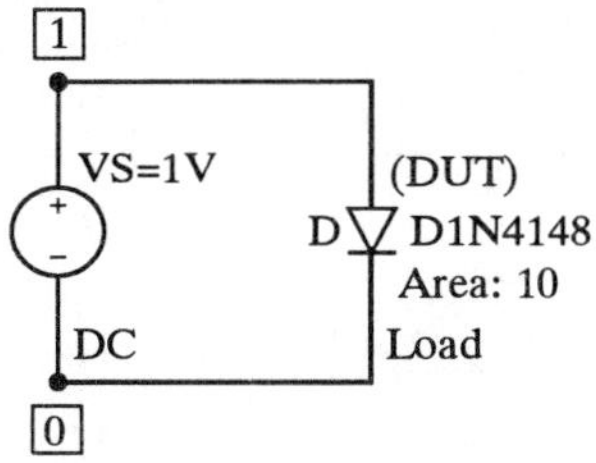

STEP 2 In order to write the circuit file we must know the diode's model name and we must know in which file the diode's model parameters reside. The general statement is

D*xxx* N+ N- model_name <area_factor>

where D*xxx* is the diode name (beginning with D); and N+ and N− are the numbers given to the anode and cathode nodes respectively.
Model_name is the diode's model name (eg. D1N4148) that is also in the .MODEL statement. In this case, the .MODEL statement is in EVAL.LIB. The area_factor is a number pertaining to the equivalent number of D1N4148 diodes in parallel, so that the current rating is increased by the area factor.

```
»»»»» EX11_2_1.CIR »»»»»»»»»»»»»»»»»»»»»»»»»»»»»»»»»»»»

STATIC I-V CHARACTERISTIC OF DIODE D1N4148 (X10)
* To determine the dc large-signal characteristic.

. INC      OPTIONS;                    Convergence aid.
. LIB      EVAL.LIB

* PARAMETER
. PARAM  AREA=10;       Area factor.  Equivalent to 10 diodes in parallel.

* SOURCE
VS    1   0   DC  10V

* DEVICE UNDER TEST
DIODE    1   0   D1N4148 {AREA};       The area_factor is 10.
* The diode's model parameters are in EVAL.LIB.

* ANALYSIS
. DC       VS -5,  1,  0.1;                A VS sweep in 0.1V increments.
. PROBE
. END

»»»»»»»»»»»»»»»»»»»»»»»»»»»»»»»»»»»»»»»»»»»»»»»»»»
```

STEP 3

A PSpice simulation can be run with the circuit file EX11_2_1.CIR. To plot the diode's *I-V* characteristic a dc sweep of the voltage is performed in the analysis.

STEP 4

Using PROBE with the file EX11_2_1.DAT, a trace of the diode current versus the diode-voltage drop can be plotted. See Fig. 11.3.

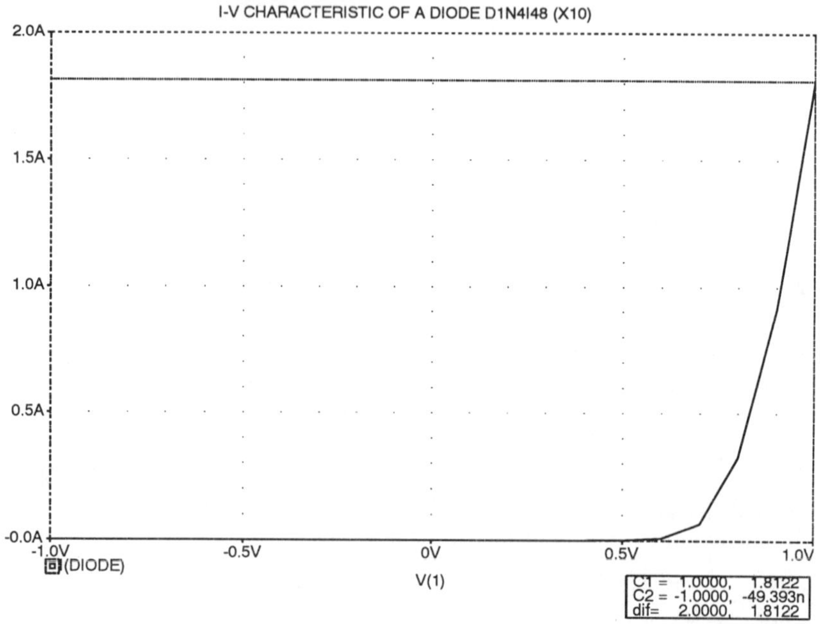

Fig. 11.3

Part (a) of Solution. For a diode current $I_D = 0.9\text{A}$, the diode-voltage drop is $V_{AK} \approx 0.9\text{V}$.

Part (b) of Solution. From the cursor box in Fig. 11.3, the current is $I_D = 1.81\text{A}$ for a voltage drop $V_{AK} = 1\text{V}$.

Part (c) of Solution. From the cursor box in Fig. 11.3, the diode's leakage current is $I_D = -49.4\text{nA}$ for a reverse bias $V_{AK} = -1\text{V}$.

END OF EXAMPLE 11.2.1

Drill Exercise D11.2.1
Do a PSpice simulation and plot the *I-V* characteristic of the diode D1N4148 with an effective area factor of 20. Plot the trace I_D versus V_{AK} over a range −106V, 2V. Determine (**a**) the value of the reverse breakdown voltage V_{BV} of the diode, (**b**) the value of the diode's reverse saturation current I_s and (**c**) the value of the diode current for a voltage drop V_{AK}=2V.
(Ans: (**a**) 100V, (**b**) ~224nA at −80V, (**c**) 33.9A.)

11.2.2 A DIODE'S MODEL PARAMETERS

PSpice characterizes the junction diode by means of parameter values. In a circuit file, the diode is described by a data statement and a control statement.

For example,

```
DIO 1 0 D_IDEAL 2;            2 is the area factor.
.MODEL D_IDEAL D(RS=1m CJO=0.1pF N=0.001)
```

The diode named DIO in the circuit file has a model name D_IDEAL to link the diode with a .MODEL statement. In this case, three parameter values are given. They are the neutral-region resistance R_s, the zero-bias junction capacitance C_{JO} (C_D for zero bias in Fig. 11.1d) and the emission coefficient *n*. The parameters, that are not given in the statement, assume default values. The parameter values R_s and C_{JO} are for an area-factor 1 (default value). For an area-factor 2, PSpice divides the value of R_s by 2 and multiplies the value of C_{JO} by 2. Most of the model parameters are listed in Table 11.1 on the next page.

Table 11.1 shows that there are many parameters that PSpice uses in the equations to analyze the diode behaviour. There are even more parameters† relating to temperature dependency. In any analysis the temperature is assumed constant (27°C by default). Some parameters are temperature dependent. Account is taken of temperature by setting the operating temperature in the .TEMP control statement such as

```
.TEMP 50deg, 75degC;          Two simulations (a sweep).
```

The parameters BV and IBV are specified in absolute terms in the .MODEL statement.

† See MicroSim PSpice A/D Circuit Analysis Reference Manual (eg. V6.2 for DOS, Chap. 4 p104).

Table 11.1 PSpice diode parameters.

Model	Parameter	Dependency[1]	Default value
IS[2]	Saturation current (A)	A, T	1E−14
N	Emission coefficient		1
RS[2]	Parasitic resistance (Ω)	A, T	0
IKF[3]	High-injection knee current (A)	A, T	∞
XTI	IS temperature exponent		3
EG	Energy gap (eV)		1.11
CJO[2]	Zero-bias junction capacitance (F)	A,T	0
M	CJ grading coefficient		0.5
VJ	Junction potential (V)	T	1
FC	CJ forward-bias coefficient		0.5
ISR[3]	Recombination current parameter (A)	A, T	0
NR[3]	Emission coefficient for ISR		2
BV[2]	Reverse breakdown (V)	T	∞
IBV[2]	Current at BV (A)	A	1E−10
TT[2]	Transit time (s)		0
NBV[3]	Reverse breakdown ideality factor		1
IBVL[3]	Low-level reverse-breakdown knee current (A)		0
NBVL[3]	IBVL ideality factor		1
KF	Flicker noise coefficient		0
AF	Flicker noise exponent		1

1 “A” relates to the area factor and “T” relates to temperature.
2 Parameters important in power electronics.
3 PSpice enhancements.

The parameters given in Table 11.1 are not given by device manufacturers in their data sheets. Data sheets are usually the only source of device characteristics available to us, if the device description is not in a PSpice library file. Consequently, there must be some means of extracting the parameter values from the data sheets. PSpice can help us with its PARTS option or we can use what is described in the literature†. There are some limitations. For example, RS, N and TT are constants for simulation. In real life, the resistance R_s of the neutral regions changes with the level of injection, as does the emission coefficient n, and the reverse recovery time t_{rr} is a function of current. Therefore, the use of the above parameters does create some error.

† See Motorola Rectifier Applications Handbook, Chap. 3, the SPICE Diode Model, by H. T. Russell, 1991.

In some analyses of power electronics circuits this may be of little concern. However, in the study of forward recovery and reverse recovery during switching, for example, there can be significant differences between simulation and real results. To combat the parameter limitations it is possible to add circuit elements† to the diode model.

In Table 11.1, some of the parameters have default values of zero. To try to avoid nonconvergence problems it is useful to have finite values of RS and CJO.

In this section we will use typical values of model parameters without going through the exercise of parameter determination by calculation.

EXAMPLE 11.2.2 (PESS EXAMPLE 3.2)

Consider the chopper circuit illustrated in Fig. 11.4. The diode D_F has the following specifications at 27°C.

Diode ratings 1000V, 100A. Parameter values IS = 3μA, N = 2.8, RS = 21mΩ, CJO = 30pF, VJ = 0.3V, BV = 1000V, IBV = 0.14A, TT = 50ns.

Junction operating temperature 50°C.

Do a PSpice simulation for steady conditions. Plot traces of the diode current and diode voltage during a diode reverse recovery interval. Determine **(a)** the diode current slew rate (di_D/dt) for turn-off, **(b)** the reverse-recovery time t_{rr}, **(c)** the softness factor $S = t_f/t_r$, **(d)** the reverse-recovery current I_{RR}, **(e)** the storage charge Q_{rr} and **(f)** the reverse-recovery voltage V_{RR}.

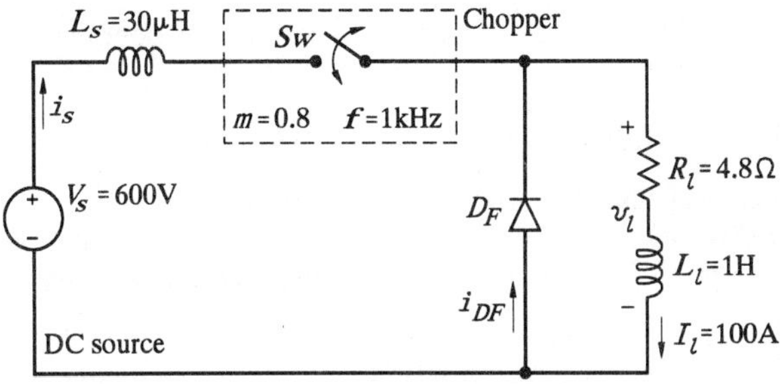

Fig. 11.4 Freewheeling diode in a chopper circuit.

† Liang. Y. and V. J. Gosbell, "Diode Forward and Reverse Recovery Model for Power Electronic SPICE Simulations", IEEE Trans Power Electronics, Vol. 5, No. 3, July 1990, pp 346-356.
F. Bertha, et al. "An Improved Power Diode Model for PSpice, Applied to Converter Simulation" European Power Electronics Conference, Brighton, September 1993, pp 249-254.

Solution

There are four steps to achieve a solution.

STEP 1 The circuit diagram in Fig. 11.4 can be transformed into a PSpice configuration as shown in Fig. 11.5b. We will not operate the switch as a chopper. Instead, we will allow the initial conditions that the switch SW is open and the load current is 100A (mV_s/R_l). At $t = 1\mu s$ the switch SW is allowed to close and at $t = 1$ms the switch is opened again. In this way we can obtain an operational picture of diode turn-off.

STEP 2 From the given specifications and from the PSpice configuration in Fig. 11.5, the circuit file can be written. Account is taken of the diode's junction temperature.

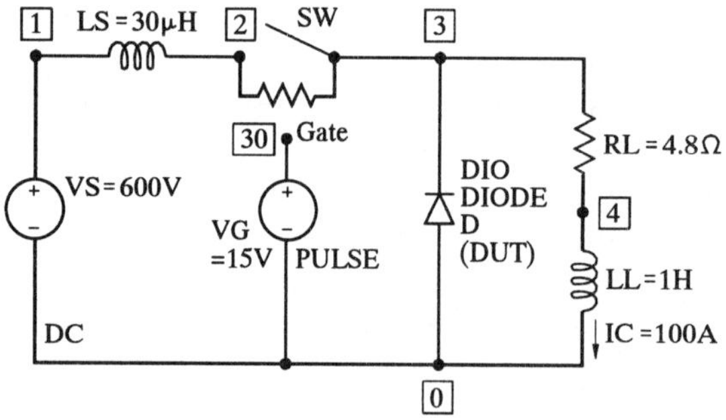

Fig. 11.5 PSpice configuration for diode under test.

```
»»»»» EX11_2_2.CIR »»»»»»»»»»»»»»»»»»»»»»»»»»»»»»»»»»»»»

DIODE FORWARD AND REVERSE RECOVERY
.INC     OPTIONS;                              Convergence aid.
.TEMP    50degC;       Resistors and the given diode parameters are modified.

* PARAMETERS
.PARAM  FREQ=1kHz  PER={1/FREQ};               Switching frequency.
.PARAM  TR=2ns     TF={TR};                    Pulse rise and fall times.
.PARAM  DEL=1us;                               Pulse delay.
.PARAM  TON={PER-DEL-2*TR};                    Pulsewidth.
.PARAM  VP=15V;                                Pulse magnitude.
.PARAM  VS=600V;                               DC source voltage.
.PARAM  RLOAD=4.8ohms
.PARAM  LLOAD=1H
.PARAM  LSOURCE=30uH
```

```
* SOURCES and ELEMENTS
VS1    1   0   DC  {VS}
VG     30  0   PULSE(0 {VP} {DEL} {TR} {TF} {TON} {PER})
* VG is the switch control voltage.
RL     3   4   {RLOAD}
LL     4   0   {LLOAD}       IC=100A
LS     1   2   {RSOURCE}  IC=0

* DEVICES
SW     2   3   30   0    SWITCH;                  Voltage-controlled switch Sw.
. MODEL   SWITCH     VSWITCH(RON=1E-3 ROFF=1E6
+                                VON=10V  VOFF=1E-3)
DIO    0   3   DIODE
. MODEL   DIODE   D(IS=3u  N=2.8  RS=21m  CJO=30pF  VJ=0.3
+                       BV=1000V  IBV=0.14  TT=50ns)

* ANALYSIS
. TRAN      0.1us 10us  0    10ns      UIC
. PROBE
. END
```

» »

STEP 3 A PSpice simulation can be run. To obtain results with reasonable resolution, the run time is kept short (TSTOP=10us) and the maximum time step is kept short (TMAX=10ns).

STEP 4 Using PROBE, we can plot traces of the diode voltage $v_D \equiv$ v(0,3), the integral of the diode current S(i(DIO)) for the storage charge Q_{rr}, and the diode current $i_D \equiv$ i(DIO)). Figure 11.6, depicted on the next page, shows the traces *zoomed* to cover the interval from about 5.9μs to 6.18μs.

Part (a) of Solution. From the diode current trace, the current falls from 100A to zero in 4.98μs. Thus the slew rate is approximately 20A/μs.

Part (b) of Solution. From cursor measurements on the diode current trace, the reverse recovery time is t_{rr} = 125ns.

Part (c) of Solution. From cursor measurements on the diode current trace, the current rise time is t_r = 55.07ns and the current fall time is t_f = 69.89ns. Therefore, the softness factor is $S = t_f/t_r$ = 69.89/55.07 = 1.27. This is not a snappy turn-off.

Part (d) of Solution. From a cursor measurement on the diode current trace, the reverse-recovery current is I_{RR} = −0.939A.

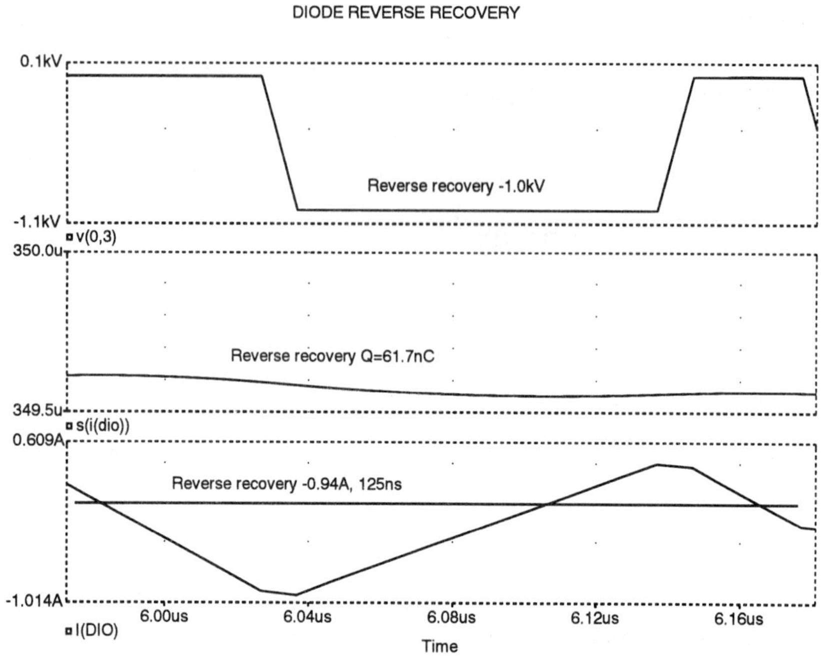

Fig. 11.6

Part (e) of Solution. The storage charge is calculated as the time integral of the diode current over the reverse-recovery interval t_{rr}. From cursor measurements on the charge trace S(i(DIO)), $Q_{rr}=61.67\text{nC}$. An approximate calculation is $Q_{rr}=\frac{1}{2}I_{RR}t_{rr}=58.7\text{nC}$.

Part (f) of Solution. From the trace of the diode voltage, the reverse-recovery voltage is $V_{RR}=-1\text{kV}$. This is the breakdown limit.

END OF EXAMPLE 11.2.2

Drill Exercise D11.2.2

Repeat EXAMPLE 11.2.2 to study the forward recovery characteristic. Make any reasonable modifications to the circuit (eg. let $L_s=0$ and add stray inductance in the diode branch). The junction temperature is 50°C. Plot traces of the diode current and voltage over the interval of turn-on for steady conditions. Determine **(a)** the diode's current rise time, **(b)** the diode's voltage recovery time for turn-on and (c) the steady forward voltage drop. (d) Comment on the diode's voltage waveshape. (Ans: (a) 0.9ns, (b) 0.9ns, (c) 3.09V, (d) no voltage overshoot, because conductivity modulation is not modelled.)

Drill Exercise D11.2.3

The circuit diagram in Fig. 11.7 represents the JEDEC reverse-recovery circuit for practical tests on a diode under test. Transform the circuit diagram to a PSpice configuration. Do a simulation for the diode D1N4148 (found in EVAL.LIB). Let the diode's forward current be 1A (area factor of 10), the slew rate 50A/μs, and a junction temperature of 100°C. Plot a trace of the reverse-recovery current of the diode and determine (a) the reverse-recovery time t_{rr} and (b) the peak reverse-recovery current I_{RR}.

Fig. 11.7

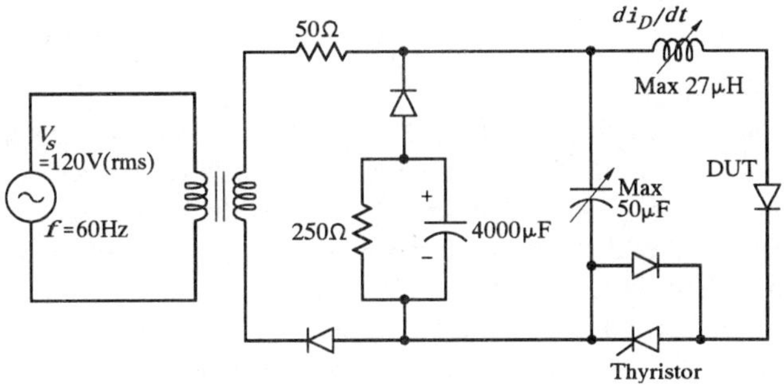

Drill Exercise D11.2.4

Consider the PSpice configuration shown in Fig. 11.8. Do a PSpice simulation for a junction temperature of 27°C to determine (a) the reverse-recovery time t_{rr} (b) the peak reverse-recovery current I_{RR} and (c) the slew rate di_D/dt. (Ans: (a) 18.8ns, (b) 0.366A, (c) 30A/μs.)

Fig. 11.8

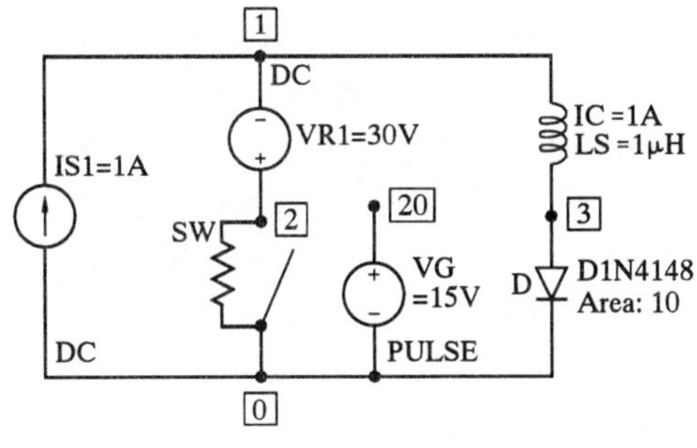

11.2.3 MODEL PARAMETERS AND PARTS

PSpice's program PARTS† transforms the terminal characteristics from data sheets to model parameters that describe the internal characteristics of a device. The model parameters are used by PSpice for analysis. The evaluation Version 6.1 of PSpice allows PARTS to be run only for diode parameters, so we will take advantage of this. The following example explains how the model parameters of a device can be obtained.

EXAMPLE 11.2.3

Use PSpice's program PARTS to determine the parameter values (listed in Table 11.1) of a power junction diode rated at 100V, 30A.

Solution

The external characteristics of this 100-V, 30-A diode are similar to Motorola's diode MUR3010PT. This commercial package comprises two diodes. If the diodes were connected in parallel the parameters of the pair would have an AREA FACTOR of 2. Here, the device is considered to be a single equivalent diode with an AREA FACTOR of 1.

It is assumed that the computer is on and that we are in the PSpice subdirectory. Check that the subdirectory contains the file PSPICE.DEV. If there is no such file, either refer to the installation manual regarding SETUPDEV, or create the file after typing SETUPDEV, hitting <ENTER>, and following the instructions that appear on the screen.

1. At the DOS prompt type PARTS <ENTER> A list of device types appears.
2. For the diode select <1> A command line appears.
3. Type in a device name such as dRAM3010 <ENTER>
   A screen appears for the diode's dc forward characteristic. Inspect this screen and explore the menu. If you get into trouble hit <ESC> and start again. We need to enter three data points.
4. From the menu, select **X_axis** by hitting <X>
5. From the submenu, select **Set_range** by hitting <S>
6. Enter a suitable range 0.4V,1.6V <ENTER>, and <ESC>
7. From the menu, select **Y_axis** by hitting <Y>

---

† For more details see MicroSim PSpice A/D Circuit Analysis Reference Manual, Version 6.2 for DOS, Chap. 10, and Circuit Analysis User's Guide, Chap. 15.

8. From the submenu, select **Set_range** by hitting <S>
9. Enter a suitable range 0.2A,100A <ENTER>, and <ESC>
   Now, we are ready to enter data-sheet values from the characteristic of forward current versus forward voltage at 25°C.
10. From the menu, select **Device_curve** by hitting <D>
11. From the data sheet enter the value of V_{FWD} 0.9V <ENTER>
12. Enter value of I_{FWD} 20A <ENTER> for $V_{FWD} = 0.9V$.
    If you have made a mistake with an entry, instead of escaping (or exiting) you can use **Change** or **Delete**
    At this point the model parameter IS changes to 99.9E-3A.
13. Select **Add**
14. Enter a new value of V_{FWD} 0.5V <ENTER>
15. Enter the value of I_{FWD} 0.2A <ENTER> for $V_{FWD} = 0.5V$.
16. Select **Add**
17. Enter 1.3V <ENTER> for V_{FWD}
18. Enter 100A <ENTER> for I_{FWD}, for $V_{FWD} = 1.3V$
19. Select **Exit**
20. Select **Fit**, and again, until there is no change. The characteristic fit gives IS = 186.9uA, N = 2.767 and RS = 3.56mΩ.

ASIDE

21. From the menu, select **Exit**
22. Select <0> to **Exit Program**
    Following this last command, PARTS creates files DRAM3010.MOD and DRAM3010.MDT. See the subdirectory listing. An inspection of the ASCII file DRAM3010.MOD will show the diode's .MODEL statement. The file DRAM3010.MDT is in code.
23. At the DOS prompt type PARTS <ENTER>
24. Select <1>
25. Enter the device name DRAM3010 <ENTER>
    PARTS searches for the file DRAM3010.MDT. The screen data that appears has been taken from DRAM3010.MDT. Now we can continue.

---

26. Select **Next_set** for the junction capacitance screen. Browse.
27. Select **X_axis**
28. Select **Set_range**

29. Enter the range 1V,100V <ENTER> for the reverse voltage.
30. Select **Y_axis,** then **Log,** then **Set_range**
31. Enter the range 10pF,800pF <ENTER>
32. Select **Exit**
33. Select **Device_curve** Now we can select three data points from the manufacturer's reverse-bias capacitance curve.
34. Enter the value 1.9V <ENTER> for V_{REV}.
35. Enter the CJ value 600pF <ENTER> Note the list of parameter values.
36. Select **Add**
37. Enter the value 12V <ENTER> from the data sheet.
38. Enter the value 300pF <ENTER> for $V_{REV}=12V$.
39. Select **Add**
40. Enter the value 100V <ENTER> from the data sheet.
41. Enter the value 150pF <ENTER> for $V_{REV}=100V$.
42. Select **Exit,** then select **Fit** a number of times. The improved fit gives CJO=1.181nF, M=0.3858, VJ=0.3905V and FC=0.5.
43. Select **Next_set** for the reverse-leakage screen. Browse.
44. Select **Y_axis, Log,** then **Set_range**
45. Enter the range 0.01uA,0.2uA <ENTER> and <ESC>
46. Select **Device_curve,** to enter three data points.
47. Enter the value 20V <ENTER> for the reverse voltage V_{REV}.
48. Enter the I_{REV} value 0.03uA <ENTER> for $V_{REV}=20V$.
49. Select **Add**
50. Enter the V_{REV} value 50V <ENTER>
51. Enter the I_{REV} value 0.08uA <ENTER> for $V_{REV}=50V$.
52. Select **Add**
53. Enter the V_{REV} value 80V <ENTER>
54. Enter the I_{REV} value 0.16uA <ENTER> and <ESC>
55. Select **Fit** a number of times.
    The model parameters become ISR=10.01E−21A, NR=4.788.
56. Select **Next_set** for reverse breakdown.
57. Select **Model_parameters,** then select **BV**
58. For BV enter 100V <ENTER>
59. Select **IBV** and enter 20uA <ENTER> from the data sheet.
60. Select **Exit,** then **Next_set,** then **Device_data** for reverse recovery.

61. Select **1)Trr** and enter 35ns <ENTER> for t_{rr}.
62. Select **2)Ifwd** and enter 2A <ENTER>
63. Select **3)Irev** and enter 3.5A <ENTER>
64 Select **4)Rl** and enter 3ohms <ENTER>
This gives a transit time TT = 77.44ns.
65. Select **Exit** to save results to DRAM30.MOD.

An inspection of the file DRAM3010.MOD reveals the model parameters

```
D(IS=186.9uA N=2.767 RS=3.56m IKF=0 XTI=3 EG=1.11
+ CJO=1.181nF M=0.3858 VJ=0.3905V FC=0.5 ISR=10.01E-21A
+ NR=4.788 BV=100V IBV=20uA TT=77.44E-9)
```

This file DRAM3010.MOD can be found in POWER.LIB for use in the following drill exercise. In a circuit file two statements are needed to call this 100V, 30A diode into the main circuit file. They are

```
. LIB  POWER.LIB
DIODE N+ N- DRAM3010
```

END OF EXAMPLE 11.2.3

Drill Exercise D11.2.5

Consider the circuit configuration in Fig. 11.9. The turn-on time of the switch SW is to be 10ns. The device under test for reverse recovery is the diode dRAM3010 whose parameters were obtained by PARTS in EXAMPLE 11.2.3 and whose model parameters are in the file POWER.LIB. For the given circuit conditions run a PSpice simulation for an interval before and after the switch is closed. Determine (**a**) the reverse-recovery current I_{RR}, (**b**) the reverse-recovery time t_{rr} and (**c**) the reverse-recovery voltage V_{RR}.
(Ans: (**a**) -60.87A, (**b**) 13.1ns, (**c**) -43.7V.)

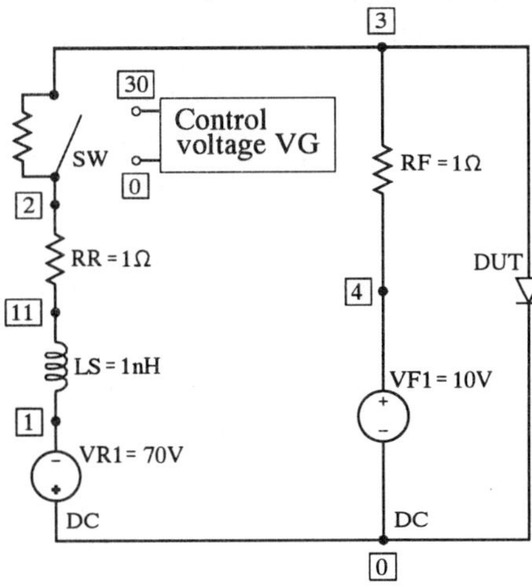

Fig. 11.9

11.3 THE THYRISTOR

In the library EVAL.LIB, PSpice has a thyristor subcircuit that begins
. SUBCIRCUIT anode gate cathode Scr PARAMS:
\+ <PARAMETER=VALUE ... > etc.
This thyristor† has parameter default values

Vdrm=400V;	Forward breakover voltage.
Vrrm=400V;	Reverse breakdown voltage.
Idrm=10uA;	Peak blocking current.
Ih=6mA;	Holding current.
dVdt=5E7V/s;	dv_{TH}/dt trigger value.
Igt=5mA;	Gate trigger current.
Vtm=1.7V;	On-state voltage $V_{TH(ON)}$.
Vgt=0.7V;	Gate trigger voltage.
Itm=24A;	Rated anode current.
Ton=1us;	Turn-on time.
Toff=15us;	Turn-off time.

Any thyristor can be modelled by using the call statements
. LIB EVAL.LIB
Xthyristor Scr N1 N2 N3 PARAMS: <PARAMETER=VALUE ... >
The parameter values must be known. They are usually obtained from the manufacturer's data sheets. See Fig. 11.10 for the symbolized subcircuit.

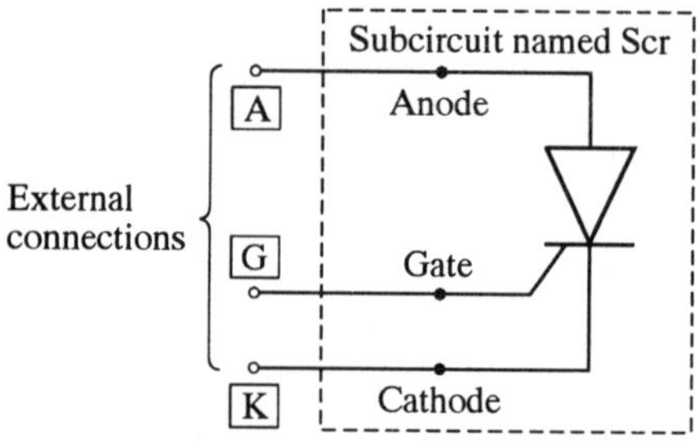

Fig. 11.10 Thyristor subcircuit configuration.

† See PESS, Chapter 5.

It is worth viewing the description of this thyristor macromodel in EVAL.LIB and drawing the PSpice configuration. It will be found that the modelling is accomplished using elements, a voltage-controlled switch for the on/off states, zero-voltage current sensors, and dependent sources for analogue behavioural modelling.

The analogue behavioural modelling of the thyristor named Scr accounts for forward breakover, for *dv/dt* turn-on, gate current i_{gt} and a positive v_{AK} for turn-on, and an anode current greater than the holding current I_h to maintain the switch on. In addition, *RC* delays are incorporated for turn-on and turn-off times, and diodes are added for reverse leakage and reverse breakdown.

There are other PSpice models†.

EXAMPLE 11.3.1 (PESS 5.1)

Consider the circuit configuration in Fig. 11.11. The inductor is to be charged from an ideal dc source, and a thyristor is to be the controller. From a PSpice simulation, plot a trace of the circuit current. Determine **(a)** the duration of the gate pulse IG1 to ensure thyristor turn-on is successful and **(b)** the thyristor turn-on time.

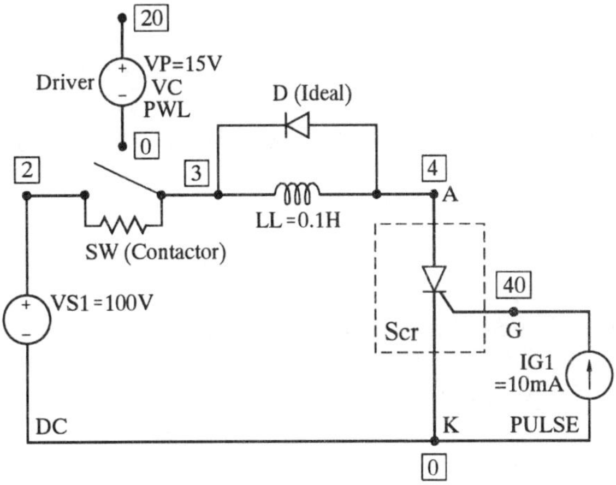

Fig. 11.11 PSpice configuration.

† For example, see Giacoletto, L. J., "Simple SCR and TRIAC PSpice Computer Models", IEEE Trans Industrial Electronics, Vol. 36, No. 3, Aug., 1989, pp451-455.

Solution

There are three steps to achieve a solution.

STEP 1 We must make a number of assumptions. The thyristor is chosen to have the default values of Scr in EVAL.LIB (see the Scr description at the beginning of Section 11.3). The contactor is closed at $t=0$, so that the dv/dt across the thyristor is less than the critical value of $50V/\mu s$. The gate current IG is applied at about $t=5\mu s$ and a multiple-run simulation is done for a sweep of the gate-current pulse time (TON). From the configuration a circuit file can be written.

```
»»»»» EX11_3_1.CIR »»»»»»»»»»»»»»»»»»»»»»»»»»»»»»»»»»»
THYRISTOR GATE TURN-ON AND LATCHING

.LIB     EVAL.LIB
.INC     OPTIONS;                                   Convergence aid.

* PARAMETERS
.PARAM  VS=100V  VP=15V   IG=10mA
.PARAM  DEL=5us  TR=10ns  TF={TR}  TON=10us  PER=1s
.PARAM  LLOAD=0.1H

* SOURCES
VS1   2  0   DC {VS};                               DC source.
VC    20 0   PWL(0,0 1us,{VP});                     Switch's control voltage.
IG1   0  40  PULSE(0 {IG} {DEL} {TR} {TF} {TON} {PER})

* CIRCUIT ELEMENT
LL    3  4  {LLOAD}

* DEVICES
SW    2  3  20  0  SWITCH;                          Contactor.
.MODEL SWITCH VSWITCH(RON=1E-3 ROFF=1E6 VON=10 VOFF=1E-3)
D     4  3  D_IDEAL
.MODEL D_IDEAL  D(RS=1m CJO=0.1pF N=0.001)
XTH   4  40 0   Scr;             Call for the thyristor from EVAL.LIB.

* ANALYSIS
.STEP    PARAM TON LIST  0.4us  0.5us  0.6us
.TRAN    0.1us   15us      0      30ns
.PROBE   i(SW),  v(4);           Thyristor current and voltage.
.END
»»»»»»»»»»»»»»»»»»»»»»»»»»»»»»»»»»»»»»»»»»»»»»»»»»»»
```

STEP 2 A PSpice simulation can be run with the circuit file EX11_3_1.CIR.

STEP 3 Using PROBE, traces can be plotted for the thyristor current and voltage for different values of the gate pulsewidth. See Fig. 11.12.

Part (a) of Solution. From the PROBE trace, the minimum pulsewidth of the gate current for thyristor turn-on appears to be $0.5\mu s$.
At $t_{ON} = 0.4\mu s$, the thyristor fails to latch on.

Part (b) of Solution. A trigger pulse is applied to the thyristor gate at $t = 5\mu s$ for $0.5\mu s$. There is a delay of $0.37\mu s$, then the thyristor voltage falls to 10% of its blocking value in $1.89\mu s$.

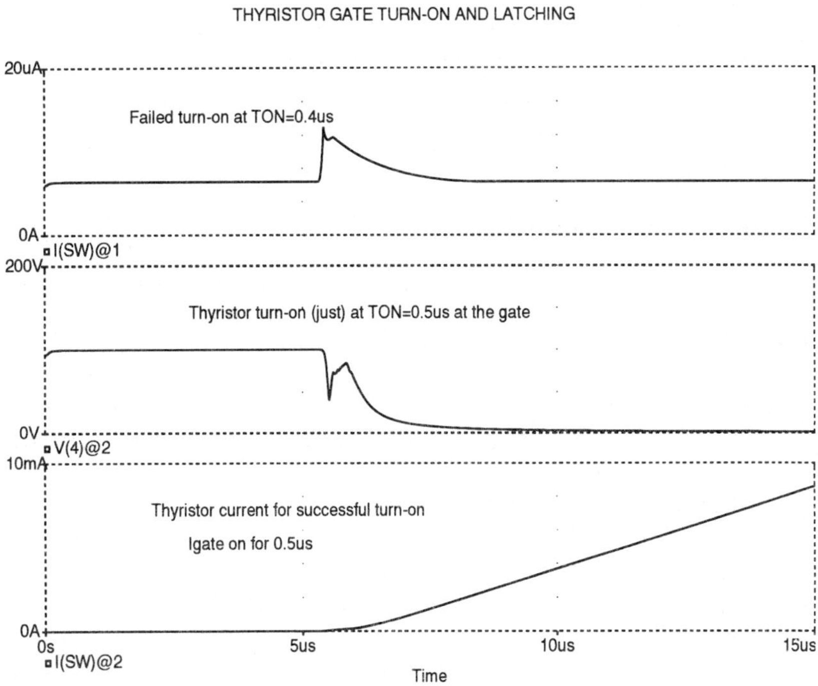

Fig. 11.12

END OF EXAMPLE 11.3.1

Drill Exercise D11.3.1
Consider the PSpice configuration in Fig. 11.11. Replace the load inductor with a 5-ohm resistor and inhibit the gating signal. Use the EVAL.LIB description of the thyristor Scr with the default parameter values. Follow the pattern of EXAMPLE 11.3.1 to find the critical *dv/dt* that turns on the thyristor. Note that *dv/dt* must be applied long enough to the PSpice thyristor so that the latching current is reached and maintains the switch on. If $V_s=100$V does not allow turn-on, try $V_s=390$V. If a voltage-controlled switch is used as a contactor, let ROFF=1E10ohms, otherwise the thyristor has a high initial voltage across it and will not turn on with *dv/dt*. (Ans: TR=15μs turns on Scr, but *dv/dt* is not constant; *dv/dt* reaches a maximum of 150V/μs.)

Drill Exercise D11.3.2
Repeat Drill Exercise D11.3.1, but replace VS1 and SW with a combined PULSE source to control the *dv/dt* across the thyristor Scr. Determine the thyristor's turn-on *dv/dt* and turn-on time.
(Ans: *dv/dt* turn-on is V_s dependent. For $V_s=390$V, critical $dv/dt=50.19$V/μs, $t_{on}=0.68\mu$s. For $V_s=60$V, critical $dv/dt=53.1$V/μs, $t_{on}=0.48\mu$s.)

11.4 THE TRIAC

The bidirectional switch, the triac (see PESS Chapter 8), is modelled as a near-ideal device in Chapter 4. In EVAL.LIB there is a PSpice macromodel of a triac. It is modelled for room temperature operation and is a modified version of the thyristor subcircuit named Scr with two PSpice voltage-controlled switches. The level of the modelling is described in terms of the given parameters in the subcircuit. With reference to Fig. 11.13, the EVAL.LIB's subcircuit for the triac has a first line that is

```
.SUBCKT Triac MT2 gate MT1 PARAMS:
+  Vdrms=400V;     Forward breakover voltage.
+  Idrm=10uA;      Peak blocking current.
+  Ih=6mA;         Holding current for MT2 positive.
+  dvdt=50E6V/s;   Critical dv/dt turn-on.
+  Igt=20mA;       Gate current for turn-on. MT2 positive, gate negative.
+  Vgt=0.9V;       Gate voltage for turn-on. MT2 positive, gate negative.
+  Vtm=1.3V;       On-state voltage, VTr(ON).
+  Itm=17A;        Rated current.
+  Ton=1.5us;      Turn-on time.
```

Fig. 11.13
EVAL.LIB's external connections for the triac.

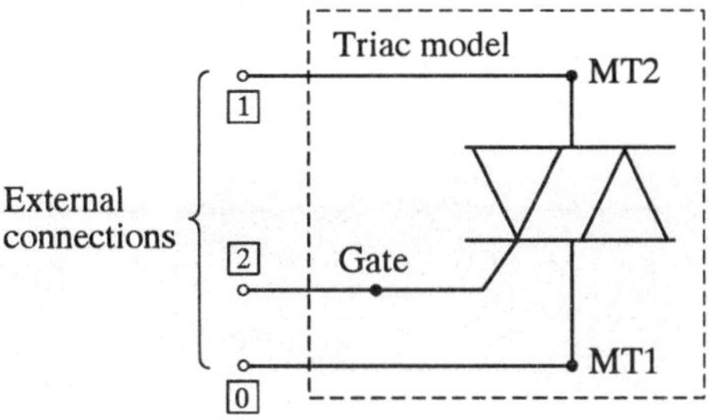

For any other triac, whose parameters can be determined from the manufacturer's data sheets, we can use PSpice's model. For example, for the commercial triac 2N5444, rated at 200V, 56A, we could use the 2N5444 in a simulation using the two statements

```
.LIB      EVAL.LIB
Xtriac 1 2 0 Triac PARAMS: Vdrm=200V Idrm=10uA Ih=70mA
+   dvdt=50E6 Ton=1us Igt=70mA Vgt=2V Vtm=1.65V Itm=56A
```

In fact, the 2N5444 triac, described in EVAL.LIB, is called by

```
.LIB   EVAL.LIB
Xtriac    1 2 0 2N5444
```

View this triac in EVAL.LIB together with the subcircuit named Triac.

EXAMPLE 11.4.1 (PESS EXAMPLE 8.1)

A 400-V, 17-A triac is used to modulate power from a 120-V(rms), 60-Hz supply to a load whose resistance is 10Ω. For the condition of a symmetrical delay trigger angle of $\alpha = 60°$ for the triac, do a PSpice simulation. Plot a trace of the load current and determine **(a)** the average power delivered to the load and **(b)** the total average power dissipated by the triac. (Ans: **(a)** 1.14kW, **(b)** 9.6W.)

Solution

We must choose a specific triac. The device described in the subcircuit named Triac is appropriate. Two steps of a solution are described.

STEP 1 Figure 11.14, depicted on the next page, shows a relevant PSpice configuration of the triac circuit, which acts as an ac-voltage regulator.

STEP 2

From the configuration in Fig. 11.14 a circuit file, EX11_4_1.CIR, can be written. Run a PSpice simulation.

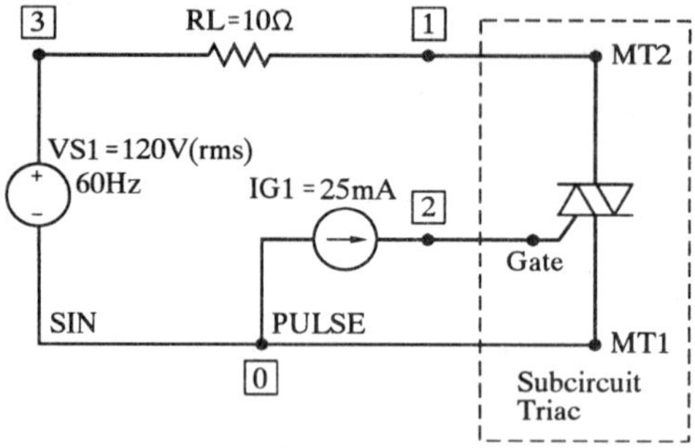

Fig. 11.14 PSpice configuration, ac power modulation.

»»»»» EX11_4_1.CIR »»

```
TRIAC POWER MODULATION AND LOSSES
* To determine the load power and triac losses.
* The triac subcircuit is in EVAL.LIB.
.LIB     EVAL.LIB
.INC     OPTIONS;                             See Section 7.5.1. Convergence aid.
* PARAMETERS
.PARAM   FREQ=60Hz    PERIOD={1/FREQ};         AC source frequency.
.PARAM   VRMS=120V    VMAX={SQRT(2)*VRMS};     Source voltage.
.PARAM   IG=25mA;                              Gate current.
.PARAM   ALFA=60deg   DEL={ALFA*PERIOD/360};   Delay angle.
.PARAM   TR=5ns  TF={TR};                      Rise and fall times.
.PARAM   TON={PERIOD/2-DEL-2*TR};              Gate pulsewidth.
.PARAM   RLOAD=10ohms
* SOURCES and ELEMENTS
VS    3  0  SIN(0 {VMAX} {FREQ})
IG1   2  0  PULSE(0 {IG} {DEL} {TR} {TF} {TON} {PERIOD/2})
RL    3  1  {RLOAD}
* DEVICE
Xtriac 1  2  0  Triac;       From EVAL.LIB. Default parameters used.
* ANALYSIS
.TRAN    0.1ms 16.6667ms 0     50us;           One cycle.
.PROBE   v(3),  i(RL),   i(IG1), v(2);         Source variables.
.END
```

»»»

END OF EXAMPLE 11.4.1

Drill Exercise D11.4.1

Consider the circuit diagram in Fig. 11.14. Change the supply voltage to 208V(rms) at 60Hz and the load to 20Ω. The trigger angle of the triac is to be $\alpha = 90°$. Follow the pattern of EXAMPLE 11.4.1, use the 400-V, 17-A triac named Triac in EVAL.LIB and do a PSpice simulation. Determine **(a)** the energy loss in the model triac due to the turn-on process and **(b)** the energy loss in the model triac during one half cycle due to the on-state conduction.

(Ans: **(a)** 0.361mJ, **(b)** 42.26mJ.)

11.5 THE MOSFET AND PERFORMANCE ANALYSIS

The circuit depicted in Fig. 11.15 is a chopper circuit using a power MOSFET and a free-wheeling diode. In practice, a gate resistor R_{gate} is often placed in series with the gate of a power device to dampen ringing and overshoots that arise during switching due to the combination of line inductance and gate capacitance. This is necessary since gate-signal transients that exceed 20volts can cause the gate oxide to break down. Gate resistors also provide a means of controlling the switching characteristics of a power device. This is necessary since the freewheeling diode has a finite reverse-recovery time and therefore a finite turn-off time. If the MOSFET switches on too fast, compared with diode turn-off, a large *shoot-through* current pulse can occur. This shoot-through current should be limited to reduce EMI and reduce switching losses. The following example demonstrates how PSpice can be used to analyze these effects. The PERFORMANCE ANALYSIS feature in PROBE is used.

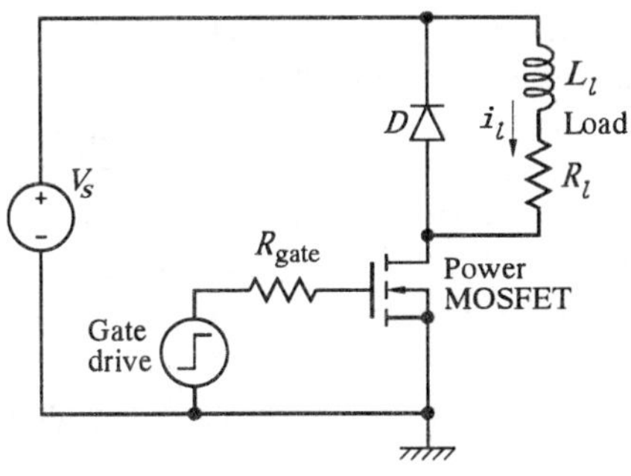

Fig. 11.15
MOSFET chopper circuit.

EXAMPLE 11.5.1

Simulate the circuit depicted in Fig. 11.15 using a range of gate resistors ranging from 1 ohm to 400ohms. The load is a 1 H inductor with an initial current of 20A before the MOSFET is turned on. The dc supply voltage is 48V. Use the detailed IRF150 MOSFET model that is included in EVAL.LIB and the diode model dRAM3010 introduced in EXAMPLE 11.2.1. Plot **(a)** the family of transient curves of the MOSFET current waveform over the switch-on interval for the range of gate-resistance values and **(b)** the relationship between peak shoot-through current versus the value of gate resistance. Determine an appropriate gate resistor to limit the peak shoot-through current in the MOSFET to 30A.

Solution

There are six steps to achieve a solution.

STEP 1

The first step is to transform the circuit diagram into a PSpice configuration. Before we proceed, a few notes should be made regarding the MOSFET model. The nature of this particular example requires an accurate MOSFET model that represents faithfully the switching behaviour of the device. For this example, we are required to use the IRF150 power MOSFET from International Rectifier. This model has been made available with the evaluation version of PSpice in the evaluation library. It uses PSpice's built-in MOSFET model and reflects reasonably the static and dynamic characteristics of the actual device. The model requires four connections for Drain, Gate, Source, and Bulk respectively. For all power MOSFETs, the bulk should be shorted to the source node. The dRAM3010 diode was chosen as a suitable free-wheeling diode for this example.

The problem specifies that an initial current of 20A is conducted by the inductive load. That is, at $t=0^{-}$, $i_l=20$A, and at $t=0$ the MOSFET is turned on. Initial conditions in PSpice can be established in a number of ways. The IC attribute is an easy way to set initial conditions and it works well for capacitors. Occasionally, difficulties can arise when it is used for inductors. Therefore, another approach for setting initial conditions will be shown in this example. The approach is to place a piece-wise linear current source in parallel with the inductor. The current in the inductor is rapidly *charged* up during the first moments of the transient simulation using the PWL current source. Almost immediately after time zero the PWL current source is turned off and it assumes a current of zero amperes. The current source remains at zero amperes throughout the remainder of the simulation and it no longer affects the circuit operation. The PSpice configuration is shown in Fig. 11.16.

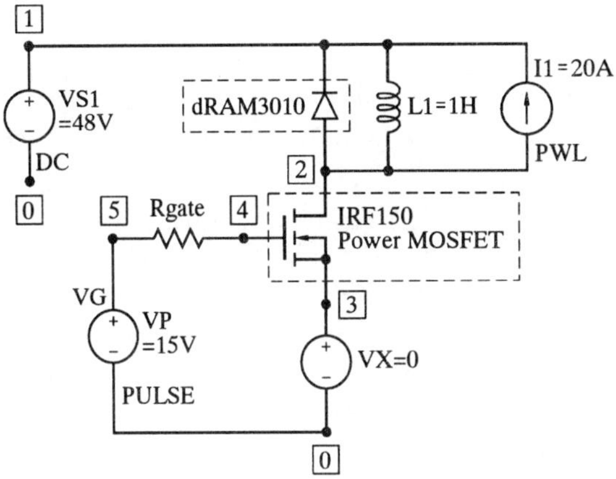

Fig. 11.16
PSpice configuration, MOSFET chopper.

STEP 2

The next step is to create the PSpice-compatible circuit file. The gate resistance is "parameterized" as Rgate, and a .STEP command is used to vary its value from 1ohm to 400ohms. The turn-on switching of the MOSFET is set by the PULSE source VG to occur at 100ns into the transient analysis. A time-step ceiling TMAX is chosen to provide sufficient detail in the output plots. The evaluation library has been added with a .LIB statement to access the IRF150 model. In addition, the OPTIONS file has also been included to aid convergence and speed up the simulation. The resulting PSpice circuit file is as follows.

```
»»»»» EX11_5_1.CIR »»»»»»»»»»»»»»»»»»»»»»»»»»»»»»»»»»»»»»
PERFORMANCE ANALYSIS OF MOSFET SWITCHING
* To analyze the peak shoot-through current in a MOSFET chopper with a
* free-wheeling diode in terms of the MOSFET gate resistance.

* The MOSFET model is in EVAL.LIB.
.INC     OPTIONS;                              Convergence aid.
.LIB     EVAL.LIB
.LIB     POWER.LIB

* PARAMETERS
.PARAM   Rgate=10ohms;                         Nominal gate resistance.
.PARAM   VS=48V;                               DC source voltage.
.PARAM   VP=15V DEL=100ns TR=10ns TF={TR} TON=60us  PER=100us
```

```
* SOURCES
VS1   1  0  {VS};                                       Power source.
VX    3  0  0;          Zero-voltage source to measure MOSFET current.
VG    5  0  PULSE(0 {VP} {DEL} {TR} {TF} {TON} {PER});        Driver.
I1    2  1  PWL(0,20A 1ns,0); Current source used to "charge" inductor.

* DEVICES
* IRF150 MOSFET from EVAL.LIB. DRAM3010 diode from POWER.LIB.
M1    2  4  3   3   IRF150;                          Chopper switch.
D1    2  1  DRAM3010;                                Free-wheeling diode.

* CIRCUIT ELEMENTS
L1     1  2  1H;                        Load inductor (constant current).
Rgate    5  4   {Rgate};                Driver gate resistance.

* ANALYSIS
* Vary the resistance of the gate resistor.
.STEP   PARAM  Rgate  LIST  1 10 25 50 75 100 200 300 400
.TRAN   2ns       1.1us   0     2ns
.PROBE  i(VX);                                   MOSFET current.
.END
```

» »

STEP 3 A PSpice simulation can be run with the circuit file EX11_5_1.CIR. The parameter sweep will cause PSpice to repeat the transient analysis for each step of the value of Rgate.

STEP 4 The next step is new, since we are doing a PERFORMANCE ANALYSIS. Before running PROBE, we must create a file containing a *goal function* definition. The goal function contains descriptions of analysis to be performed on the output data points. Since we wish to find the peak shoot-through current of the MOSFET at turn-on, we must create a goal function that locates peaks in the transient output data. The goal function, like the circuit file, is composed using a standard text editor. The goal function listing, which was used for this example to locate transient peaks, is as follows.

```
Peak(1) = y1                                  (1)
{                                             (2)
        1|                                    (3)
              search forward MAX !1           (4)
              ;                               (5)
}                                             (6)
```

The explanation of the goal function listing is as follows.

Line (1) of the goal function defines the *function name* of the goal function as it will appear in the **Add_trace** box in PROBE. Here the goal function name is called **Peak**. The (1) is the *expression argument list*. For this particular example it indicates that the function will use the results of the search identified by the number "1". The equals sign assigns the return value for the function. The return value consists of a *marked point expression*. A marked point expression is a variable or mathematical expression comprising X and Y coordinates of specific points on a curve. The specific points are found using *search commands*. In this case it is the variable **y1** which is the y-value of the data point marked by the search command in line (4).

Line (2) is an opening curly brace that signifies the beginning of the goal function definition. The complete goal function definition is always enclosed in matching curly braces.

Line (3) The **1|** syntax means the subsequent search command is identified by the number "1".

Line (4) This line contains the actual *search command*. Search commands begin at the origin of a curve and seek specific data points. Functions performed by search commands include: find the X value for a specified Y value, find the Y value for a specified X value, locate a peak or trough, or locate the maximum or minimum over a specified range. Search commands begin at the origin of the curve and also specify the direction of the search as either forward or reverse. The search command on the line (4) is searching forward from the origin for the maximum point. The exclamation mark followed by a one at the end of the search command marks and saves the data point as **(x1,y1)**. The **y1** point is the variable which appears on line (1) in the marked point expression list.

Line (5) is a semicolon used to denote the end of the search command.

Line (6) is a closing curly brace used to mark the end of the goal function definition.

If you are using a version of PSpice which is prior to Version 6.2 then the goal function can be saved as "PROBE.GF". This will make the goal function automatically available when PROBE is loaded. The goal function may also be saved under any other name with a .GF extension. If this is the case, the goal function file must be explicitly specified using the "-g" switch when invoking PROBE. For example

```
PROBE EX11_5_1.DAT -g EX11_5_1.GF
```

The extensions do not have to be included. They will be assumed by PSpice. With version 6.2 and later of PSpice, many PROBE-related files were consolidated into one file. This file has the same name as the circuit file with a .PRB extension and should be located in the current working directory. This .PRB file is automatically read in when PROBE is loaded. This .PRB file includes macros, goal functions, and PROBE settings. The goal functions should be placed within the .PRB file under the heading of [GOAL FUNCTIONS]. In addition, Version 6.2 of PROBE and later comes with a helpful file called MSIM.PRB. (In Version 6.1 with DOS view MSIM.GF.) These files contain a wide variety of useful predefined goal functions that may be used. For more information on goal functions consult the MicroSim PSpice *Circuit Analysis User's Guide.*

STEP 5

Now that the goal function has been defined we may load PROBE. When loading the simulation results, a box will appear showing the family of curves for the parameter sweep. For this example we will select all the curves. The plots for part (**a**) of the solution are quite straightforward. The current through the MOSFET is the same as the current through the dummy voltage source given by i(VX). The family of transient curves for each value of gate resistance is shown in Fig. 11.17. These plots show that the MOSFET current during turn-on has a distinctive peak due to the reverse-recovery current of the freewheeling diode. It is also shown that the turn-on time of the MOSFET increases nonlinearly as the value of the gate resistance is increased.

STEP 6

Next we will obtain the plot for part (**b**) of the solution. PROBE is invoked with the .GF file as above. PERFORMANCE ANALYSIS is enabled through the **X_axis** setting. Once **Performance_analysis** has been selected the X-axis label should change from the variable time to the name of the parameter under consideration. In this case the X-axis variable will become **Rgate** over the range 0 to 400ohms.

We have selected **All_transient_analysis**, then **X_axis**, then **Performance_analysis**. Now we select **Add_trace** and hit F4 (or click on the right mouse button). In the box, the goal function should be the **Peak**() goal function that we established in STEP4. If the **Peak**() goal function does *not* appear it may not have been found. If this occurs then return to STEP4 and check that the goal function was created properly. Next the **Peak**() goal function must be selected with appropriate input arguments specified within the parentheses. In this case we wish to find the peak MOSFET current. The MOSFET current is given by i(VX), therefore, the complete expression to type in is

```
Peak(i(VX)) <ENTER>
```

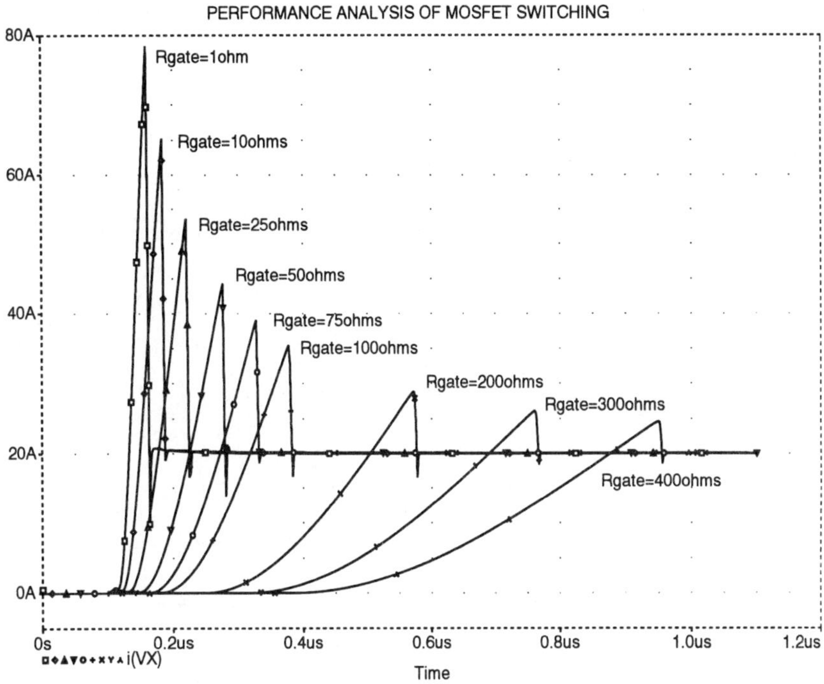

Fig. 11.17

The final plot of part (**b**) of the solution is shown in Fig. 11.18 on the next page. The peak MOSFET current is shown as a function of the gate resistor. Using the cursor in PROBE, the peak shoot-through current is kept to 30A with a gate resistance of about 183 ohms.

From the results of part (b) we can see that increasing the value of Rgate has the highly desirable effect of decreasing the peak shoot-through current. However, it should be noted, in the plot from part (a), that increasing Rgate also has the undesirable effect of increasing the switch's turn-on time. An increase in the switching time directly increases the switching losses in the MOSFET. The selection of Rgate is therefore a compromise between competing requirements. A conclusion is that the PERFORMANCE ANALYSIS feature within PSpice is a useful tool for the engineer who is faced with determining critical component values.

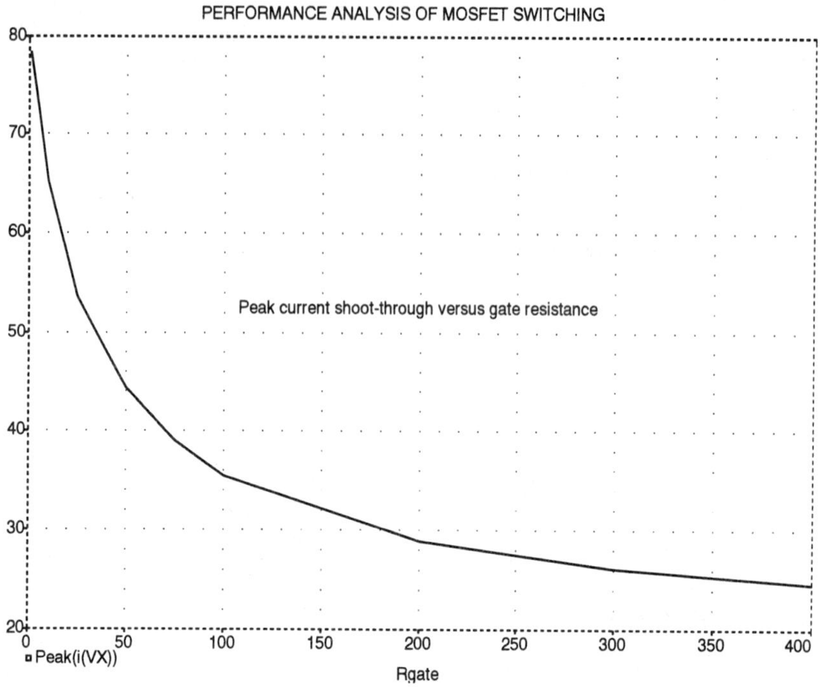

Fig. 11.18

END OF EXAMPLE 11.5.1

Drill Exercise D11.5.1

Repeat EXAMPLE 11.5.1 with the value of Rgate fixed at 10 ohms. Vary the risetime of the gate-driver voltage from 10ns to 500ns. Simulate and use PERFORMANCE ANALYSIS to plot the relationship between rise-time and peak shoot-through current.

11.6 THE BJT AND PERFORMANCE OPTIMIZATION

In some instances a design must not only be analyzed it must also be optimized. Previously, this has been a process of running numerous simulations whilst varying a key component value. This process is time consuming and difficult since circuit performance can be a complex function of numerous interdependent parameters. For these reasons MicroSim has created the PSpice OPTIMIZER program† (which was formerly called PARAGON).

The PSpice OPTIMIZER performs its optimization by computing the matrix of partial derivatives (the "Jacobian") for the circuit parameters. These derivatives are calculated by performing small perturbations of the parameters around the operating point. These partial derivatives are then used to formulate a new set of parameter values. This optimization process then repeats with the new parameters and a new matrix of corresponding partial derivatives is obtained. Each iteration produces new parameter values which will hopefully converge to a set of optimal parameter values.

The PSpice OPTIMIZER can perform a *minimization* on a single target or perform *least squares* optimization on any number of targets. Targets are essentially circuit parameters to be optimized. The optimization is guided by comparing a set of *specifications* with the actual performance of a circuit. Performance can be measured using a PROBE trace, a PROBE goal function, or evaluating a PSpice OPTIMIZER expression. The specifications are provided by the user in the form of *goals* or *constraints*. Every problem must have at least one goal along with any number of constraints. A goal describes a requirement that the circuit should come as close as possible to meeting. A constraint, on the other hand, is a strict requirement that must be met. It should be noted that the evaluation version of the PSpice OPTIMIZER is limited to one parameter optimization with a maximum of two specifications.

The PSpice OPTIMIZER is well suited for making adjustments to key circuit parameters and improving performance. It can even be used to more accurately fit a parameterized device model with measured data points. However, the PSpice OPTIMIZER cannot take a circuit that does not work and make it work. The PSpice OPTIMIZER is also no substitute for a proficient designer although it is an important tool for the designer.

The use of the PSpice OPTIMIZER will be demonstrated in the following example.

† See MicroSim's PSpice manual 6.2a for Windows.

EXAMPLE 11.6.1

Use the PSpice OPTIMIZER to calculate the value of the base resistor to minimize conduction losses in a power BJT. The configuration is shown in Fig. 11.19. The power BJT is a 2N6251 at a temperature of 25 °C. Use a typical value for the dc current gain. The load resistance is 25 ohms and the dc supply is 250 volts. The base drive power supply is 12 volts. Use an initial value of 10 ohms for the base-resistor value.

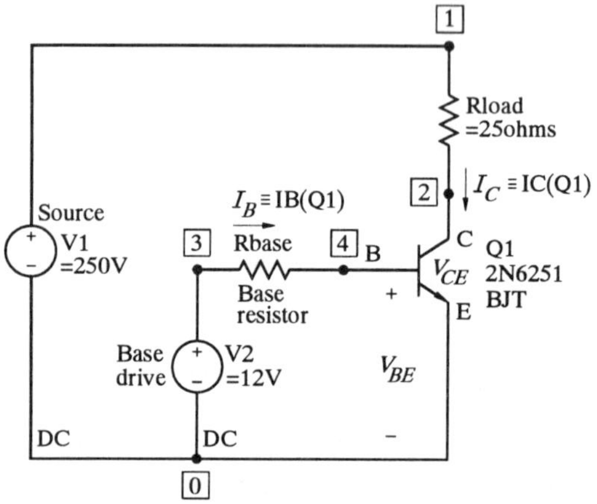

Fig. 11.19
PSPICE configuration of a BJT chopper.

Solution

The use of the PSpice OPTIMIZER requires nine steps for a solution.

STEP 1 The first step to transform a circuit diagram into a PSpice configuration has been done. See Fig. 11.19.

STEP 2 Before the PSpice circuit file can be constructed, we need to obtain the details of the model of the 2N6251 power BJT. For this particular example we need to analyze the static on-state conduction losses of the BJT. This can be accomplished using dc analysis. Some simulation problems, like this one, do not require extensive and elaborate device models. In these cases model parameters can be roughly approximated from a manufacturer's data sheet with a few simple calculations†. The major dc parameters for forward operation of a BJT include IS, BF, IKF and NE.

† S. Natarajan, "An Effective Approach to Obtain Model Parameters for BJTs and FETs from Data Books", IEEE Trans. Education, Vol. 35, No. 2, May 1992.

The IS parameter represents the saturation current. This parameter forms part of the transistor characteristics† and is given by eq. (11.2.2). Using the manufacture's data sheet†† for the 2N6251, we can get an operating point $I_C = 2A$, $V_{BE} = 0.8V$. Substituting these values into eq. (11.2.2) we obtain the result IS = 8.67E−14A. Next we need to determine BF, which represents the forward dc current gain of the BJT. This value is often represented by the symbols β or h_{fe}. From the manufacturer's data sheet we are given that the forward dc current gain varies from a minimum of 6 to a maximum of 50. Taking the geometric mean of these two extremes gives a typical value of BF $= \sqrt{6 \times 50} \approx 17$.
The IKF parameter is defined as the corner for forward beta high-current roll-off. The forward current gain in BJTs decreases at high current levels so this parameter is included to account for this effect. In the device data sheet there is a curve illustrating dc current gain versus collector current. This curve shows that at 25°C the dc current gain is reduced by one-half at a current of roughly 7.0A. Therefore, we set IKF = 7.0.
The NE parameter is the base-emitter leakage emission coefficient, we will use the default value of 1.5.
The simplified PSPICE model statement for this example becomes

```
.MODEL 2N6251 NPN (IS=8.67E-14A BF=17 IKF=7.0A)
```

STEP 3 The next step is to create a parameter file containing the circuit parameters to be optimized. This file has the same name as the circuit file except that it has a .PAR extension. The .PAR file should be referenced by an INCLUDE statement in the main circuit file. This .PAR file forms the link between the PSpice OPTIMIZER and the PSpice simulator. As new parameter values are determined by the OPTIMIZER, the .PAR file is updated as the simulations are run. Initially the .PAR file should contain the initial guess for the parameter value and finally it should contain the final optimal parameter value. The file in this case is called EX11_6_1.PAR and contains just one line as follows.

```
.PARAM Rbase=10
```

STEP 4 We want to create the PSpice compatible circuit file. The model statement from STEP 2 and the .PAR file from STEP 3 will be included. In addition, the OPTIONS file has been included to aid convergence. Since the temperature has been given in the problem we have specified it in the circuit file using a .TEMP statement.

---

† A. S. Sedra, K. C. Smith, "Microelectronic Circuits", Holt, Rinehart and Winston, 1987.

†† "Motorola Bipolar Power Transistor Data Book", 1992. Motorola Inc., PO Box 20912, Phoenix, Arizona.

In STEP 1 we determined that the dc analysis is most appropriate for this problem. Since all the power supplies are fixed, the analysis command for a single-point dc solution is used. The results of the single-point dc analysis will be used by the PSpice OPTIMIZER to judge the performance of the circuit. The PSpice circuit file is named EX11_6_1.CIR.

»»»»» EX11_6_1.CIR »»»»»»»»»»»»»»»»»»»»»»»»»»»»»»»»»»»

```
PSPICE OPTIMIZER EXAMPLE FOR A BJT CIRCUIT
* An example to use with PSpice OPTIMIZER to demonstrate the
* optimization of the value of the base resistance in a BJT chopper.

. LIB      EVAL.LIB
. INC      OPTIONS;                          Convergence aid.
* Parameters to be optimized are included in an external file.
. INC      EX11_6_1.PAR
. TEMP     25degC

* PARAMETER
. PARAM Rbase=10ohms;                        The BJT base resistance.

* SOURCES
V1  1   0   DC  250V;                        Power supply.
V2  3   0   DC  12V;                         Base drive.

* DEVICE, the BJT chopper switch in EVAL.LIB.
Q1  2   4   0   0   2N6251
. MODEL 2N6251  NPN  (1S=8.67E-14  BF=17  IKF=7.0)

* CIRCUIT ELEMENTS
Rbase   3   4   {Rbase}
Rload   1   2   25ohms;                      The load resistance.

* ANALYSIS
* Define the simulation parameters for single point dc analysis.
. DC       V1  LIST  250V
. PROBE
. END
```

»»

STEP 5 At this point it is a good idea to do a simulation with the circuit file to ensure that it is free of errors.

STEP 6 The next step is to run the PSpice OPTIMIZER in Windows. Once it is loaded, select **FILE/NEW** from the menu to begin a new optimization. All optimizations must begin by defining all the circuit parameters that are to be optimized. This is done by selecting **EDIT/PARAMETERS** from within the PSpice OPTIMIZER. A dialogue box will appear. Click on the "Add" button to add a new parameter. In the dialogue box enter **Rbase** for the name and enter 10 as the current value and the initial value. Within the same dialogue box set the upper limit to 100, and the lower limit to 1, and the tolerance to 0. When this has been entered click on the "OK" button. The new parameter name should now appear in the original parameters dialogue box. Since this example has only one parameter we are done. Click on the "CLOSE" button to proceed to the next step.

STEP 7 Now that the circuit's parameters are defined, we must set up the optimization specifications. Proceed by selecting **EDIT/SPECIFICATIONS** from within the PSpice OPTIMIZER. A dialogue box will appear. Click on the "Add" button to add a specification. We will begin by defining our goal since all optimizations must have at least one goal. The problem states that we must minimize the conduction losses of the power BJT; this is our goal. A new dialogue box appears requesting specific information about the specification to be added. To begin, enter "Power" as the name of the specification. Since power is to be minimized enter 0W as the target and set the range to 1 watt. Zero power dissipation is impossible to achieve and this goal actually represents the ideal case. This does not pose a problem, since goals do not have to be met and the PSpice OPTIMIZER will proceed to try to get as close as possible to meet its goal. Make sure that the "Constraint" box is turned off so that this specification is recognized properly as a goal. Under the "Analysis" section of the dialogue box enter the circuit file name from STEP 4 and set the analysis type to DC. Finally, we must provide an expression under the "Evaluate" field which will indicate the power dissipation in the BJT. The on-state losses in a power BJT comprise the conduction loss, collector to emitter, and the base drive loss, base to emitter. The total dissipative power P_d is expressed as

$$P_d = V_{CE} \times I_C + V_{BE} \times I_B . \tag{11.6.1}$$

Using the node numbers assigned in Fig. 11.19, the on-state losses for the BJT in this example are expressed in PSpice notation for BJTs as

```
V(2)*IC(Q1)+V(4)*IB(Q1)
```

Note the special notation for the transistor currents. This expression is to be entered in the "Evaluate" field. The goal definition is now complete so click on "OK" to proceed.

We must define one constraint to prevent the OPTIMIZER from seeking solutions which place the BJT in the active region. We want the BJT to operate like a switch, so it must operate in saturation. Therefore, we will create a constraint that requires the collector to emitter voltage to be less than 1 volt. The constraint will be entered in a manner similar to the previous goal. Click on the "Add" button and enter "VCE" as the name. Set the target to 1V with a range of 0.1V. Next, click on the constraint box to turn it on. Select the type of constraint as "**<=target**". Finally, under the analysis section of the dialogue box, enter the circuit file name as before and set the analysis type to DC. In the "Evaluate" box enter the PSpice expression "**V(2)**" which corresponds to the collector-emitter voltage. Click on "OK" and proceed to the next step.

STEP 8 If more than one goal function is defined, the PSpice OPTIMIZER always uses a least-squares optimization. However, in this example we have only one goal defined, so we are required to select whether the PSpice OPTIMIZER uses *minimization* or *least-squares* optimization. This is done by selecting the **OPTIONS/DEFAULTS** dialogue box. From within this dialogue box select "Advanced Options". This particular example involves reducing the power to the smallest value possible so the "Minimize" option should be selected.

The definition of the optimization is now complete and should be saved using the **FILE/SAVE** menu item before proceeding to the next step.

STEP 9 Select **TUNE/AUTO/START** and the optimization should begin to run automatically. The user should be able to observe as the OPTIMIZER iteratively calls PSpice and PROBE to perform multiple simulations. Throughout this process the new specifications and current parameter values will be periodically updated in the OPTIMIZER window. A "Thermometer" indicator will show how close each target is being met. The optimization process will continue until the specifications are met or no more progress can be made. The goal, which was defined, is impossible to achieve so the specifications will never be met. As a result the optimization continues until a message appears indicating that "Optimization has ended, no more progress". At this point the power dissipation has been minimized as low as possible and the final value for Rbase should appear in the parameter box.

The evaluation Version 6.2 of the PSpice OPTIMIZER quits after 11 iterations with a final Rbase value of 6.475 ohms. This value appears reasonable and can be rounded to the nearest practical resistor value by selecting the **EDIT/ROUND VALUES** menu item. The nearest rounded value of resistance is given as 6.2 ohms.

It should be noted that this BJT model is simplistic and in reality the optimal value of base resistance is highly dependent on the dc current gain (BF). In practice, the dc current gain can vary considerably between different BJT devices. Consequently, this base resistance is only optimal for a transistor which possesses the exact parameter values which were used in this example. Designers using the PSpice OPTIMIZER should always be aware of the effect that component tolerances have on their *optimal* designs.

This example demonstrates that, when used properly, the PSpice OPTIMIZER can be a useful tool for the design engineer. For more information on the operation of the PSpice OPTIMIZER consult the MicroSim PSpice OPTIMIZER *Analog Performance Optimization User's Guide.*

END OF EXAMPLE 11.6.1

Drill Exercise D11.6.1

Solve the problem in EXAMPLE 11.6.1 using PERFORMANCE ANALYSIS only. Plot the conduction power dissipation in the BJT versus the Rbase parameter and locate the minimum loss graphically. Does this agree with the PSpice OPTIMIZER results? (Ans: Yes.)

Drill Exercise D11.6.2

Repeat EXAMPLE 11.6.1 and find the optimal value of the base resistor if the BF parameter of the BJT is reduced to the minimum value of 6.

Drill Exercise D11.6.3

Use the PSpice OPTIMIZER to calculate the value of the optimal source resistance Rsource that maximizes power transfer to the load for the circuit illustrated in Fig. 11.20. (Ans: 5Ω.)

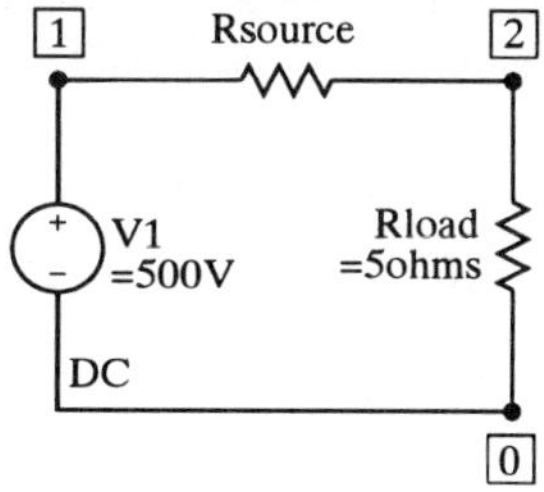

Fig. 11.20

11.7 THE IGBT, GTO AND MCT

PSpice has no built-in models for the IGBT, the GTO and the MCT. However, macromodels of these devices have been described in the literature. We give an arbitrary list below for some cases that provide circuit files so that the models can be used.

IGBT

“A SPICE Model for IGBTs” by A. F. Petrie and C. Hymowitz, Tenth Annual Applied Electronics Conference Proceedings, 1995, pp 147-152.

“SPICE Model Accurately Simulates IGBT Parameters” by A. F. Petrie and C. Hymowitz, PCIM, Jan. 1996 pp 40-51.

“IGBT SPICE Model” by F. Mihalic, K. Jezernic, K. Kreschan, M. Rentmeisler IEEE Trans Industrial Electronics. Vol. 42, No. 1, Feb. 1995, pp 98-104.

GTO

“A GTO Modular Macromodel for Use in Circuit Simulation with SPICE”, by V. M. Guzmán and M. I. Giménez de Guzmán, Ninth Annual Applied Power Electronics Conference Proceedings, 1994, pp 1059-1065.

MCT

“SPICE MODEL USER’S GUIDE” Harris Corp. Power Marketing Dept., 724 Route 202, PO Box 591, Somerville, N. J. 08876-0591, USA. Associated with this is a PSpice model disk that contains subcircuit descriptions of power MOSFETs and an MCT.

Appendix A

Library Files

In the evaluation version of PSpice there are the library files NOM.LIB and EVAL.LIB. It is worth browsing through these files.

We have created two user-defined library files called POWER.LIB and DRIVER.LIB. These files can be copied from the WEB. See the Preface and Appendix E of this text. The files comprise subcircuit descriptions of devices and drivers. Descriptions not referenced in detail in the text will be found on the WEB. The contents of these files are as follows.

POWER.LIB SUBCIRCUIT NAMES

SW_PWM	AC-DC single-phase bridge driver.	p57
V_CONT	Triac driver.	
DIO_V	A near-ideal diode model (VSWITCH).	p101
DIO_I	A near-ideal diode model (ISWITCH).	p103
DIO_GD	A near-ideal diode model (VCCS).	p108
D_IDEAL	A near-ideal PSpice diode model (D).	p112
BJ_IDEAL	A near-ideal BJT model (ISWITCH).	p116
TH_IDEAL	A near-ideal thyristor model + driver (ISWITCH).	p121
TR_IDEAL	A near-ideal triac model + driver (ISWITCH).	p124
GT_IDEAL	A near-ideal GTO + driver (ISWITCH).	p132
MG_IDEAL	A near-ideal MOSFET/IGBT model (VSWITCH).	p135
MG_GATE	Gate driver for MG_IDEAL.	p135
MT_IDEAL	A near-ideal MCT + driver (VSWITCH).	p140
MCT_IDEAL	A near-ideal MCT (VSWITCH).	p142
MCT_DRIVER	An MCT driver.	p142
. MODEL dRAM3010	Diode parameters.	p319
OPTIONS	Convergence aid. See next page.	p210
. MODEL SWITCH	Voltage-controlled switch parameters.	p146

DRIVER.LIB SUBCIRCUIT NAMES

VCHOPPER	A PWM driver for a chopper.	p146
VSAWTOOTH	A sawtooth generator.	p149
COMPARATOR	A comparator with hysteresis (Schmitt trigger).	p149
COMPARATOR2	A near-ideal comparator.	p149
VPWM	Adjustable duty-cycle driver for a chopper.	p152
VPWM2	Adjustable duty-cycle driver for a chopper.	p152
HALFWAVE_DRVA	Alpha point-on-wave driver for a rectifier.	p160
HALFWAVE_DRVAB	Alpha/beta point-on-wave driver.	p163
FULLWAVE_DRV	Single-phase bridge driver for rectification.	p164
INV_DRV	Driver for a single-phase bridge inverter.	p167
INV_PWM_DRV	Multiple-pulse inverter driver.	p169
MOD_INV_PWM_DRV	Inverter driver with modulating pulsewidth.	p174
TH_GAT	A thyristor gate driver.	p223
TH_GAT6	Double-pulse driver for a three-phase rectifier.	p238
INV_MLT_PLS	Uniform multiple-pulse inverter driver.	p245
SPWM_555_DRV	Digital driver for a single-phase inverter.	p247
PWM_555_DRV	Uniform PWM inverter driver.	p247
MPLS_TRI_INV	Multiple-pulse inverter driver.	p260
SPWM_TRI_INV	Driver for a single-phase bridge inverter.	p268
SPWM_FULL_TRI	Driver for a single-phase bridge inverter.	p278
SPWM_FULL_TRI3	Driver for a three-phase inverter.	p280
COMPARATOR3	Comparator without hysteresis.	

OPTIONS FILE FOR CONVERGENCE

An OPTIONS file can be introduced into a circuit-file listing with the . INC statement to aid convergence. The file comes from page 216. The file description follows.

```
* OPTIONS
. OPTIONS   RELTOL=0.01
. OPTIONS   ABSTOL=1nA
. OPTIONS   VNTOL=1mV
. OPTIONS   ITL1=500
. OPTIONS   ITL2=200
. OPTIONS   ITL4=40 ;    ITL5=0 by default
. OPTIONS   GMIN=1E-10
```

Appendix B

List of Circuit-file Examples

Appendix C

Quick Reference to PSpice Statements

In this text we have met some of the symbols, element, devices and dot commands in PSpice statements that make up a circuit file. The following is a partial listing taken from the commands chapter of MicroSim PSpice A/D Circuit Analysis Reference Manual.

SYMBOLS

<option>
{expression}
(parentheses for grouping)
* comment line.
, optional separator for clarity.
; comment on rest of line.
+ continuation line of a statement.
... further optional values to be included in a statement.
LIN linear base.
OCT sweep by octaves.
DEC sweep by decades.
LIST use the following list of values.
TIME is the time variable.

VOLTAGE VARIABLES

v(4), v(3, 1), v(N+ N−), VS, v(0, 2)

CURRENT VARIABLES

I(RL), i(VS), i(C2)

DATA STATEMENTS (Case insensitive)

C **Capacitor (linear)**

C*name* N+ N- *value* <IC=*initial_value*>

C*name* N+ N- {*parameter_name*}

Examples: C20 2 0 0.1uF IC=10.1V

CL 2 3 {CLOAD}

References: 25, 27, 99, 156, 298

There is a model form of the capacitor. (See PSpice manual.)

D **Diode**

D*name* N+ N- *model_name* <*area_factor*>

. MODEL *model_name* D(*parameter*=*value* ...)

Examples: DIO 3 5 DIODE

. MODEL DIODE D(RS=1m CJO=1nF TT=70ns)

References: 110, 115, 179, 309, 311, 321

E **Voltage-controlled voltage source** (analogue behavioural modelling)

E*name* N+ N- CN+ CN- *gain*

E*name* N+ N- POLY(*value*) (CN+CN-) ... *polynomial_coeffs* ...

E*name* N+ N- VALUE={*expression*}

E*name* N+ N- TABLE {*expression*} = (*input_value*, *output_value* ...)

E*name* N+ N- LAPLACE {*expression*} = {*transform*}

E*name* N+ N- FREQ {*expression*} = (*freq*, *mag*, *phase* ...)

E*name* N+ N- CHEBYSHEV {*expression*} = *type*, (*cutoff*, *atten.* ...)

Examples: Eamp 6 0 1E6

Eopamp 5 6 VALUE={1E6*v(1)}

E 2 0 VALUE={LIMIT(V(0,1)*1E6, -15V, 15V)}

Eout 3 4 TABLE {v(1,2) + i(VS)} = (0,0 1,10V)

E 5 6 POLY(1) 1,0 3 5 7 ; {v(5,6) = 3 + 5*v(1) + 7*v(1)*v(1)}

References: 32, 33, 40, 77, 89, 141

F **Current-controlled current source**

F*name* N+ N- V_*control_name* *gain*

F*name* N+ N- POLY(*value*) V_*control_name* ... *polynomial_coeffs* ...

Example: F 10 20 Vsense 20

Reference: 205

G **Voltage-controlled current source** (analogue behavioural modelling)

G*name* N+ N− CN+ CN− *gain*
G*name* N+ N− POLY(*value*) (CN+ CN− ...) *polynomial_coeffs* ...
G*name* N+ N− VALUE={*expression*}
G*name* N+ N− TABLE {*expression*} = (*input_value, output_value* ...)
G*name* N+ N− LAPLACE {*expression*} = {*transform*}
G*name* N+ N− FREQ {*expression*} = (*freq, mag, phase* ...)
G*name* N+ N− CHEBYSHEV {*expression*} = *type*, (*cutoff, atten.* ...)

Examples: Gamp 5 7 10
G 2 3 VALUE={2*PWR(rms(v(1,2)),2)}
Gbase 100 101 TABLE {v(4) −v(50,51)} = (0,0 0,1ma 1,10A)
GNOM 7 8 POLY(2) 1,0 2,0 4.0 3.0 5.0 6.0 9.0
That is, $i_{78} = 4 + 3v_1 + 5v_2 + 6v^2_1 + 9v_1 v_2$

References: 107, 119, 125, 126

H **Current-controlled voltage source**

H*name* N+ N− V_*sense gain*
H*name* N+ N− POLY (*value*) V_*sense* ... *polynomial_coeffs* ...

Example: Hamp 1 0 VX 5
That is, v(1) = 5i(VX)

Reference: MicroSim's manual (Command chapter).

I **Independent current source**

I*name* N+ N− <<DC> *value*> <AC *magnitude* <*phase*>> <*transient_form*>
Transient_form is EXP PULSE PWL SFFM or SIN with parameters. (See index.)

Examples: I1 1 0 DC 12A
I1 1 0 12A
I2 3 4 SIN(2A 10A 60Hz 0 0 120deg)

References: 115, 119, 131, 185, 253, 324, 332

K **Inductor coupling**

K*name* inductor1_*name* inductor2_*name* ... *coupling_coeff*
(There is a model form for nonlinear inductors. See PSpice manual.)

Example: K12 L11 L22 0.99
That is, $L_{12} = 0.99(L_{11} L_{22})^{½}$

References: 191, 192, 200, 203, 206

L **Inductor**

L*name* N+ N− *value_or_parameter* <IC=*value*>
(There is a model form for nonlinear inductors. See PSpice manual.)

Examples: LINDUCT 5 6 25.7mH
L11 9 10 {LLOAD} IC=2.1A

References: 17, 19

M **MOSFET**

M*name* Ndrain Ngate Nsource Nbulk *model_name* <*parameter_values* ... >
. MODEL *model_name* NMOS <*parameter=value* ... >

Example: MOS 1 2 0 0 NNOM L=30u W=10u

Reference: 332

Q **Bipolar transistor (BJT)**

Q*name* Ncollector Nbase Nemitter <Nsubstrate> *model_name* <*area_factor*>
. MODEL *model_name* NPN <*model_parameters*>

Examples: Q1 14 13 0 PMOD
. MODEL PMOD NPN ; Default parameters.
Q2 16 17 0 Q2N2222A ; From EVAL.LIB

References: 340

R **Resistor**

R*name* N+ N- *value_or_parameter*
(The resistor has a model form. See PSpice manual.)

Examples: Resistance 1 0 20kohms
RES 2 3 {RLOAD}

References: 6, 10, 63, 66

S **Voltage-controlled switch**

S*name* N+ N- CN+ CN- *model_name*
. MODEL *model_name* VSWITCH(RON=1 ROFF=1MEG VON=1 VOFF=0)
The above parameters are default values.

Example: Switch 1 0 2 0 IGBT
. MODEL IGBT VSWITCH(RON=1E-3 ROFF=1E6
+ VON=10 VOFF=0)

References: 63, 141, 179

T **Transmission line**

T*name* NA+ NA- NB+ NB- ZO=*value* <TD=*value*> <F=*value*> <NL=*value*>

Example: T1 4 0 10 0 ZO=1 TD=1.4

Reference: MicroSim's manual (Command chapter).

V **Independent voltage source**

V*name* N+ N- <<DC> *value*> <AC *magnitude* <*phase*>> <*transient_form*>
Transient_form is EXP PULSE PWL SFFM or SIN with parameters. (See index.)

Examples: Vsource 1 0 AC 1
V1 3 2 PWL(0,0 1ns,2V) ; A step input.
V2 6 0 PULSE(-1V 2V 1us 1ns 1ns 1ms 2ms)

References: 6, 8, 17, 18, 25, 26, 32, 36, 40, 115, 203, 261, 263

W **Current-controlled switch**

W*name* N+ N- V_*sense* *model_name*
. MODEL *model_name* ISWITCH(RON=1 ROFF=1MEG ION=1A IOFF=0)
The above parameters are default values.

Example: Wtriac 1 0 VA TRIAC
. MODEL TRIAC ISWITCH(RON=1E-3 ROFF=1E6
+ ION=10mA IOFF=0)

References: 115, 119, 131, 185

X **Subcircuit call**

X*name* <*nodes* ... > *subcircuit_name* <PARAMS: *parameter=value* ... >

Example: XTH1 1 2 3 Scr PARAMS: dvdt=60E6V/s

References: 37, 63, 65, 102, 151, 156, 174, 322

CONTROL STATEMENTS (DOT COMMANDS) (Case insensitive)

. AC — AC analysis

. AC LIN_or_DEC_or_OCT *points_value start_frequency end_frequency*

Examples: . AC LIN 1 50Hz 50Hz
. AC LIN 10 10kHz 100kHz
. AC DEC 50 1kHz 1MEGHz

References: 200, 203

. DC — DC analysis

. DC LIN *variable start_value end_value increment <nested_sweep>*
. DC DEC_or_OCT *variable start_value end_value points_value <nested_sweep>*
. DC *variable* LIST *values ... <nested_sweep>*
. DC PARAM *parameter_name* LIST *values ...*

Examples: . DC LIN VS 2V 10V 1V I1 0.1A 1A 0.05A
. DC PARAM RLOAD LIST 2 10 40 100ohms

References: 6, 11

. END — End of circuit file

. END

References: 6, 14

. ENDS — End of subcircuit description

. ENDS *<subcircuit_name>*

Example: . ENDS Scr

Reference: 57, 61

. FOUR — Fourier analysis

. FOUR *frequency <number_of_harmonics> variables ...*

Example: . FOUR 60Hz 15 i(RL) v(1)
Results are in the output file.

References: 87, 88, 243

. FUNC — Function definition

. FUNC *name* (*<arguments ... >*) {*definition*}

Example: . FUNC G(y) {exp(y) + 1}

Reference: 91

. IC — Initial bias point condition

. IC V(N+,N-) = *value ...*
. IC I(*inductor_name*) = *value ...* ; PSpice release 6.2.

Examples: . IC V(1) = 100V V(2,1) = 10V
. IC I(L1) = 2.6A

References: 179, 180

. INC — Include file

. INC *file_name*

Examples: . INC A:\FILE\D1_2_1.CIR
. INC C:\PSPICE\FUNC.DEF ; Function definitions.

References: 91, 93, 340

. LIB **Library file**

. LIB *<file_name>*

Examples: . LIB

(By default this is NOM.LIB, the reference to individual library files.)

. LIB EVAL.LIB

. LIB POWER. LIB

. LIB DRIVER. LIB

References: 35, 36, 63, 64, 156

. LOADBIAS **Load bias point file**

. LOADBIAS *file_name*

Example: . LOADBIAS EXAMPLE.SAV ; File from .SAVEBIAS statement.

Reference: MicroSim's manual (Commands chapter).

. MODEL **Model statement**

Describes a device named in a data statement.

. MODEL *model_name model_type (parameter=value ...)*

Examples: . MODEL BJT NPN(IS=1E-10 BF=15)

. MODEL DIO D(IS=1E-14 CJO=1pF RS=1)

References: 48, 50, 57, 58, 73, 99, 110, 111, 115, 119, 311, 339

See Model types page 64.

See the MicroSim manual for a more comprehensive definition.

. NODESET **Nodeset statement**

An aid to calculate the bias point.

. NODESET V(N+,N-) = *value ...*

. NODESET I(*inductor_name*) = *value ...* ; PSpice release 6.2.

Examples: . NODESET V(2) = 3V V(5,6) = 10V

. NODESET I(L1) = 20.1A

Reference: 212

. OP **Bias point (operating point)**

Causes bias point data to be printed to output file.

. OP

Reference: See MicroSim's manual (Command chapter).

. OPTIONS **Options**

. OPTIONS *<option_name ... > <option_name = value ... >*

Example: . OPTIONS NOLIST RETOL=0.01

References: 156, 179, 210, 213, 216

See PSpice manual for list of options.

. PARAM **Parameters**

. PARAM *element_or_source_name = value ...*

. PARAM *parameter = {expression}*

Example: . PARAM VS=100V RLOAD={VS/10.2}

References: 63, 65, 115

. PLOT **Plot**

. PLOT *analysis_type variable* <(*lower_limit, upper_limit*)> ...
Example: . PLOT TRAN v(1) (1mV,100mV) i(RL)
References: 17, 21

. PRINT **Print**

. PRINT AC_or_DC_or_TRAN *variables* ...
Example: . PRINT AC VM(10) VP(10) VR(10) VI(10)
References: 6, 13, 200, 203

. PROBE **Graphical postprocessor**

. PROBE *variables* ...
Examples: . PROBE ; All variables available.
. PROBE i(RL) v(1) ; Values of two variables available.
. PROBE/CSDF ; Creates a data file in text format.
References: 17, 21, 40

. SAVEBIAS **Save bias point to a file**

Used with .LOADBIAS
. SAVEBIAS *filename analysis_type* <TIME=*value*> <*options*>
Example: . SAVEBIAS EXAMPLE.SAV TRAN TIME=50ms REPEAT
Reference: MicroSim's manual (Commands chapter).

. SENS **Sensitivity analysis**

. SENS *variables* ...
Reference: MicroSim's manual (Commands chapter).

. STEP **Parametric analysis (sweep)**

. STEP <LIN> *variable start_value end_value increment*
. STEP DEC_or_OCT *variable start_value end_value number_points*
. STEP PARAM *name* LIST *values* ...
Examples: . STEP V(1) 2V 10V 1V
. STEP DEC NPN Q1(IS) 1E-16 1E-6 10
. STEP PARAM RLOAD LIST 2 4 10 120ohms
References: 63, 67

. SUBCKT **Subcircuit definition**

. SUBCKT *name* <*nodes* ... > <PARAMS: *parameter=value* ... >
Example: . SUBCKT MOSFET 10 11 12 PARAMS: dvdt=1E8
References: 57, 58, 101, 102, 326

. TEMP **Temperature**

. TEMP <*temperature_values* ... >
Example: . TEMP 25 50 100degC ; Three consecutive simulations (sweeps).
References: 311, 314, 340

. TF **Transfer function**

. TF *output_variable input_source_name*
Example: . TF V(10) VS ; Results in output file.
Reference: MicroSim's manual (Commands chapter).

. TRAN **Transient analysis**

. TRAN</OP> TSTEP TSTOP<TSTART<TMAX>> <UIC> ; Release 6.1.

. TRAN</OP> TSTEP TSTOP<TSTART<TMAX>> <SKIPBS> ; Release 6.2.

TSTEP: print step size. TSTOP: final time. TSTART: start printing.
TMAX: step ceiling value. UIC: use initial conditions.
SKIPBS: skip bias-point calculation.

Examples: . TRAN 1ms 200ms
. TRAN 0.1ms 2ms 1ms 50us

References: 17, 20, 25, 27

. WATCH **Watch analysis results**

. WATCH <DC> <AC> <TRAN> *variables* <*low_limit, high_limit*> ...

Example: . WATCH TRAN i(L1) vBE(Q1) (0V,4V)

Reference: MicroSim's manual (Commands chapter).

. WIDTH **Output width control**

. WIDTH OUT=*value* ; See .OPTIONS reference.

Examples: . WIDTH OUT=80 ; Default columns in output file.
. WIDTH OUT=132 ; Only other choice.

Reference: MicroSim's manual (Commands chapter).

Appendix D

Bibliography

Computer-Aided Circuit Analysis Using SPICE
W. Banzhaf
Prentice Hall, 1989 (ISBN 0-13-162579-9)

Spicey Circuits
R. Chattergy
CRC Press, 1992 (ISBN 0-8493-7173-2)

Power Electronics: Computer Simulation, Analysis, and Education Using PSpice
N. Mohan
Minnesota Power Electronics Research and Education 1992
PO Box 14503, Minneapolis, MN 55414, USA

Engineering Circuit Analysis with PSpice and Probe
(Describes the use of PROBE in detail)
R. Conant
McGraw-Hill, Inc., 1993

Introduction to PSpice (Supplement to Electric Circuits, 4th ed.)
J. W. Nilsson, S. A. Riedel
Addison-Wesley, 1993 (ISBN 0-201-51318-8)

SPICE for Power Electronics and Electric Power
M. H. Rashid
Prentice Hall, 1993 (ISBN 0-13-030420-4)

Semiconductor Device Modelling with SPICE
G. Massobrio, P. Antognetti
McGraw-Hill, Inc., 1993 (ISBN 0-07-002469-3)

The SPICE Book
A. Vladimirescu
John Wiley & Sons, Inc., 1994 (ISBN 0-471-60926-9)

Inside SPICE. Overcoming the Obstacles of Circuit Simulation.
R. Kielkowski
McGraw-Hill, Inc., 1994 (ISBN 0-07-911525-X)

SPICE: Practical Device Modeling
R. Kielkowski
McGraw-Hill, Inc., 1995 (ISBN 0-07-911524-1)

SPICE. A Guide to Circuit Simulation & Analysis Using PSpice
P. W. Tuinenga
Prentice Hall, 1994, 3rd ed. (ISBN 0-13-747270-6)

Schematic Capture with PSpice
M. E. Herniter
Merrill (Macmillan Publishing Company), 1994 (ISBN 0-02-353771-X)

The Art of Simulation Using PSpice
B. Al-Hashimi
CRC Press, 1995

Hands on PSpice
J. G. Gottling
Houghton Mifflin Co., 1995 (ISBN 0-395-69916-94)

SPICE for Circuits and Electronics Using PSpice
M. H. Rashid
Prentice Hall, 1995, 2nd ed. (ISBN 0-13-124652-6)

PSpice for Windows
E. Brumgnach
Delmar Publishers, Inc., 1995 (ISBN 0-8273-6821-6)

PSpice for Windows - A Circuit Simulation Primer
R. W. Goody
Prentice Hall, 1995 (ISBN 0-02-345022-3)

The Illustrated Guide to PSpice
R. Lamey
Delmar Publishers, Inc., 1995 (ISBN 0-8273-6524-1)

Power Electronics Semiconductor Switches
R. S. Ramshaw
Chapman & Hall, 1993 (ISBN 0-412-28870-2)
(This text is the focus for PSpice simulation in this guide.)

MicroSim PSpice A/D Reference Manuals
MicroSim Corp.
20, Fairbanks, Irvine, California 92718
Fax: (714) 455-0554
Internet: Sales@MicroSim.com
Tech. Support@MicroSim.com

Appendix E

WEBsite

The WEBsite can be accessed by using the URL

http://www.ece.uwaterloo.ca/~pwr_elec

The material that can be viewed and downloaded from this WEBsite is as follows. WEB EXAMPLES and Drill Exercises have the prefixes W and WD respecively. Extra Drill Exercises have the prefix ED. The numbers relate to the text section.

GENERAL MATERIAL

OPTIONS (a file to aid convergence. See Appendix A.)
POWER .LIB (a file containing subcircuits. See Appendix A.)
DRIVER .LIB (a file containing driver circuits. See Appendix A.)
CIRCUIT FILES of all EXAMPLES in the text.
(See Appendix B.) Found in relevant chapters.

Instructors who have adopted the text may access the circuit files of the Drill Exercises and Problems by contacting the publisher.

EXTRA MATERIAL

Chapter 1 PSpice and Electric Circuits

Section 1.4.4 PSpice Comparator
Worked EXAMPLE, Drill Exercises
Extra Drill Exercises

Chapter 2 PSpice Switches and Power Modulation

Section 2.4 AC-AC Modulation
Worked EXAMPLE, Drill Exercises

Chapter 3 Waveform Distortion

Extra Drill Exercises

Chapter 4 Ideal Semiconductor Switches

Worked EXAMPLE (diode DIODE_I), Drill Exercises

Extra Drill Exercises

Chapter 5 Simulation of Driver Circuits

Sawtooth Generator

Worked EXAMPLE

Comparator

Worked EXAMPLE

Section 5.3.2 Full-wave AC-DC Conversion

Worked EXAMPLE, Drill Exercises

Section 5.4.4 Digital Driver (SPWM)

Worked EXAMPLE, Drill Exercise

Extra Drill Exercises

Chapter 6 DC-DC Converter Circuits

Extra Drill Exercises

Chapter 7 Diode Rectifiers

Section 7.4.2 Transformer Test

Worked EXAMPLES, Drill Exercises

Extra Drill Exercises

Chapter 8 Controlled Rectification

Extra Drill Exercises

Chapter 9 Switch-mode Inverters

Section 9.2.3 SPWM Inverter Digital Driver

Worked EXAMPLE, Drill Exercise

Section 9.3.1 Multipulse Inverter

Worked EXAMPLE, Drill Exercise

Section 9.3.3 Current Source Inverter

Worked EXAMPLE

Section 9.3.4 Cycloconverter

Worked EXAMPLE, Drill Exercises

Extra Drill Exercises

Chapter 10 Resonant Converters

Section 10.4 Integral-cycle Converter

Worked EXAMPLES, Drill Exercises

N.B. Please view this WEBsite often to note any up-dated material.

Appendix F

Problems

Chapter 6

P6.1 (PESS[†] Problem 2.1) Consider the primitive chopper shown in Fig. P6.1. For the conditions shown determine by simulation (**a**) the average current in the switch and (**b**) the average power absorbed by the load. Use the model of the switch provided in EXAMPLE 6.2.1. (Ans: (**a**) 59.94A, (**b**) 5.988kW.)

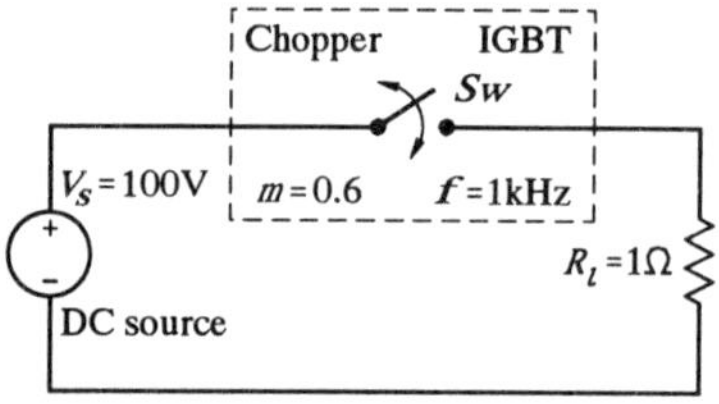

Fig. P6.1

P6.2 A MOSFET chopper regulates power from a 12-V dc source to a 10-Ω resistive load. The switching frequency is 25kHz. The circuit contains an inductor $L = 2.1$mH in series with the load resistor and a freewheeling diode across both. For a duty cycle of $m = 0.7$ determine by simulation (**a**) the load-current excursion ($\Delta I = I_{l\max} - I_{l\min}$), (**b**) the load-voltage excursion and (**c**) the average power absorbed by the load. Use the switch and diode models from EXAMPLE 6.2.1. (Ans: (**a**) (840.696−789.284)mA = 51.4mA, (**b**) 0.514V, (**c**) 6.648W.)

P6.3 A MOSFET chopper (buck converter) regulates power from a 12-V dc source to a 10-Ω resistive load. The switching frequency is 25kHz. The circuit contains elements, inductor $L = 0.1$mH, $C = 2000\mu$F and a diode. For a duty cycle $m = 0.7$, determine by simulation (**a**) the inductor's current excursion ($\Delta I = I_{L\max} - I_{L\min}$) and (**b**) the average power absorbed by the load. (**c**) What is the current rating of the MOSFET for all possible conditions? Use the model switch and diode found in EXAMPLE 6.2.1.
(Ans: (**a**) (796.27 − 745.38)mA = 50.9mA, (**b**) 6.64W, (**c**) 1.2A.)

P6.4 (PESS Problem 2.2) An IGBT buck converter (see Fig. 6.1) modulates power from a 100-V dc source to a resistive load of 2Ω. The inductance of the chopper circuit has a value of 0.4mH and the capacitance is 1000μF. If the switching frequency is f = 1kHz and if the duty cycle is $m = 0.7$, determine by simulation (**a**) the transient peak current in the switch at the initial

† PESS refers to "Power Electronics Semiconductor Switches" by Raymond Ramshaw.

turn-on, **(b)** the average power absorbed by the load, **(c)** the inductor current excursion, **(d)** the load-voltage excursion and **(e)** the average current in the converter switch. Use the model switch and diode from EXAMPLE 6.2.1. (Ans: **(a)** 143A (maximum average 75A), **(b)** 2.53kW, **(c)** (62.36 - 7.25) = 55.11A, **(d)** (73.46 - 66.58) = 6.88V, **(e)** 26.2A.)

P6.5 A basic circuit diagram of a battery charger is shown in Fig. P6.5. The buck converter operates at a switching frequency of 100kHz and a duty cycle of 0.68. Determine **(a)** the useful average power delivered to the battery being charged and **(b)** the charging current excursion (ripple). Use the MOSFET subcircuits MG_IDEAL and MG_GATE and the diode subcircuit DIO_GD in POWER.LIB. (See EXAMPLES 4.7.1 and 4.2.2.) (Ans: **(a)** 2.131kW, **(b)** $\Delta I = I_{b\max} - I_{b\min} = 4.42$A.)

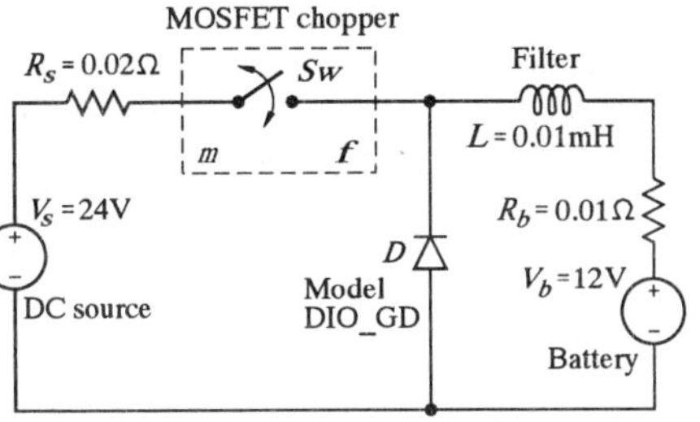

Fig. P6.5

P6.6 A buck converter has a circuit diagram similar to that shown in Fig. 6.1. The chopper operates at a switching frequency of 6kHz to modulate power from a 36-V direct-voltage source to a resistive load of 5Ω. The inductance value is 0.5mH and the filter capacitor is 600μF. If the chopper duty cycle is $m = 0.4$, do a PSpice simulation and determine **(a)** the time it takes to reach steady-state operation, **(b)** the inductor-current excursion, ($\Delta I = I_{L\max} - I_{L\min}$), **(c)** the capacitor voltage excursion and **(d)** the average power delivered to the load. Use any of the appropriate near-ideal models of Chapter 4 for the switches. Comment on the results. (Note. The transient load current is a damped oscillation with a period ~3ms, a 5A overshoot and a steady average value of about 3A.) (Ans: **(a)** about 20ms, **(b)** $\Delta I_L = 2.885$A, **(c)** $\Delta V_C = \Delta I_l \times R = 18.8 \times 10^{-3} \times 5 = 94$mV, **(d)** 41.49W.)

P6.7 Consider the circuit diagram of the buck converter in Fig. 6.1. The dc supply voltage is 48V, the load resistance is 4Ω, the inductance is 0.6mH, the capacitance is 1000μF, the chopper frequency is 8kHz and the duty cycle is $m = 0.55$. Do a PSpice simulation to find the average load voltage in the steady state. The ideal theoretical load voltage is $mV_s = 26.4$V. Use any of the appropriate near-ideal models of switches that are described in Chapter 4. (Ans: 26.35V.)

P6.8 Consider the circuit diagram of the buck converter in Fig. 6.1. The dc supply voltage is $V_s = 24$V, the load resistance is 1Ω, the inductance is 40μH, the capacitance is 1500μF, the chopper frequency is $f = 10$kHz and the duty cycle is $m = 0.8$. In the steady state the ideal theoretical excursion of the capacitor voltage is $\Delta V_C = \Delta I_L/(8fC)$, where ΔI_L is the inductor current excursion. From a PSpice simulation determine **(a)** the excursion ΔI_L, **(b)** the excursion ΔV_C and **(c)** the difference between the theoretical and simulation results of the excursion ΔV_C. Use any of the appropriate near-ideal models of the switches from Chapter 4. (Ans: **(a)** 9.64A, **(b)** 79.86mV, **(c)** theoretical, 80.3mV.)

P6.9 Consider the circuit diagram of the buck converter in Fig. 6.1. The dc supply voltage is 48V, the load resistance is 4Ω, the inductance is 12μH, the capacitance is 1000μF, the chopper frequency is $f = 20$kHz and the duty cycle is $m = 0.4$. Such operation provides discontinuous inductor current. From a PSpice simulation determine for steady conditions **(a)** the time to reach steady state, **(b)** the peak inductor current, **(c)** the capacitor voltage excursion ($\Delta V_C = I_{C\max} - I_{C\min}$) and **(d)** the average load voltage (theoretically 19.2V). Use any of the appropriate near-ideal models of switches from Chapter 4.
(Ans: **(a)** ~5ms, **(b)** 26.69A, **(c)** 194.8mV, **(d)** 32.03V.)

P6.10 Consider the circuit diagram of the buck converter in Fig. 6.1. The dc supply voltage is $V_s = 48$V with an internal resistance of 0.1Ω, the load resistance is 4Ω, the inductor has inductance of 0.6mH and resistance of 0.1Ω, the capacitance is 1000μF, the chopper frequency is 8kHz and the duty cycle is 0.55. From a PSpice simulation find **(a)** the transient interval and **(b)** the average load voltage in the steady state. Compare with the results of P6.7. Use any of the appropriate switches from Chapter 4.
(Ans: **(a)** ~18ms, **(b)** 25.41V, a drop of about 1V.)

P6.11 A near ideal boost converter (see Fig. 6.5) modulates power from a 100-V dc source to an 11.09-Ω resistive load. The nominal rating of the load is 5kW for which the theoretical value of the duty cycle is $m = 0.575$. Theoretically, the minimum inductance is $L = 0.575$mH if its current is to be finite for the above conditions. Practical design values for L and C are $L = 1.0$mH and $C = 200\mu$F. From a PSpice simulation for a switch frequency $f = 1$kHz and a duty cycle $m = 0.575$ determine **(a)** the time to reach steady conditions (modify L or C if necessary) and **(b)** the peak transient inductor current and capacitor voltage. **(c)** Initialize the inductor current and capacitor voltage for a steady-state simulation. **(d)** Determine the steady, periodic average value of the load voltage (235V theoretically). **(e)** Find, using PROBE, the average power absorbed by the load resistance.
(Ans: **(a)** ~ 20ms, **(b)** 135A, 360V, **(c)** $I_{L\min} \approx 17.75$A, $V_{C\max} = 252$V, **(d)** 226.9V (theoretical 235.3V), **(e)** 4.67kW.)

P6.12 A near-ideal boost converter has a circuit diagram depicted in Fig. 6.5a. Parameter values are $V_s = 48$V, $L = 0.3$mH, $C = 800\mu$F and $R_l = 8\Omega$. The BJT operates at a frequency $f = 5$kHz with a duty cycle $m = 0.4$. An ideal theoretical analysis ($C \to \infty$) gives the inductor current excursion, $\Delta I = I_{L\max} - I_{L\min} = mV_s/(Lf) = 12.8$A. From a PSpice simulation plot the steady-state traces of i_L, i_D, i_C and i_l. Compare the simulation waveforms with those in Fig. 6.5b. Determine **(a)** from the trace of i_L the inductor current excursion and **(b)** from the trace of i_l the average load current. Use the models of the BJT and the diode given in Chapter 4. (Ans: **(a)** $\Delta I_L = 12.73$A (theoretically 12.8A), **(b)** 9.945A (theoretically 10A.)

P6.13 Consider the circuit diagram in Fig. P6.13. The switch *Sw*boost is allowed to remain open. The switch *Sw*buck is switched at a frequency of 1kHz with a duty cycle $m = 0.7$. Determine from a PSpice simulation the average steady-state power delivered to the load source whose voltage is 100 volts. Use any appropriate unidirectional switch described in Chapter 4. (Ans: 12.5kW.)

Fig. P6.13

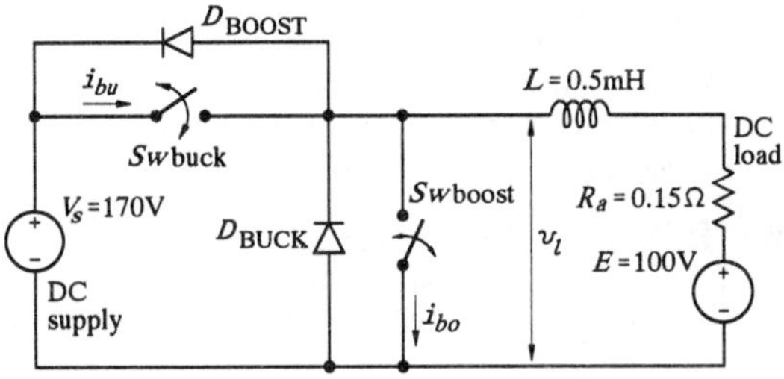

P6.14 Consider the circuit diagram in Fig. P6.13. The switch *Sw*buck is allowed to remain open. The switch *Sw*boost operates at a frequency $f = 1.5$kHz and duty cycle $m = 0.9$. From a PSpice simulation determine the steady-state value of the average power that is returned from the load to the 170-V dc supply. Use any appropriate unidirectional switch described in Chapter 4. (Ans: 845W.)

P6.15 Consider the circuit diagram of a boost converter in Fig. P6.15. An ideal theoretical analysis gives the average load voltage to be 30V. From a PSpice simulation determine for steady conditions **(a)** the average load voltage, **(b)** the average load current, **(c)** the inductor current excursion $\Delta I = I_{L\max} - I_{L\min}$ and **(d)** the capacitor voltage excursion, $\Delta V = V_{C\max} - V_{C\min}$.
(Note: The transient interval is about 15ms.)
(Ans: **(a)** 28.4V, (theoretically ~30V), **(b)** 2.368A, **(c)** 1.744A, **(d)** 0.237V.)

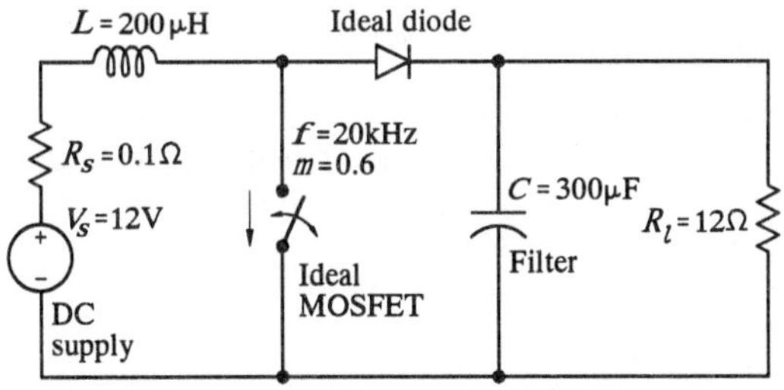

Fig. P6.15

P6.16 In an ideal boost converter (see Fig. 6.5) the 12-V dc supply is converted to a 48-V output voltage ($m = 0.75$). The dc supply ideally delivers 100W (load $R_l \approx 23\Omega$). In a practical design the 12-V supply has a source resistance $R_s = 0.1\Omega$. The converter inductor has a value $L = 20\mu$H and $R = 0.1\Omega$ and the filter capacitor has a value $C = 350\mu$F. Determine from a PSpice simulation, for steady-state operation at 25kHz with a duty cycle $m = 0.75$, **(a)** the value of the average power absorbed by the load and **(b)** the load current excursion. Use the MOSFET and diode models from Chapter 4. (Ans: **(a)** 77.89W, **(b)** 7.047mA.)

P6.17 The boost converter described in problem P6.16 is redesigned to have a new inductor $L = 160\mu$H, $R = 0.1\Omega$ and a new capacitor $C = 500\mu$F. The chopper is operated at 25kHz with a duty cycle $m = 0.5$. Ideally the output voltage would be 24V. Determine from a PSpice simulation for steady conditions **(a)** the average load voltage, **(b)** the capacitor voltage excursion, $\Delta V = V_{C\max} - V_{C\min}$, **(c)** the diode current excursion, $\Delta I = I_{D\max} - I_{D\min}$ and **(d)** the average diode current. (Ans: **(a)** 23.18V, **(b)** 39.76mV, **(c)** 1.416A, **(d)** 1.0126A.)

P6.18 Consider the Ćuk chopper diagram in Fig. 6.9. Change C_1 to 100μF, C_2 to 350μF, L_1 to 500μH, L_2 to 500μH and the duty cycle to $m=0.7$. Plot a transient trace of the source current i_s over an interval of 1ms starting at $t=49$ms. Determine **(a)** the source current excursion and **(b)** the average source current (ideally 5.444A). (Ans: **(a)** 1.642A, **(b)** 5.313A.)

P6.19 A flyback converter (see EXAMPLE 6.5.1) is to modulate voltage from a 100-V dc supply (0.1mΩ source resistance) to a 6-ohm resistive load at a nominal voltage of 30V ($m=0.6$). The IGBT is switched at a frequency of 30kHz. Assume perfect coupling for the transformer windings which have a turns ratio $n=N_1/N_2=5$. Let the primary winding of the transformer have a nominal inductance of 160μH. The filter capacitor has a value of 200μF. Do a PSpice simulation and determine **(a)** the transient interval, **(b)** the maximum IGBT current during the transient interval, **(c)** the steady-state average current in the IGBT and **(d)** the average load voltage. (Ans: **(a)** ~0.3ms, **(b)** 40.34A, **(c)** 4.68A, **(d)** 50.58V.)

Chapter 7

P7.1 A single-phase, half-wave rectifier has a 100-V(rms), 60-Hz supply and a dc load of inductance $L_l=6.2$mH and resistance $R_l=6.8\Omega$. Carry out a PSpice simulation using a diode model D_IDEAL that is described in Section 4.2.4. Plot traces of the diode current and voltage over an interval of 35ms. Determine from a trace the current rating of the diode. (Note the voltage spikes due to the numerical analysis.) (Ans: $I_{Dav}=6.41$A.)

P7.2 A single-phase, half-wave rectifier has a 120-V(rms), 50-Hz supply and a dc load of inductance $L_l=7.6$mH and resistance $R_l=11.1\Omega$. Carry out a PSpice simulation using the subcircuit named D_IDEAL from the library file POWER.LIB for the diode. Plot traces of the diode current and voltage. From a trace determine the average power absorbed by the load. (Ans: $I_{l\text{rms}}=7.486$A, average power 622W.)

P7.3 A single-phase, half-wave diode rectifier modulates power from a 240-V(rms), 50-Hz ac supply, whose source resistance is $R_s=0.3\Omega$, to a resistive load of value $R=10\Omega$. Across the load a capacitor, $C=10{,}000\mu$F, is used to smooth the rectifier output. From a PSpice simulation, determine **(a)** the output-voltage, steady-state excursion ΔV (peak-to-peak deviation from the average load voltage V_{lav}) and **(b)** the current rating of the diode. Use the diode model DIO_I, described in Section 4.2.2. (This problem is analysed in PESS EXAMPLE 3.6.) (Ans: **(a)** 44.5V (theoretically 53V), **(b)** 42.1A (theoretically 30.7A). This does not account for the initial transient of 550A (peak).)

P7.4 A single-phase, half-wave diode rectifier modulates power from a 240-V(rms), 60-Hz ac supply, whose source inductance is $L_s=60\mu$H, to a load, whose resistance is $R=1.0\Omega$ and whose filter inductance is $L=500$mH. See Fig. P7.4. Determine **(a)** the average load current and **(b)** the commutation overlap u in steady state. Use the diode model D_IDEAL, described in Section 4.2.4. and written as a subcircuit in POWER.LIB. (This problem is analysed in PESS EXAMPLE 3.7.) (Ans: **(a)** 106.2A, **(b)** $\Delta t=1.8051$ms, $u=0.68$rad (39°) measured from maximum current $i_s=107.1$A to zero in steady state at $t=2.9575$s.)

Fig. P7.4

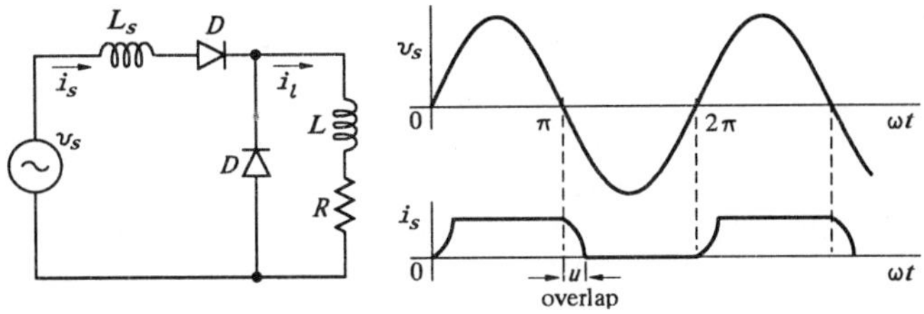

P7.5 A single-phase, full-wave, diode bridge converts power from an ac source of 100V(rms) at 50Hz to a dc load whose resistance is $R_l = 3.2\Omega$. See Fig. 7.3a. From a PSpice simulation, incorporating the diode model DIO_V, plot traces of the diodes' voltage and current. Determine **(a)** the value of the peak current of each diode and **(b)** the current rating of each diode. (Notes: (1) The model DIO_V is described in EXAMPLE 4.2.1 and the subcircuit for DIO_V is in POWER.LIB. (2) The current in a diode, modelled by DIO_V, cannot be measured directly, because DIO_V is not a two-terminal device. Use a zero-voltage source to measure the diode current.) (Ans: **(a)** 44.115A, **(b)** 14A.)

P7.6 A single-phase, full-wave, diode bridge converts power from an ac source of 120V(rms) at 60Hz to a dc load whose resistance is $R_l = 8.6\Omega$ and whose inductance is $L_l = 4.3$mH. See Fig. 7.3a. Do a PSpice simulation incorporating the diode model DIO_GD (see EXAMPLE 4.2.3 and POWER.LIB). Plot the traces of the ac input current i_s and the load current i_l. Determine from PROBE traces **(a)** the time to reach steady-state conditions, **(b)** the periodic average power input (P_s) in steady state, **(c)** the periodic average output power (P_l) in steady state and **(d)** the ac input power factor (PF $= P_s/(V_s I_s)$). From a Fourier analysis (13 harmonics) determine **(e)** the value of the rms fundamental component of current, **(f)** the displacement power factor ($\cos\theta_1$), **(g)** the average power input ($P_s = V_s I_{s1} \cos\theta_1$), **(h)** the THD (total harmonic distortion) and **(i)** the rms value of the input current i_s. (There are *numerical* spikes.)
(Ans: **(a)** ~100ms, **(b)** 1.63kW, **(c)** 1.63kW, **(d)** 0.9866, **(e)** 13.73A, **(f)** 0.9899, **(g)** 1.63kW, **(h)** 6.7%, **(i)** 13.76A.)

P7.7 For the biphase, full-wave, diode rectifier illustrated in Fig. P7.7 carry out a PSpice simulation similar to that in EXAMPLE 7.4.2. Find **(a)** the time to reach steady-state conditions. Determine **(b)** the load voltage excursion (peak-to-peak ripple), **(c)** the average value of the load voltage, **(d)** the diode current rating and **(e)** the total harmonic distortion (THD) of the transformer input current.
(Ans: **(a)** ~8ms, **(b)** 0.3V, **(c)** ~17.1V, **(d)** >9.1A **(e)** 42.2%.)

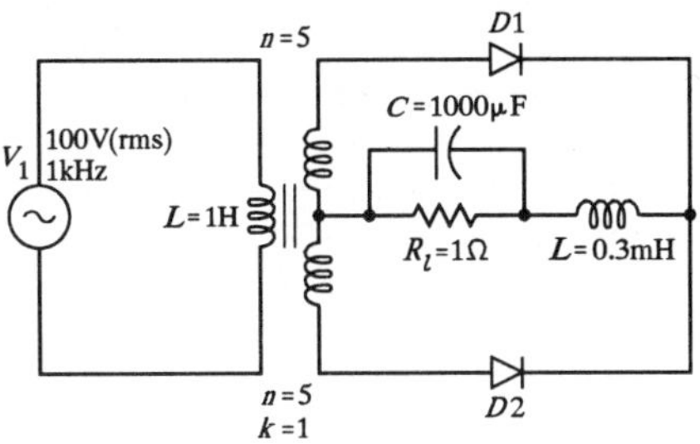

Fig. P7.7

P7.8 A biphase, full-wave, diode rectifier converts power from a 120-V(rms), 60-Hz ac supply which has an equivalent source impedance of $(0+j1.131)\Omega$. The dc load has an inductance whose value is large enough that the load current is virtually constant at 55A. The centre-tapped transformer has a turns ratio $n=N_1/N_2=4$. The circuit diagram is similar to Fig. 7.5. In this case, the load can be modelled by an independent current source. Carry out a PSpice simulation using the diode model named D_IDEAL as a subcircuit from POWER.LIB. Use PROBE to plot traces of the diode currents and determine **(a)** the resistance of the load and **(b)** the total harmonic distortion THD of the ac input current.
(Ans: **(a)** 0.46Ω, **(b)** 32.2%.)

P7.9 Consider the same rectifier and source as in problem P7.8. In this case let the load have a resistance 0.46Ω and an inductance of 200mH. Use PROBE to plot a trace of the load current. Determine **(a)** the average value of the load current in the steady state and the current excursion, and **(b)** the total harmonic distortion of the ac input current.
(Ans: **(a)** 54.975A, 0.21A, **(b)** 37.9%. Note: Steady state is reached after about 2.2s.)

P7.10 Consider the circuit diagram in Fig. 7.11. The load of the rectifier draws a constant current of 500A. The supply voltage has a magnitude of 600V(rms line-to-line). The three-phase transformer has a rating 600V/60V(rms line-to-line), 50kVA at 60Hz. Simulate the circuit. From PROBE plot traces of a diode's voltage and current and determine the average load voltage. (Ans: 69.66V.)

P7.11 A three-phase bridge rectifier is depicted in Fig. 7.13. Specifications are

$$V_{sa}=600\text{V(rms) at 50Hz},\ L_s=0,\ R_s=0,\ L_f=0,\ R_f=0,$$
$$C=300\mu\text{F},\ L_l=28.1\text{mH and } R_l=3\Omega.$$

Do a PSpice simulation using any diode models from Section 4.2. Plot traces of the input line current, and the running average of the input phase power for steady conditions. Determine **(a)** the time to reach steady state, **(b)** the peak and rms values of the ac line currents, **(c)** the total average power input, **(d)** the ac-line power factor, **(e)** the displacement power factor and **(f)** the total harmonic distortion of the ac line currents.
(Ans: **(a)** ~56ms, **(b)** 623.1A, 384.1A, **(c)** 657.9kW, **(d)** 0.952, **(e)** 0.9997, **(f)** 28.26%.)

P7.12 Consider the circuit diagram of a three-phase rectifier in Fig. 7.13, with data

$$V_{sa}=600\text{V(rms) at 50Hz},\ L_s=0,\ R_s=0,\ L_f=24\text{mH},\ R_f=3\text{m}\Omega,$$
$$C=400\mu\text{F},\ L_l=28.1\text{mH and } R_l=3\Omega.$$

Do a PSpice simulation using any of the diode models from Chapter 4 or POWER.LIB. Plot traces of the input line current, and the running average of the input power per phase for steady conditions. Determine, **(a)** the time to reach steady state, **(b)** the maximum and rms values of the ac line currents, **(c)** the total average power input, **(d)** the ac-line power factor, **(e)** the displacement power factor and **(f)** the total harmonic distortion of the ac line current.
(Ans: **(a)** ~300ms, **(b)** 468.8A, 379.37A, **(c)** 651.3kW, **(d)** 0.954, **(e)** 1, **(f)** 27.9%.)

P7.13 A three-phase diode bridge is illustrated in Fig. 7.13, with data

$$V_{sa}=600\text{V(rms) at 50Hz},\ L_s=0.5\text{mH},\ R_s=0.4\text{m}\Omega,\ L_f=30\text{mH},\ R_f=3\text{m}\Omega,$$
$$C=500\mu\text{F},\ L_l=28.1\text{mH and } R_l=3\Omega.$$

Do a PSpice simulation using any of the diode models from Section 4.2 and POWER.LIB. Plot traces of the input line current, and the running average of the input power per phase for steady conditions. Determine **(a)** the time to reach steady state, **(b)** the load-current excursion, **(c)** the average load current, **(d)** the rms value of the line currents, **(e)** the total average power input, **(f)** the ac-line power factor, **(g)** the displacement power factor, **(h)** the total harmonic distortion of the ac line current and **(i)** the commutation overlap angle in degrees.
(Ans: **(a)** 160ms, **(b)** 0.088A, **(c)** 444.63A, **(d)** 352.78A, **(e)** 595.8kW, **(f)** 0.938, **(g)** 0.957, **(h)** 21.3%, **(i)** 23.13°.)

P7.14 Consider the circuit diagram of a three-phase diode rectifier in Fig. 7.13, with data

$$V_{sa} = 600\text{V(rms) at 50Hz}, \quad L_s = 0.5\text{mH} \quad \text{and} \quad R_s = 0.4\text{m}\Omega.$$

The load has a filter such that the load current is virtually constant at 450A. Do a PSpice simulation using any of the diode models from Section 4.2. Plot traces of the load voltage, the ac-line current and the diode currents. Determine **(a)** the load-voltage excursion, **(b)** the average load voltage, **(c)** the average power absorbed by the load, **(d)** the rms input current, **(e)** the input power factor, **(f)** the commutation overlap angle, **(g)** the displacement power factor and **(h)** the total harmonic distortion of the ac-line current. (Note: DIO_V has nonconvergence problems, DIO_I works only with UIC, DIO_GD works only without UIC.)
(Ans: **(a)** 317.7V, **(b)** 1.336kV, **(c)** 601.2kW, **(d)** 356.9A, **(e)** 0.9357, **(f)** 28.27°, **(g)** 0.957, **(h)** 21.3%.)

Chapter 8

P8.1 A single-phase, half-wave, controlled rectifier modulates power from a rectangular-wave, ac source of 50V(peak) at 60Hz to a dc load whose resistance is $R_l = 0.5\Omega$ and whose inductance is $L_l = 4.6$mH. See the circuit diagram in Fig. 8.1a. Do a PSpice simulation of the circuit to find the current rating of the thyristor. Use the subcircuit named TH_GAT in DRIVER.LIB for the thyristor and its driver. (Ans: 24.19A.)

P8.2 (PESS EXAMPLE 2.5) Consider the circuit diagram in Fig. 8.1a. The single-phase, half-wave converter has a power supply consisting of an ideal 115-V, 60-Hz source. The load, $R_l = 8\Omega$, $L_l = 20$mH, has a freewheeling diode. Do a PSpice simulation for a firing angle $\alpha = 60°$. Determine **(a)** the average value of the load voltage, **(b)** the average value of the load current and **(c)** the thyristor conduction angle each cycle. A model of a thyristor and driver (TH_GAT) is found in DRIVER.LIB and a model of a diode (D_IDEAL) is found in POWER.LIB. (Ans: **(a)** 38.78V, **(b)** 4.789A, **(c)** current continuous.)

P8.3 (PESS Problem 2.7) A single thyristor modulates power from a single-phase, 115-V, 10-Hz source to a virtually inductive load of 10mH. Do a PSpice simulation. **(a)** Do a sweep of the thyristor firing angle to determine α for a peak energy of 750J to be delivered to the load. **(b)** What are the voltage and current ratings of the thyristor?
(Ans: **(a)** 59.9° (theoretically 60°), **(b)** 170V, 257.2A (theoretically 170V, 260A).)

P8.4 (PESS Problem 2.11) A single thyristor modulates power from a single-phase, 120-V, 60-Hz source to a load, across which there is connected a freewheeling diode. See Fig. 8.1. The load resistance is $R_l = 0.8\Omega$ and the load inductance is $L_l = 1\text{mH}$. The thyristor firing angle is set to a value $\alpha = 60°$. Do a PSpice simulation. Determine **(a)** whether the load current is continuous, **(b)** the average value of the load current, **(c)** the load current at $\omega t = \pi$ radians, **(d)** the average power dissipated in the load resistance during the freewheeling intervals and **(e)** the total average power delivered by the source. (Ans: **(a)** discontinuous, **(b)** 50.51A (theoretically 50.6A), **(c)** 80.22A (theoretically 80.56A), **(d)** 193W (theoretically 194.7W), **(e)** 5.48kW (theoretically 5.49kW).)

P8.5 Simulate the converter illustrated in Fig. 8.4 for the given specifications.
$V_s = 120\text{V}$ at 60Hz, $R_s = 1\text{m}\Omega$, $L_s = 4\text{mH}$, no diode, $R_l = 8\Omega$, $C = 0$, $L_l = 20\text{mH}$, $\alpha = 30°$. Use the thyristor model described as a subcircuit named TH_GAT in DRIVER.LIB. Plot traces of the input current i_s, the thyristor currents i_{TH1}, i_{TH3} and the load current i_l. Determine **(a)** the total harmonic distortion of the input current, **(b)** the displacement power factor $\text{DPF} = \cos\theta_1$ of the input current, **(c)** the form factor FF of the load current and **(d)** the commutation overlap angle u (from i_{TH1}, i_{TH3}). (Ans: **(a)** 12.79%, **(b)** 0.788, **(c)** 1.076, **(d)** 8.64°.)

P8.6 Consider the circuit diagram in Fig. 8.4. The converter specifications are as follows.
$V_s = 100\text{V(rms)}$ at 50Hz, $R_s = 1\text{m}\Omega$, $L_s = 7\text{mH}$, $R_l = 5\Omega$,
a freewheeling diode connected, $C = 0$, $L_l = 30\text{mH}$, $\alpha = 45°$,
Do a PSpice simulation with the diode model D_IDEAL from POWER.LIB and the thyristor and driver model TH_GAT from DRIVER.LIB. Plot traces of the input current i_s, the thyristor currents i_{TH1}, i_{TH3} the diode current i_D and the load current i_l. Determine **(a)** the peak load current and the load current excursion and **(b)** the total harmonic distortion THD of the input current. (Ans: **(a)** 15.419A, 5.03A, **(b)** 19.19%.)

P8.7 (PESS Problem 2.14) Consider the ac-dc converter shown in Fig. 8.4. Assume that the load-inductance value is high enough that the load current is virtually constant at 46.78A. The system specifications are as follows.
$V_s = 120\text{V(rms)}$ at 60Hz, $R_s = 1\text{m}\Omega$, $L_s = 1\text{mH}$, no diode, $C = 0$, $L_l \approx \infty$, $\alpha = 30°$.
From a PSpice simulation, plot traces of load voltage v_l, the thyristor currents i_{TH1}, i_{TH3} and the source current i_s. Account for the fact that over the range $0 \le \omega t \le \alpha$, the load current is zero. Determine **(a)** the output voltage form factor FF, **(b)** the output voltage ripple factor RF, **(c)** the rectification ratio RR, **(d)** the average output power, **(e)** the input harmonic distortion THD, **(f)** the input power factor PF, **(g)** the input displacement power factor DPF and **(h)** the commutation overlap u. (Ans: PROBE values **(a)** 1.54, **(b)** 1.171, **(c)** 0.65, **(d)** 3959W, **(e)** 39.7%, **(f)** 0.74, **(g)** 0.766, **(h)** 22.7°. Theoretical values **(a)** 1.28, **(b)** 0.8, **(c)** 0.78, **(d)** 4377W, **(e)** 48.3%, **(f)** 0.78, **(g)** 0.866, **(h)** 22.7°.)

P8.8 (PESS Problem 2.17) Consider the single-phase, full-wave, ac-dc converter, that is illustrated in Fig. 8.4. The converter has the following specifications.
$V_s = 240\text{V}$ at 50Hz, $R_s = 1\text{m}\Omega$, $L_s = 1\text{mH}$, diode connected, $R_l = 2\Omega$, $C = 0$, $L_l = 2\text{mH}$, $\alpha = 90°$.
From a PSpice simulation, determine **(a)** whether the load current is continuous, **(b)** the minimum load current, **(c)** the output voltage form factor FF, **(d)** the average power absorbed by the load and **(e)** the average value of the diode current and the peak value.
(Ans: **(a)** Continuous, **(b)** 0.437A, **(c)** 1.36, **(d)** 9.272kW, **(e)** 2.14A, 24.25A.)

P8.9 Consider the converter, illustrated in Fig. 8.4. The load is a 96-V battery with a series resistance $R_b = 0.2\Omega$. The converter specifications are as follows.

$V_s = 240\text{V}$ at 50Hz, $R_s = 1\text{m}\Omega$, $L_s = 0$, no diode, $C = 0$, $L_l = 0$, $\alpha = 30°$.

Do a PSpice simulation to determine **(a)** the maximum and minimum values of the battery current, **(b)** the energy dissipated in the battery resistance in the time taken to charge the battery 15As. Use PROBE's integration facility (S). (Note: This circuit has non-convergence problems if the source and load have inductances. Try the simulation separately with $L_s = 1\text{mH}$, $L_l = 0$, and $L_s = 0$, $L_l = 1\text{mH}$). (Ans: **(a)** 1.2kA, 0, **(b)** 2.876kJ.)

P8.10 Repeat P8.9 with *TH*2 replaced by a diode *D*2, and *TH*3 replaced by a diode *D*3 to form a half-controlled rectifier. Compare the results of the solutions of P8.10 and P8.9. The results are sensitive to L_s and L_l and TMAX. (Ans: **(a)** 750A, 212A, **(b)** 1.654kJ.)

P8.11 A biphase controlled rectifier has a circuit diagram as shown in Fig. 8.6a. For a condition that the load current is virtually constant, the circuit parameters are as follows.

$V_{s2} = 120\text{V(rms)}$ at 60Hz, $R_{s2} = 0$, $L_{s2} = 0$, no diode, $L_l \approx \infty$, $C = 0$, $I_l = 5.4\text{A}$, $\alpha = 60°$.

From a PSpice simulation, plot traces of i_{TH1}, i_{TH2} and v_l. Determine **(a)** the value of the load resistance and **(b)** the average value of the load power. (Ans: **(a)** 9.95Ω, **(b)** 290W.)

P8.12 A biphase controlled converter (see Fig. 8.6b) has the following specifications.

$V_{s2} = 120\text{V(rms)}$ at 60Hz, $R_{s2} = 1\text{m}\Omega$, $L_{s2} = 1\text{mH}$, no diode, $L_l \approx \infty$, $C = 0$, $I_l = 5.4\text{A}$, $\alpha = 60°$.

From a PSpice simulation, plot traces of i_{TH1}, i_{TH2} and v_l. Determine **(a)** the commutation overlap angle u of the thyristor currents, **(b)** the value of the load resistance and **(c)** the average value of the load power. (Ans: **(a)** 0.79°, **(b)** 11.08Ω, **(c)** 323W.)

P8.13 Consider the circuit diagram in Fig. 8.6b. This biphase controlled rectifier has the following specifications. $V_{s2} = 120\text{V(rms)}$ at 60Hz, $R_{s2} = 1\text{m}\Omega$, $L_{s2} = 1\text{mH}$, no diode, $R_l = 10\Omega$, $C = 0$, $L_l = 25\text{mH}, 45\text{mH}, 60\text{mH}$, $\alpha = 60°$. From a PSpice simulation, plot traces of i_{TH1}, i_{TH2} and i_l for each of the converter loads. Determine **(a)** the average value of the load current and **(b)** the average value of the load power.
(Ans: **(a)** 6.31A, 5.03A, 5.03A, **(b)** 570W, 339W, 305W.)

P8.14 Consider a biphase converter as illustrated in Fig. 8.6b, with data

$V_{s2} = 120\text{V(rms)}$ at 60Hz, $R_{s2} = 1\text{m}\Omega$, $L_{s2} = 2\text{mH}$, no diode, $R_l = 10\Omega$,
$C = 100\mu\text{F}$, $L_l = 25\text{mH}, 45\text{mH}$, $\alpha = 60°$.

From a PSpice simulation, plot traces of i_{TH1}, i_{TH2}, i_C and i_l for each of the loads. Determine **(a)** the average value of the load current and **(b)** the average value of the power in the load. (Ans: **(a)** 6.996A, 5.322A, **(b)** 662.8W, 358.4W.)

P8.15 Consider a biphase converter with a freewheeling diode, as illustrated in Fig. 8.6b. The circuit specifications are as follows. $V_{s2} = 120\text{V(rms)}$ at 60Hz, $R_{s2} = 1\text{m}\Omega$, $L_{s2} = 1\text{mH}$, diode connected, $R_l = 10\Omega$, $C = 0$, $L_l = 25\text{mH}, 45\text{mH}, 60\text{mH}$, $\alpha = 60°$. From a PSpice simulation, plot traces of i_{TH1}, i_{TH2}, i_D and i_l for each of the loads. Determine **(a)** the average value of the diode current, **(b)** the average value of each thyristor current, **(c)** the average value of the load current, **(d)** the average value of the load power and **(e)** the load current excursion. (Ans: **(a)** 1.25A, 1.54A, 1.58A, **(b)** 2.99A, 2.47A, 2.218A, **(c)** 7.36A, 6.82A, ~6.45A, **(d)** 678.5W, 551.3W, ~490.6W, **(e)** 9.12A, 5.71A, 2.22A.)

P8.16 A biphase converter (see Fig. 8.6b) is used to charge a superconducting coil. The circuit conditions are as follows.

$V_{s2} = 10\text{V(rms)}$ at 10Hz, $R_{s2} = 1\mu\Omega$, $L_{s2} = 0$, no diode, $R_l = 0$, $C = 0$, $L_l = 10\text{mH}$.
Determine **(a)** the load current increment in the first half cycle for $\alpha = 0$ and **(b)** the average load current for $\alpha = 120°$. Note. The analysis is not stable for all α.
(Ans: **(a)** 219.8A, **(b)** 21.7A.)

P8.17 A biphase controlled rectifier (see Fig. 8.6b) has as its load a dc source V_d. This battery delivers power to the ac supply. The specifications of the circuit are as follows.

$V_{s2} = 120\text{V(rms)}$ at 60Hz, $R_{s2} = 1\text{m}\Omega$, $L_{s2} = 1\text{mH}$, no diode, $R_l = ?$, $C = 0$,
$I_l \approx 5\text{A}$ (constant in steady state), $L_l \approx \infty$, $V_d = 210\text{V}$, $\alpha = 150°$.

From a PSpice simulation, plot traces of i_{TH1}, i_{TH2}, the voltage across the entire load (V_D, R_l and L_l), and v_{TH1}. Determine **(a)** the value of the load resistance R_l, **(b)** the average value of the power delivered to the ac source, **(c)** the power absorbed by the load resistance, **(d)** the peak thyristor voltage and **(e)** the commutation overlap angle u.
(Ans: **(a)** 25.37Ω, **(b)** 415.8W, **(c)** 634.2W, **(d)** 337.4V, **(e)** 1.29°.)

P8.18 Consider the controlled rectifier in Fig. 8.7. The circuit specifications are as follows.

$V_s = 120\text{V(rms, phase)}$ at 60Hz, $R_s = 0$, $L_s = 0$, diode connected, $R_l = 10\Omega$,
$C = 0$, $L_l = 10\text{mH}$, $\alpha = 35°$.

Do a PSpice simulation. Use PROBE to plot traces of i_{TH1}, i_D, v_l and i_l. Determine **(a)** the average load current and current excursion ΔI, **(b)** the average power consumed by the load, **(c)** the average thyristor current, **(d)** the total rms value of the input line current, **(e)** the fundamental rms value of the input line current, **(f)** the displacement power factor DPF at the input, **(g)** the total harmonic distortion THD at the input, **(h)** the input power factor PF and **(i)** the average diode current. (Ans: **(a)** 11.52A, 10.76A, **(b)** 1434W, **(c)** 3.775A, **(d)** 6.888A, **(e)** 4.668A, **(f)** 0.8517, **(g)** 71.95%, **(h)** 0.577, **(i)** 0.207A.)

P8.19 Consider a three-phase, half-wave, controlled rectifier, whose circuit diagram is shown in Fig. 8.7. The load inductance is high enough so that the load current is virtually ripple free. The circuit specifications are as follows.

$V_s = 120\text{V(rms, phase)}$ at 60Hz, $R_s = 1\text{m}\Omega$, $L_s = 3\text{mH}$, no diode, $L_l \approx \infty$,
$C = 0$, $I_l = 11.52\text{A}$ (constant), $\alpha = 35°$.

Do a PSpice simulation with the load current modelled by a PWL current source. Plot traces of i_{TH1}, i_{TH2} and v_l. Determine **(a)** the average load voltage, **(b)** the average value of the load power, **(c)** the value of the load resistance, **(d)** the total rms value of the input line current, **(e)** the average value of the input line current, **(f)** the displacement power factor DPF at the input, **(g)** the total harmonic distortion THD at the input, **(h)** the input power factor PF and **(i)** the commutation overlap angle u. (Ans: **(a)** 108.6V, **(b)** 1251W, **(c)** 9.43Ω, **(d)** 6.58A, **(e)** 3.84A, **(f)** 0.775, **(g)** 63.57%, **(h)** 0.529, **(i)** 7.96°.)

P8.20 A three-phase, half-controlled rectifier, depicted in Fig. 8.8, has data
$V_s = 120\text{V(rms, phase)}$ at 60Hz, $R_s = 0$, $L_s = 0$, no diode, $R_l = 10\Omega$, $C = 0$, $L_l = 15\text{mH}$, $\alpha = 35°$.
Do a PSpice simulation. Plot traces of i_{sa}, v_l and i_l. Determine **(a)** the average value of the load voltage, **(b)** the average value of the load current, **(c)** the load-current excursion, **(d)** the average power consumed by the load, **(e)** the total rms value of the input line current, **(f)** the

fundamental rms value of the input line current, **(g)** the displacement power factor DPF at the input, **(h)** the total harmonic distortion THD at the input and **(i)** the input power factor PF. (Note: Nonconvergence problem with diode D_IDEAL. Use diode DIO_I model.)
(Ans: **(a)** 254.37V, **(b)** 25.44A, **(c)** 6.02A, **(d)** 6503W, **(e)** 20.91A, **(f)** 18.97A, **(g)** 0.954, **(h)** 42.51%, **(i)** 0.866.)

P8.21 Consider the three-phase, fully-controlled rectifier whose circuit diagram in shown in Fig. 8.9a. The circuit specifications are as follows.

$V_s = 120$V(rms, phase) at 60Hz, $R_s = 0$, $L_s = 0$, $R_l = 10\Omega$, $C = 160\mu$F, freewheeling diode connected, $L_l = 20$mH, $\alpha = 90°$.

From a PSpice simulation, plot traces of i_s, i_D, i_L, i_l and v_D. Determine **(a)** the average value of the load voltage, **(b)** the average power absorbed by the load, **(c)** the load current excursion, **(d)** the load form factor FF, **(e)** the average inductor current, **(f)** the average diode current, **(g)** the rms value of the input line current, **(h)** the total harmonic distortion THD at the input and **(i)** the input power factor PF. (Ans: **(a)** 3.743A, **(b)** 140.8W, **(c)** 0.687A, **(d)** 1.002, **(e)** 3.75A, **(f)** 1.656A, **(g)** 2.47A, **(h)** 99.92%, **(i)** 0.158.)

P8.22 A three-phase rectifier, depicted in Fig. 8.9, has the following specifications.

$V_s = 120$V(rms, phase) at 60Hz, $R_s = 1$mΩ, $L_s = 5$mH, no diode, $R_l = 10\Omega$, $C = 0$, $L_l = 0$, $\alpha = 35°$, $90°$ (sweep).

From a PSpice simulation, plot traces of i_{sa}, i_{TH1}, i_{TH3} and v_l. Compare the results with EXAMPLE 8.7.1. Determine **(a)** the average value of the load voltage, **(b)** the load voltage form factor FF, **(c)** the average power absorbed by the load, **(d)** the average value of the thyristor current, **(e)** the angle u of commutation overlap, **(f)** the total harmonic distortion THD at the input, **(g)** the fundamental rms value of the input line current, **(h)** total rms value of the input line current and **(i)** the input power factor PF.
(Ans: **(a)** 196.6V, 27.63V, **(b)** 1.002, 1.27, **(c)** 3.878kW, 123.5W, **(d)** 6.55A, 0.92A, **(e)** 18.9°, 0, **(f)** 22.97%, 80.03%, **(g)** 15.29A, 2.22A, **(h)** 15.71A, 2.87A, **(i)** 0.687, 0.117.)

P8.23 Consider a fully-controlled, three-phase rectifier, whose circuit diagram is shown in Fig. 8.9a. The circuit has the following specifications.

$V_s = 120$V(rms, phase) at 60Hz, $R_s = 1$mΩ, $L_s = 5$mH, no diode, $R_l = 10\Omega$, $C = 0$, $L_l = 15$mH, $\alpha = 35°$, $90°$ (sweep).

From a PSpice simulation, plot traces of i_{sa}, i_{TH1}, i_{TH3}, i_l and v_l. Compare with the results of problem P8.22. Determine **(a)** the average value of the rectifier output voltage, **(b)** the average value of the load current, **(c)** the form factor FF of the load current, **(d)** the average power absorbed by the load, **(e)** the average value of the thyristor current, **(f)** the angle u of commutation overlap, **(g)** the total harmonic distortion THD at the input, **(h)** the fundamental rms value of the input line current, **(i)** total rms value of the input line current and **(j)** the input power factor PF. (Ans: **(a)** 195.1V, 17.57V, **(b)** 19.5A, 1.73A, **(c)** 1.0004, 1.21, **(d)** 4.059kW, 44.1W, **(e)** 6.5A, 0.577A, **(f)** 19.64°, 0, **(g)** 21.92%, 67.86%, **(h)** 15.15A, 1.38A, **(i)** 15.54A, 1.714A, **(j)** 0.683, 0.070.)

P8.24 Consider a fully-controlled, three-phase rectifier, whose circuit diagram is shown in Fig. 8.9a. The circuit has the following specifications.

$V_s = 120$V(rms, phase) at 60Hz, $R_s = 1$mΩ, $L_s = 5$mH, no diode, $R_l = 10\Omega$, $C = 160\mu$F, $L_l = 15$mH, $\alpha = 35°$, $90°$.

From a PSpice simulation, plot traces of i_{sa}, i_{TH1}, i_{TH3}, i_L, i_C, i_l and the rectifier output voltage. Determine **(a)** the average value of the rectifier output voltage, **(b)** the average value of the load current and the current excursion, **(c)** the average power absorbed by the load, **(d)** the form factor FF of the load current, **(e)** the load inductor current excursion, **(f)** the capacitor current excursion, **(g)** the average value of the thyristor current, **(h)** the angle u of commutation overlap, **(i)** the total harmonic distortion THD of the input, **(j)** the input power factor PF. (Note: Did not run .STEP successfully, but ran individual programs.)
(Ans: **(a)** 194.8V, 19.0V, **(b)** 19.47A, 0.35A; 1.864A, 0.841A, **(c)** 3.789kW, 35.6W, **(d)** 1, 1.012, **(e)** 1.537A, 3.319A, **(f)** 1.45A, 3.35A, **(g)** 6.5A, 0.622A, **(h)** 23.59°, 0, **(i)** 21.97%, 67.23%, **(j)** 0.682, 0.0578.)

P8.25 A fully-controlled, three-phase rectifier, whose circuit diagram is shown in Fig. 8.9a, has the following specifications.

$V_s = 120$V(rms, phase) at 60Hz, $R_s = 1\text{m}\Omega$, $L_s = 5$mH, no diode, $R_l = ?$,
$C = 0$, $I_l = 30\text{A} = \text{constant}$, $L_l \approx \infty$, $\alpha = 35°, 90°$.

From a PSpice simulation, plot traces of i_{sa}, i_{TH1}, i_{TH3} and v_l. Determine **(a)** the average value of the load voltage, **(b)** the load voltage excursion, **(c)** the form factor FF of the output voltage, **(d)** the average power absorbed by the load, **(e)** the value of the load resistance, **(f)** the average value of the thyristor current, **(g)** the angle u of commutation overlap, **(h)** the total rms value of the input line current, **(i)** the fundamental rms value of the input line current, **(j)** the total harmonic distortion THD of the input current and **(k)** the power factor PF at the input.
(Ans: **(a)** 175.56V, −53.99V, **(b)** 129.5V, 181.1V, **(c)** 1.022, 1.352, **(d)** 5.267kW, −1620W (generating), **(e)** 5.85Ω, generating, **(f)** 10A, 10A, **(g)** 30.12°, 23.38°, **(h)** 23.53A, 23.77A, **(i)** 23.15A, 23.25A, **(j)** 17.3%, 20.22%, **(k)** 0.623, 0.19.)

Chapter 9

P9.1 Consider the circuit diagram of the centre-tapped inverter in Fig. 9.1. The circuit has the following specifications.

$V_s = 200$V, $L_l = 50\mu$H, $C_l = 0$, $R_l = 2\Omega$, IGBT switches, antiparallel diodes connected,
$m = 0.4$, $f = 400$Hz. Single-pulse output per half cycle. Gate voltage 15V.

Following the pattern of EXAMPLE 9.2.1 do a PSpice simulation. Plot traces of i_{s1}, i_{sw1}, i_{D1} and i_l. Determine **(a)** the total average power delivered by the sources, **(b)** the average value of the current in the switch $Sw1$, **(c)** the average value of the current in the diode $D1$, **(d)** the rms value of the load current, **(e)** the fundamental rms value of the load current and **(f)** the total harmonic distortion of the load current.
(Ans: **(a)** 15.46kW, **(b)** 38.98A, **(c)** 0.336A, **(d)** 87.98A, **(e)** 84.94A, **(f)** 24.41%.)

P9.2 A centre-tapped source inverter is illustrated in Fig. 9.1. The data are

$V_s = 300$V, $L_l = 90$mH, $C_l = 0$, $R_l = 18\Omega$, $Sw1$ and $Sw2$ are IGBTs, diodes connected,
$m = 0.5$, $f = 60$Hz. Single-pulse output per half cycle. Gate voltage 15V.
Note. Use initial conditions that $i_l = -11.372$A for steady state.

From a PSpice simulation plot traces of the load current i_l, the load voltage v_l and the source 1 current i_{s1}. Determine **(a)** the average power delivered to the load, **(b)** the average current delivered by each source, **(c)** the total harmonic distortion THD of the output current and **(d)** the fundamental rms value of the load current. (Note: The real IGBT has a built-in reverse diode that shares reverse current with diode D. This can be observed in traces of i_{sw1} and i_{D1}, since the switch $Sw1$ is on for t_{ON} seconds each cycle. An ideal IGBT would be modelled by a PSpice switch in series with a diode for blocking reverse current. We have not added a series diode in this text.) (Ans: **(a)** 909.5W, **(b)** 1.51A, **(c)** 13.5%, **(d)** 7.035A.)

P9.3 A centre-tapped source inverter is illustrated in Fig. 9.1. For multiple-pulse operation the circuit has the following specifications.

$V_s = 300\text{V}$, $L_l = 0$, $C_l = 0$, $R_l = 18\Omega$, $m = 0.3$, $f = 60\text{Hz}$. Gate voltage 15V.

Following the pattern of Drill Exercise D9.2.3 do a PSpice simulation for 1, 2, 3 and 4 pulses per half cycle of the output-voltage waveform. Plot traces of the output voltage for each case. Determine for each case **(a)** the rms value of the load voltage, **(b)** the total harmonic distortion THD of the output waveform, **(c)** the amplitude of the fundamental component of the load voltage, **(d)** the amplitude of the third harmonic of the load voltage, **(e)** the amplitude of the fifth harmonic of the load voltage and **(f)** the amplitude of the seventh harmonic of the load voltage. (Ans: **(a)** 232.37V for all cases, **(b)** 32.9%, 84.5%, 90.9%, 90.5%, **(c)** 309.2V, 246.0V, 234.0V, 232.8V, **(d)** 39.0V, 176.6V, 102.4V, 90.98V, **(e)** 76.4V, 77.11V, 152.8V, 74.45V, **(f)** 17.21V, 12.12V, 89.56V, 144.0V.)

P9.4 Consider the centre-tapped source inverter, whose circuit diagram is shown in Fig. 9.1. The circuit specifications are as follows.

$V_s = 300\text{V}$, $L_l = 90\text{mH}$, $C_l = 0$, $R_l = 18\Omega$, IGBT switches,
$m = 0.3$, $f = 60\text{Hz}$. Three-pulse output per half cycle. Gate voltage 15V.

Do a PSpice simulation following the pattern of Drill Exercise D9.2.3 to find the initial value of current in the inductance for steady conditions. Plot the steady-state traces† of v_l, i_l and i_{s1}. Determine **(a)** the rms value of the load current, **(b)** the average power absorbed by the load, **(c)** the total harmonic distortion THD of the load current and **(d)** the average value of a source current. (Ans: IC = −2.7882A, **(a)** 2.848A, **(b)** 146W, **(c)** 66.59%, **(d)** 0.231A.)

P9.5 A centre-tapped source inverter, whose circuit diagram is shown in Fig. 9.1, has the following specifications.

$V_s = 100\text{V}$, $L_l = 10\text{mH}$ and 90mH, $C_l = 0$, $R_l = 10\Omega$, $m = 0.3$, $f = 50\text{Hz}$.
Two-pulse output per half cycle. IGBT gate voltage 15V.

Do a PSpice simulation following the pattern of Drill Exercise D9.2.3, but use the driver subcircuit named INV_MLT_PLS in DRIVER.LIB. Plot traces of the gate voltage and the output variables v_l and i_l for steady conditions. Determine, for steady state, **(a)** the rms value of the output voltage, **(b)** the average power absorbed by the load, **(c)** the total harmonic distortion of the output current, **(d)** the rms value of the fundamental component of the output current and **(e)** the initial value of the inductor current each cycle. (Ans: **(a)** 89.35V, 99.58V, **(b)** 358.9W, 27.5W, **(c)** 85.5%, 65.54%, **(d)** 4.51A, 1.362A, **(e)** 0, −1.539A.)

P9.6 Consider the centre-tapped source inverter, whose circuit diagram is shown in Fig. 9.1. The circuit has the following specifications.

$V_s = 100\text{V}$, $L_l = 10\text{mH}$ and 60mH, $C_l = 0$, $R_l = 10\Omega$, $Sw1$ and $Sw2$ are IGBTs,
diodes connected, SPWM, $f = 50\text{Hz}$. Gate voltage 15V.
Modulation index $M = 1$. Sawtooth carrier frequency 800Hz.

---

† True traces of IGBT and diode currents cannot be obtained unless more accurate models are used.

Do a PSpice simulation using the power circuit from EXAMPLE 9.2.1 and the driver from the subcircuit named MOD_INV_PWM_DRV in DRIVER.LIB. Plot the steady-state traces of a gate voltage, the load voltage and the load current. Determine, for steady conditions, **(a)** the rms value of the load voltage, **(b)** the average power absorbed by the load and **(c)** the total harmonic distortion THD of the load current.
(Ans: **(a)** 89.94V, 97.84V, **(b)** 244.9W, 85.35W, **(c)** 27.85%, 16.57%.)

P9.7 Consider the centre-tapped source inverter whose circuit diagram is shown in Fig. 9.1. The power circuit specifications are as follows.

$V_s = 100$V, $L_l = 6$mH, $C_l = 0$, $R_l = 2\Omega$, *Sw*1 and *Sw*2 are IGBTs, diodes connected, $f = 50$Hz. SPWM with eight pulses per half cycle.

The driver specifications are as follows.

Subcircuit named SPWM_555_DRV in DRIVER.LIB
Timer source VCC = 12V, 555 sinusoidal reference signal 10V(max), 2V(min).
Carrier signal 12V, 800Hz.

See EXAMPLE W9.2.1 on the WEBsite to carry out a PSpice simulation. Plot traces of the output voltage and current and plot the frequency specta of the output current and voltage waveforms. Determine **(a)** the rms value of the load voltage, **(b)** the average power absorbed by the load and **(c)** the total harmonic distortion THD of the output voltage and current waveforms. (Ans: **(a)** 87.64V, **(b)** 326.1W, **(c)** 169.7%, 49.2%.)

P9.8 (PESS EXAMPLE 2.10) Consider the single-phase inverter illustrated in Fig. 9.11. The circuit has the following specifications.

$V_s = 600$V, $L_l = 0$, $R_l = 20\Omega$, IGBT switches,
$m = 0.347$, $f = 500$Hz. Single-pulse output per half cycle. Gate voltage 15V.

Do a PSpice simulation following the pattern of EXAMPLE 9.3.1. Plot traces of the source current and the load voltage. Plot the frequency spectrum of the output voltage waveform. Determine **(a)** the average power absorbed by the load, **(b)** the average value of the source current, **(c)** the average current in each switch, **(d)** the total harmonic distortion THD of the output-voltage waveform, **(e)** the amplitude of the ninth harmonic of the load-voltage waveform and **(f)** the harmonic factor of the fifth harmonic of the load voltage. (Ans: **(a)** 12.49kW (theoretically 12.5kW), **(b)** 20.82A (theoretically 20.83A), **(c)** 10.41A, **(d)** 26.7% (theoretically 30%), **(e)** 33.64V (theoretically 32V), **(f)** 0.165 (theoretically 0.15.)

P9.9 (PESS Problem 2.28) A single-phase, bridge inverter, whose circuit diagram is shown in Fig. 9.11, has the following specifications.

$V_s = 600$V, $L_l = 0$, $R_l = 10\Omega$, IGBT switches, maximum power transfer,
$f = 60$Hz. Single-pulse output per half cycle. Gate voltage 15V.

Follow the pattern of EXAMPLE 9.3.1 to do a PSpice simulation. Plot traces of the source current and the load current. Determine **(a)** the instantaneous value of the load current as a Fourier series, **(b)** the total harmonic distortion THD of the load-current waveform, **(c)** the harmonic factor of the ninth-harmonic component of the load current and **(d)** the IGBT current rating. (Ans: **(a)** $i_l = 76.4\cos\omega t + 25.5\cos 3\omega t + 15.3\cos 5\omega t + 10.9\cos 7\omega t + 8.5\cos 9\omega t + \ldots$, **(b)** 45.0%, **(c)** 0.11, **(d)** > 30A.)

P9.10 A single-phase, bridge inverter, depicted in Fig. 9.11, has the following specifications.

$V_s = 300V$, $L_l = 90mH$, $R_l = 10\Omega$, IGBT switches, diodes connected,
$m = 0.5$, $f = 60Hz$. Single-pulse output per half cycle. Gate voltage 15V.

Adapt EXAMPLE 9.3.1 to do a PSpice simulation. Plot traces of the source current i_s, the gate voltage of driver1, the load current i_l and the load voltage in the steady state. Plot the frequency spectra of the load current and voltage waveforms. Determine **(a)** the total harmonic distortion THD of the load current and voltage waveforms and **(b)** the average power absorbed by the load. (Ans: IC = -13.046A, **(a)** 12.54%, 45.01%, **(b)** 599W.)

P9.11 Repeat problem P9.10 with the change $m = 0.3$. Note, especially, the load voltage waveform during the interval that a pair of switches turns off and a pair of diodes conducts. This waveform does not follow the driver output waveform.
(Ans: IC = -12.8A, **(a)** 13.21%, 47.5%, **(b)** 529W.)

P9.12 (PESS Problem 2.29) A single-phase, bridge inverter, as depicted in Fig. 9.11, has the following specifications.

$V_s = 600V$, $L_l = 2mH$, $R_l = 20\Omega$, IGBT switches, diodes connected,
maximum power, $f = 500Hz$. Single-pulse output per half cycle. Gate voltage 15V.

Follow the pattern of EXAMPLE 9.3.1 and do a steady-state PSpice simulation. Determine **(a)** the peak current from the dc source, **(b)** the current ratings of the unidirectional switches, **(c)** the average power absorbed by the load, **(d)** the rms value of the load current, **(e)** the average current from the dc source and **(f)** the average value of a diode current.
(Ans: **(a)** 29.98A, **(b)** 12.46A for a unidirectional switch (the switches can be bidirectional and share current with the diodes), **(c)** 14.55kW, **(d)** 26.97A, **(e)** 24A, **(f)** 0.46A.)

P9.13 (PESS EXAMPLE 2.11) A single-phase, bridge inverter, is depicted in Fig. 9.11. The circuit specifications are as follows

$V_s = 600V$, $L_l = 10mH$, $R_l = 0$, diodes connected,
$m = 0.5$, $f = 200Hz$. Single-pulse output per half cycle. Gate voltage 15V.

Use EXAMPLE 9.3.1 as a pattern to do a PSpice simulation. Plot traces of the source current and load current. Adjust the initial conditions to obtain waveform symmetry. Determine **(a)** the total harmonic distortion THD of the load-current waveform, **(b)** the average current in the IGBT proper and **(c)** the true average diode current.
(Ans: **(a)** 12.11%, **(b)** 9.2A, **(c)** 9.2A.)

P9.14 (PESS Problem 2.31) Consider a single-phase, bridge inverter, whose circuit diagram is shown in Fig. 9.11 with the exception that the load elements are in parallel. The circuit has the following specifications.

$V_s = 600V$, $L_l = 5mH$, $R_l = 20\Omega$, IGBT switches, diodes connected. Gate voltage 15V.
Maximum power to the load, $f = 500Hz$. Single-pulse output per half cycle.

Follow the pattern of EXAMPLE 9.3.1 and do a PSpice simulation. Plot traces of the source current, the load current, the current in R_l and the current in L_l, and the IGBT current (as though it conducted unidirectionally). Determine **(a)** the peak to peak value and average value of the dc source current (the former gives the peak to peak load current), **(b)** the average value of the currents in the switches, **(c)** the average power delivered by the dc source and **(d)** the total harmonic distortion THD of the load current.
(Ans: **(a)** 120A, 30A, **(b)** 16.8A, **(c)** 3kW, **(d)** 29.26%.)

P9.15 A single-phase, bridge inverter, depicted in Fig. 9.11, has the data
$V_s = 100V$, $L_l = 2mH$, $R_l = 1\Omega$, IGBT switches, diodes connected,
$m = 0.4$, $f = 100Hz$. Single-pulse output per half cycle. Gate voltage 15V.
The gate-driver output signals v_g are depicted in Fig. 9.15. Do a PSpice simulation and plot traces of the source current, the gate voltages, and the load voltage and current. Determine **(a)** the average power delivered by the source and **(b)** the total harmonic distortion THD of the load-voltage and current waveforms. **(c)** Why is there an interval during which there is no source current? (Ans: **(a)** 2.85kW, **(b)** 27.0%, 8.8%.)

P9.16 (PESS Problem 2.32) Consider the single-phase bridge inverter, illustrated in Fig. 9.11. The circuit has the following specifications.
$V_s = 200V$, $L_l = 0$, $R_l = 2\Omega$, IGBT switches, $m = 0.4$, $f = 60Hz$. Gate voltage 15V.
Use the gate driver MPLS_TRI_INV, described in DRIVER.LIB and WEBsite EXAMPLE W9.3.1 and do PSpice simulations for single-pulse and double-pulse output waveforms of the inverter. Determine **(a)** the ratio of the third harmonic voltages across the load for the two cases and **(b)** the total harmonic distortion THD for both cases.
(Ans: **(a)** 0.34, **(b)** 22.62%, 52.55%.)

P9.17 A single-phase, bridge inverter, depicted in Fig. 9.11, has the data
$V_s = 50V$, $L_l = 0.6mH$, $R_l = 2\Omega$, IGBT switches, diodes connected,
$m = 0.3$, $f = 1kHz$. Three-pulse output per half cycle. Gate voltage 15V.
Use the gate driver MPLS_TRI_INV, from Problem P9.16 and do a PSpice simulation. Plot traces of the source current, the load voltage, a gate driver signal and the load current over an interval of two cycles for steady conditions. Determine **(a)** the initial condition for steady state, **(b)** the average power delivered by the source and **(c)** the total harmonic distortion THD of the load-current waveform. (Ans: **(a)** IC = -2.3222A, **(b)** 25.65W, **(c)** 94.49%.)

P9.18 A single-phase bridge inverter, depicted in Fig. 9.11, has the data
$V_s = 96V$, $L_l = 10mH$, $R_l = 1\Omega$, IGBT switches, diodes connected,
$m = 0.3$, $f = 40Hz$. Four-pulse output per half cycle. Gate voltage 15V.
Using the PWM gate driver INV_MLT_PLS, described in DRIVER.LIB, do a PSpice simulation. Plot traces of the dc source current, the load voltage, a gate signal and the load current for steady state. Determine **(a)** the steady-state initial condition, **(b)** the average power delivered to the load, **(c)** the fundamental rms value of the load current, **(d)** the peak value of the load current and **(e)** the total harmonic distortion THD of the load-current waveform.
(Ans: **(a)** IC = -13.913A, **(b)** 118.4W from source, **(c)** 9.84A, **(d)** 21.0A, **(e)** 51.12%.)

P9.19 Consider the circuit diagram of a single-phase bridge inverter in Fig. 9.11. The circuit has the following specifications. $V_s = 48V$, $L_l = 12mH$, $R_l = 2\Omega$, IGBT switches, diodes connected, $m = 0.3$, $f = 100Hz$. Four-pulse output per half cycle. Gate voltage 12V. Use the 555-timer driver named PWM_555_DRV from DRIVER.LIB to obtain equal, uniform pulses. Do a PSpice simulation and plot traces of the source current, the load voltage, a gate voltage and the load current in the steady state. Determine **(a)** the initial conditions for steady state, **(b)** the total harmonic distortion THD of the load current and the load voltage, **(c)** the fundamental rms value of the load current, **(d)** the peak value of the load current and **(e)** the average power delivered to the load.
(Ans: **(a)** IC = -2.6641A, **(b)** 46.14%, 296.1%, **(c)** 1.80A, **(d)** 3.82A, **(e)** 7.96W.)

P9.20 Consider the circuit diagram of a single-phase bridge inverter in Fig. 9.11. The circuit has the following specifications.

$V_s = 48\text{V}$, $L_l = 10\text{mH}$, $R_l = 2\Omega$, IGBT switches, diodes connected, $f = 100\text{Hz}$. Output waveform as in Fig. 9.16 with $\alpha_1 = 17.8°$ and $\alpha_2 = 38.0°$. Gate voltage 15V. Adapt Fig. 9.17 to suit this problem and do a PSpice simulation. Plot traces of the source current, a gate voltage, the load voltage and the load current over two cycles for steady conditions. Determine **(a)** the initial conditions for steady state, **(b)** the average power delivered by the source, **(c)** the harmonic factors of the third, fifth and seventh harmonic components in the load-current waveform ($\text{HF} = I_{ln}/I_{l1}$), **(d)** the total and fundamental rms values of the load current and **(e)** the total harmonic distortion THD of the load-voltage and current waveforms. (Ans: **(a)** IC=−7.7284A, **(b)** 69.34W, **(c)** 0.0145, 0.0551, 0.0567 (low third), **(d)** 5.89A, 5.84A, **(e)** 66.56%, 9.57%.)

P9.21 A single-phase, bridge inverter is depicted in Fig. 9.11. A harmonic-cancellation strategy is to be employed by including *notches* in the waveform as shown in Fig. 9.16. The circuit has the following specifications.

$V_s = 100\text{V}$, $L_l = 10\text{mH}$, $R_l = 5\Omega$, IGBT switches, diodes connected, $f = 100\text{Hz}$.
Gate waveform has $\alpha_1 = 29.8°$ and $\alpha_2 = 57.6°$. Gate voltage 15V.

Follow the pattern of EXAMPLE 9.3.2 and do a PSpice simulation. Plot traces of the source and load currents over two cycles for steady conditions. Determine **(a)** the average power delivered to the load, **(b)** the rms values of the third, fifth and seventh harmonic components of the load current, **(c)** the peak load current and **(d)** the load power factor and the displacement power factor.
(Ans: **(a)** 183.1W, **(b)** 1.18A, 1.54A, 1.08A, **(c)** 11.87A, **(d)** 0.32 (low), 0.62.)

P9.22 An SPWM gating strategy is to be employed on a single-phase, bridge inverter that is depicted in Fig. 9.11. The circuit has the following specifications.
$V_s = 48\text{V}$, $L_l = 5\text{mH}$, $R_l = 2\Omega$, IGBT switches, diodes connected, $f = 50\text{Hz}$. Gate voltage 13V. SPWM modulation index $M = 0.8$. Eight-pulse gating per half cycle of the output waveform. Follow the pattern of EXAMPLE 9.3.3 to do a PSpice simulation. Plot traces of the inverter output voltage, a gate signal and the output current. Determine **(a)** the initial conditions for steady state, **(b)** the average power absorbed by the load, **(c)** the total harmonic distortion THD of the output-current waveform, **(d)** the peak value of the load current, **(e)** the harmonic factors ($\text{HF} = I_{ln}/I_{l1}$) of the third and fifth harmonic components of the output current and **(f)** the output power factor PF.
(Ans: **(a)** 0, **(b)** 87.9W, **(c)** 33.41%, **(d)** 12.67A, **(e)** 0.283, 0.049, **(f)** 0.298.)

P9.23 The circuit diagram of a single-phase cycloconverter is shown in Fig. 9.3.3 on the WEB. The circuit has the following specifications.

$V_s = 120\text{V(rms)}$ at 60Hz, $L_l = 0$, $R_l = 10\Omega$, output frequency $f_o = 20\text{Hz}$,
delay angle α modulated as shown in Fig. P9.23 on the next page.

Follow the pattern of EXAMPLE W9.3.3 on the WEB and do a PSpice simulation. Plot traces of the source voltage and current, and the load current. Determine **(a)** the rms value of the source current, **(b)** the average power delivered to the load, **(c)** the power factor at the input, **(d)** the total harmonic distortion THD of the load-current waveform, **(e)** the fundamental rms value of the load current and **(f)** the harmonic factors HF of the third, fifth, seventh and ninth

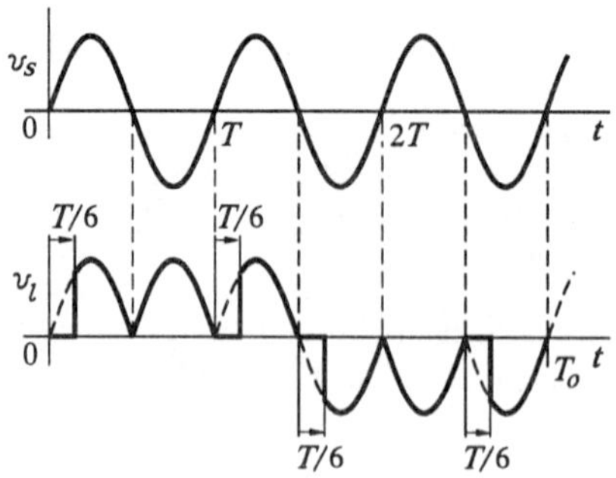

harmonics of the load-current waveform. (Ans: (a) 12.194A, (b) 1033W, (c) 0.706, (d) 78.23%, (e) 7.91A, (f) 0.356, 0.573, 0.303, 0.219.)

Fig. P9.23

P9.24 (PESS Problem 2.34) Consider the three-phase, bridge inverter illustrated in Fig. 9.23. The circuit has the following specifications.

$V_s = 600$V, $L_l = 6.67$mH, $R_l = 0$, $f = 200$Hz. Gate voltage 15V.

Inverter mode of operation 180°, six steps per cycle.

Follow the pattern of EXAMPLE 9.4.1 and do a PSpice simulation. Plot traces of the source current and the load line currents. Determine **(a)** the initial conditions for steady-state operation, **(b)** the peak value of the source current, **(c)** the average value of the source current, **(d)** the total harmonic distortion THD of the ac line current, **(e)** the rms value of the line current and **(f)** the maximum energy stored in any one phase.
(Ans: **(a)** $i_a = -50$A, $i_b = 25$A, $i_c = 25$A, **(b)** 25A, **(c)** 0, **(d)** 4.61%, **(e)** 32.39A, **(f)** 3.5J.)

P9.25 A three-phase, bridge inverter, depicted in Fig. 9.23, has the data

$V_s = 600$V, $L_l = 6.67$mH per phase, $R_l = 20\Omega$ per phase, $f = 100$Hz.

Gate voltage 15V. 180°, six-step mode of operation.

Follow the pattern of EXAMPLE 9.4.1 and do a PSpice simulation. Plot traces of the source current, a phase voltage, a line voltage and a line current for steady conditions. Determine **(a)** the initial conditions for steady state, **(b)** the rms value of an output phase voltage, **(c)** the rms value of an output line voltage, **(d)** the average power absorbed by the load, **(e)** the average source current and **(f)** the total harmonic distortion THD of the line-current waveform.
(Ans: **(a)** $i_a = -10$A, $i_b = -10$A, $i_c = 20$A, **(b)** 282.8V, **(c)** 489.9V, **(d)** 10.8kW, **(e)** 18A, **(f)** 16.71%.)

Chapter 10

P10.1 (PESS Problem 2.26) Consider the resonant inverter depicted in Fig. 10.1. The circuit has the following specifications.

$V_s = 100$V, $R = 1\Omega$, $L = 1$mH, $C = 25.0\mu$F, inverter frequency $f = 930$Hz.

Follow the pattern of EXAMPLE 10.2.1 and do a PSpice simulation. Plot the steady-state traces of the input current, the load voltage and the load current. Determine **(a)** the initial conditions for steady state, **(b)** the average power delivered to the load, **(c)** the current ratings of the thyristors and the diodes, **(d)** the phase difference between the load voltage and current and

(e) the total harmonic distortion THD of the load current. (Ans: (a) IC = −1.266kV, IC = 29.89A, (b) 3.96kW, (c) 44.2A, 4.6A, (d) current leads 23.6°, (e) 5.5%.)

P10.2 A parallel resonant inverter is illustrated in Fig. 10.4a. The specifications are $V_s = 50$V, $L_s = 1$H, $R = 20\Omega$, $L = 20.13$mH, $C = 125.8\mu$F, inverter frequency $f = 100$Hz. Follow the pattern of EXAMPLE 10.2.2 and do a PSpice simulation. Plot traces of the source current, and the currents in the load elements for steady conditions, over an interval of two cycles. Determine (a) the transient time to reach steady state, (b) the initial conditions for steady state, (c) the average power delivered to the load and (d) the total harmonic distortion THD of the load voltage. (e) Comment on the distortion. (Ans: (a) 500ms, (b) IC = 3.053A (i_s), IC = −6.1855A (i_L), (c) 152.1W, (d) 8.32%. (e) Distortion could be reduced if the inverter frequency equalled the damped frequency (94.9Hz).)

P10.3 Consider the full-wave, series-loaded ZCS flyback converter depicted in Fig. 10.10. The circuit has the following specifications.

$V_s = 24$V, $L_r = 0.6\mu$H, $C_r = 0.4\mu$F, $L = 20\mu$H, $C = 3\mu$F, $R_l = 10\Omega$, $f = 200$kHz.

Follow the pattern of EXAMPLE 10.3.2 and do a PSpice simulation. Plot traces of the source current, and the load current over the transient interval. Determine for steady conditions (a) the peak values of the source current, the switch current, the switch's diode current and the load current, (b) the load-current excursion, (c) the switch and resonant diode current ratings and (d) the average power absorbed by the load.
(Ans: (a) 11.66A, 57.61A, 39.39A, 3.84A, (b) 0.35A, (c) 12.3A, 6.7A, (d) 134W.)

P10.4 Consider the resonant zero-voltage switching (ZVS) converter depicted in Fig. 10.13. The circuit has the following specifications.

$V_s = 36$V, $L_r = 4\mu$H, $C_r = 0.5\mu$F, $L = 10\mu$H, $C = 10\mu$F, $R_l = 1\Omega$, $f = 800$kHz.

Follow the pattern of EXAMPLE 10.3.3 and do a PSpice simulation. Plot traces of the source current, the gate voltage and clock voltage, the switch voltage and the load voltage. Determine (a) the time to reach steady state and (b) the peak switch voltage and load voltage during the transient interval. Find for steady conditions (c) the switching duty cycle $m = (T - t_{OFF})/T$, (d) the peak value of the switch voltage, (e) the values of the average source current and rms load voltage, (f) the average power delivered to the load and (g) the load voltage excursion. (Ans: (a) 80μs, (b) 110.7V, 16.5V, (c) 0.4, (d) 75.7V, (e) 3.0A, 10.37V, (f) 108W, (g) 1.3V.)

INDEX

A

B

C

D

J

K

L

M

N

O

P

Q

R